Holm Altenbach · Xiao-Wei Gao ·
Stavros Syngellakis · Alexander H.-D. Cheng ·
Piotr Lampart · Anton Tkachuk
Editors

Advances in Mechanical and Power Engineering II

Selected Papers from The International Conference on Advanced Mechanical and Power Engineering (CAMPE 2023), October 16–19, 2023

Springer

Editors
Holm Altenbach
Fakultät für Maschinenbau
Otto-von-Guericke-Universität
Magdeburg, Germany

Xiao-Wei Gao
School of Aeronautics and Astronautics
Dalian University of Technology
Dalian, China

Stavros Syngellakis
Wessex Institute of Technology
Southampton, UK

Alexander H.-D. Cheng
School of Engineering
University of Mississippi
Oxford, MS, USA

Piotr Lampart
The Szewalski INSTITUTE OF
FLUID-FLOW MACHINERY Polish
Academy of Sciences
Gdansk, Poland

Anton Tkachuk
Department of Engineering and Physics
Karlstad University
Karlstad, Sweden

ISSN 2195-4356 ISSN 2195-4364 (electronic)
Lecture Notes in Mechanical Engineering
ISBN 978-3-031-82978-9 ISBN 978-3-031-82979-6 (eBook)
https://doi.org/10.1007/978-3-031-82979-6

This Springer imprint is published by the registered company Springer Nature Switzerland AG
The registered company address is: Gewerbestrasse 11, 6330 Cham, Switzerland

Lecture Notes in Mechanical Engineering

Lecture Notes in Mechanical Engineering (LNME) publishes the latest developments in Mechanical Engineering—quickly, informally and with high quality. Original research or contributions reported in proceedings and post-proceedings represents the core of LNME. Volumes published in LNME embrace all aspects, subfields and new challenges of mechanical engineering.

To submit a proposal or request further information, please contact the Springer Editor of your location:

Europe, USA, Africa: Leontina Di Cecco at Leontina.dicecco@springer.com
China: Ella Zhang at ella.zhang@cn.springernature.com
India, Rest of Asia, Australia, New Zealand: Swati Meherishi at swati.meherishi@springer.com

Topics in the series include:

- Engineering Design
- Machinery and Machine Elements
- Mechanical Structures and Stress Analysis
- Automotive Engineering
- Engine Technology
- Aerospace Technology and Astronautics
- Nanotechnology and Microengineering
- Control, Robotics, Mechatronics
- MEMS
- Theoretical and Applied Mechanics
- Dynamical Systems, Control
- Fluid Mechanics
- Engineering Thermodynamics, Heat and Mass Transfer
- Manufacturing Engineering and Smart Manufacturing
- Precision Engineering, Instrumentation, Measurement
- Materials Engineering
- Tribology and Surface Technology

Indexed by SCOPUS, EI Compendex, and INSPEC.

All books published in the series are evaluated by Web of Science for the Conference Proceedings Citation Index (CPCI).

To submit a proposal for a monograph, please check our Springer Tracts in Mechanical Engineering at https://link.springer.com/bookseries/11693.

Preface

This volume of Lecture Notes in Mechanical Engineering gathers selected papers presented at the 15th International Conference on Advanced Mechanical and Power Engineering (CAMPE-2023), held online from Kharkiv, Ukraine, on October 16–19, 2023. Between 1972–2023, XXII CAMPE Editions were carried out. The first 7 conferences did not include international authors' (1972–1993), the other 15 conferences held in 1994–2023 were international events. CAMPE–2023 was organized by the National Academy of Sciences of Ukraine, the Anatolii Pidhornyi Institute of Mechanical Engineering Problems of the National Academy of Sciences of Ukraine, the Engineering Academy of Ukraine, and the Gas Institute of the National Academy of Sciences of Ukraine.

CAMPE-2023 Keynote Speaker at Plenary Session - Imran Ali, Jamia Millia Islamia University, New Delhi, India ("Recent advances in hydrogen energy storage").

The aim of the conference was to present the latest achievements in mechanical and power engineering and to provide an occasion for discussion and exchange of views and opinions. The conference covered the following topics:

- Development and Improvement of Numerical Methods of Calculation of Fluid Flow and/or Heat Transfer
- Numerical Analysis of Fluid Flow and/or Heat Transfer in Power Equipment
- Identification and Optimization of Fluid Flow and/or Heat Transfer in Power EquipmentProfiling of Flow Parts of Turbomachines
- Development of Thermal Circuits and Improvement of Thermodynamic Characteristics of Power Equipment
- Efficient Fuel Combustion
- Renewable Energy Sources
- Resource-Saving and Energy Efficient Technologies
- Energy Storage
- Environmental Safety Issues of Power Engineering
- Hydrogen Technologies
- Theoretical and Applied Mechanics
- Mechanical Structures and Stress Analysis
- Engineering Design
- Materials Engineering
- Nanotechnology and Microengineering
- Methods and Technologies for Additive Manufacturing
- Engine Technology
- Aerospace Technology and Astronautics

The organizers received 64 contributions from 233 authors and 17 countries around the world.

After a thorough peer review process carried out in collaboration between the conference committee and the Springer series editors, 31 papers were accepted for publication in this book, prepared by authors from 17 countries (acceptance rate around 49%).

We would like to thank the members of the International Program Committee for their hard work during the review process. Their involvement and commitment were crucial for both the success of CAMPE-2023 and for the realization of this book.

The book "Advances in Mechanical and Power Engineering" has been organized in three parts, according to the main topic covered by the respective contributions:

1. Fluid Mechanics and Heat Transfer in Power Engineering;
2. Energy Saving Technologies and Environmental Safety;
3. Dynamics and Strength of Power Equipment".

The first part "Fluid Mechanics and Heat Transfer in Power Engineering" covers recent developments of advanced computational and analytical methods for applications in mechanical engineering, turbomachinery, heat and power engineering. It reports on both theoretical and experimental research, and the use of modern computer technology and effective mathematical methods for the analysis of fluid mechanics and heat transfer problems, as well as for the identification and optimization of fluid flow. It highlights new strategies for implementing intelligent solutions at the purpose of improving compressor and turbine stages and effective methods for profiling flow parts of turbomachines.

The second part "Energy Saving Technologies and Environmental Safety" presents some comparative analysis of computational methods adopted for renovation projects in mechanical engineering, such as those relating to the development of thermal circuits, and the improvement of thermodynamic characteristics of power equipment. The works belonging to this part are also devoted to methods for ensuring ecological and human safety and for reducing material costs and environmental pollution. New advancements in energy saving technologies, renewable energysources, topical issues in environmental safety, resource-saving and hydrogen technologies are presented.

The third part "Dynamics and Strength of Power Equipment" reports on recent developments in mathematical models and methods to simulating the dynamics characteristics and durability of machines, and engineering designs applications. It describes advanced mathematical models for simulating the behavior of power equipment under static and dynamic loads, and for making qualitative predictions on reliability and durability. It also covers topics related to stress-strain state, stability, structure optimization, and dynamic processes of mechanical systems and objects. It shows applications in nanotechnology, microengineering, engine and aerospace technologies, astronautics, materials engineering, and additive manufacturing.

CAMPE-2023 programme comprises an impressive range of topics addressing important issues relating to energy production and its effects on the environment.Solutions are proposed that would enhance the performance of powerengineering systems for the benefit of society.There are high quality and innovative contributions on modelling andexperimentation in material science, structural engineering as well as heat and mass transfer.

We would like to thank the organisers for their invitation and congratulate them for their efforts towards a very ambitious, topical and technically advanced content.Our best wishes to all participants for a most enjoyable and rewarding conference.

August 2024

Holm Altenbach
Alexander H.-D. Cheng
Xiao-Wei Gao
Piotr Lampart
Stavros Syngellakis
Anton Tkachuk

Organization

Honorary Chair

Olexandr Kyrylenko	National Academy of Sciences of Ukraine, Kyiv, Ukraine
Victor Subotin	Joint Stock Company «Ukrainian Energy Machines», Kharkiv, Ukraine
Igor Kravchenko	SE Ivchenko-Progress, Zaporizhzhia, Ukraine
Anatolii Boiko	National Technical University «Kharkiv Polytechnic Institute», Kharkiv, Ukraine
Oleksandr Shubenko	Anatolii Pidhornyi Institute of Mechanical Engineering Problems of the National Academy of Sciences of Ukraine, Kharkiv, Ukraine
Anatolii Tarelin	Anatolii Pidhornyi Institute of Mechanical Engineering Problems of the National Academy of Sciences of Ukraine, Kharkiv, Ukraine

General Chair

Andrii Rusanov	Anatolii Pidhornyi Institute of Mechanical Engineering Problems of the National Academy of Sciences of Ukraine, Kharkiv, Ukraine

Co-chair

Andrii Kostikov	Anatolii Pidhornyi Institute of Mechanical Engineering Problems of the National Academy of Sciences of Ukraine, Kharkiv, Ukraine

Steering Committee

Program Committee

Holm Altenbach	Otto-von-Guericke-Universität Magdeburg, Magdeburg, Germany
Svitlana Alyokhina	Fachhochschule Technikum Wien Vienna, Austria
Ali Abdulateef Kareem Alzubadi	Physics department, College of Science, Baghdad University, Iraq
Marco Amabili	McGill University, Montreal, QC, Canada
Igor Andrianov	Rheinisch-Westfälische Technische Hochschule Aachen, Aachen, Germany
Andrii Avramenko	Anatolii Pidhornyi Institute of Mechanical Engineering Problems of the National Academy of Sciences of Ukraine, Kharkiv, Ukraine
Kostiantyn Avramov	Anatolii Pidhornyi Institute of Mechanical Engineering Problems of the National Academy of Sciences of Ukraine, Kharkiv, Ukraine
Jan Awrejcewicz	Lodz University of Technology, Lodz, Poland
Jean-Michel Bergheau	École Centrale de Lyon, Ecully, France
Omar El Bounagui	Université Mohammed V de Rabat Faculté des Sciences in Rabat, Morocco
Ivan Breslavsky	McGill University, Montreal, Quebec, Canada
Weiqiu Chen	State Key Lab of CAD&CG, Zhejiang University, Hangzhou, China
Alexander Cheng	School of Engineering, University of Mississippi, Oxford, USA
Giuseppe Ciaburro	Università degli Studi della Campania Luigi Vanvitelli, Naples, Italy
Olga Chernousenko	National Technical University of Ukraine «Igor Sikorsky Kyiv Polytechnic Institute»
Neelam Choudhary	Bennett University, Greater Noida, India
Dmitriy Dikin	Temple University, Philadelphia, USA
Ney Augusto Dumont	Pontifical Catholic University of Rio de Janeiro, Brazil
Alaa Abd El-Bary	Arab Academy for Science, Technology and Maritime Transport, Alexandria, Egypt
Igor Emri	Faculty of Mechanical Engineering, University of Ljubljana, Slovenia
Hamid Ez-Zahraouy	Université Mohammed V de Rabat Faculté des Sciences in Rabat, Morocco
Chia-Ming Fan	National Taiwan Ocean University, Keelung, Taiwan
Alexander Fidlin	Karlsruhe Institute of Technology, Germany
Andreas Fischer	Technische Universität Dresden, Dresden, Germany

Yevgen Levchenko	Joint Stock Company «Ukrainian Energy Machines», Kharkiv, Ukraine
Mykhaylo Lototskyy	University of the Western Cape, Bellville, South Africa
Martin Luxa	Institute of Thermomechanics Academy of Sciences of the Czech Republic, Praha, Czech Republic
Gennadiy Lvov	National Technical University «Kharkiv Polytechnic Institute», Kharkiv, Ukraine
Kyrylo Maksymenko-Sheiko	Anatolii Pidhornyi Institute of Mechanical Engineering Problems of the National Academy of Sciences of Ukraine, Kharkiv, Ukraine
José-Antonio Marmolejo-Saucedo	Universidad Panamericana, Ciudad de Mexico, Mexico
Gennadii Martynenko	National Technical University «Kharkiv Polytechnic Institute», Kharkiv, Ukraine
Oleksandr Menshykov	University of Aberdeen, Aberdeen, UK
Minavar Mir-Salim-Zada	Institute of Mathematics and Mechanics Azerbaijan National Academy of Sciences, Baku, Azerbaijan
Francisco Monticeli	Department of Aeronautical Engineering, Technological Institute of Aeronautics (ITA), São José dos Campos, Brazil
Darkhan Myrzaliyev	M. Auezov South Kazakhstan University, Shymkent, Kazakhstan
Jose Fernando da Costa Oliveira	University of Porto, Porto, Portugal
Ernian Pan	National Yang Ming Chiao Tung University, Hsinchu, Taiwan
Chitaranjan Pany	Vikram Sarabhai Space Centre, Trivandrum, India
Anatoliy Pavlenko	Kielce University of Technology, Kielce, Poland
Dimitrios Pavlou	University of Stavanger, Norway
Francesco Pellicano	University of Modena and Reggio Emilia, Modena, Italy
Evgeny Petrov	University of Sussex, Brighton, UK
Valery Pilipchuk	Wayne State University, Detroit, USA
Christophe Pierre	Stevens Institute of Technology, Hoboken, USA
Ján Piteľ	Technical University of Kosice, Slovakia
Sergiy Plankovskyy	O.M. Beketov National University of Urban Economy, Kharkiv, Ukraine
Viktor Popov	Ascend Technologies Limited, Eastleigh, UK
Robertas Poškas	Lithuanian Energy Institute, Kaunas, Lithuania
Jure Ravnik	University of Maribor, Slovenia

Łukasz Witanowski	Robert Szewalski Institute of Fluid-Flow Machinery of the Polish Academy of Sciences, Gdansk, Poland
Luiz Wrobel	Brunel University London, Uxbridge, UK
Ihor Yanchevskii	National Technical University of Ukraine «Igor Sikorsky Kyiv Polytechnic Institute»
Chuanzeng Zhang	Universität Siegen, Siegen, Germany
Krzysztof Kamil Zur	Bialystok University of Technology, Bialystok, Poland

Contents

Fluid Mechanics and Heat Transfer in Power Engineering

Parametric Effects on Advanced Exergy and Exergoeconomics of a Gas-Fired Steam Power Plant Boiler

Uchenna Azubuike, Howard Njoku(✉), and Onyemaechi Ekechukwu

Applied Renewable & Sustainable Energy Research Group, Department of Mechanical Engineering, University of Nigeria, Nsukka 410001, Nigeria
howard.njoku@unn.edu.ng

Abstract. The achievable potentials for thermal system components improvement and the interactions of the system components cannot be obtained solely by the conventional exergy and exergoeconomic analysis. This information is imperative in the design and operation of installed thermal systems. However, advanced exergy and exergoeconomic analyses employ exergy destruction splitting to reveal the limitations associated with the conventional exergy analysis. Parametric advanced exergy analyses have been previously performed to assess the performances of some thermal systems, limited however to gas-fired steam turbine power plants. The goal of this study was to apply the advanced exergy and exergoeconomic analysis to the boiler of a gas-fired steam power plant to assess the performances of the superheater under different operating conditions (temperature and pressure). The operating temperature of the superheater were found to significantly affect the splits of exergy destruction and the sum of cost rates of avoidable endogenous exergy destruction and avoidable endogenous investment. Increasing the current plant operating temperature of the superheater decreases the avoidable, endogenous, avoidable endogenous and avoidable exogenous exergy destruction rates by 2.6% (1.4 MW), 4% (11.5 MW), 3% (1.2 MW) and 1.4% (0.2 MW), respectively. However, decreasing the current plant operating temperature of the superheater decreases the sum of cost rates of its avoidable endogenous exergy destruction and avoidable endogenous investment by 7% (350 $/h).

Keywords: Advanced Thermodynamics · Exergy Analysis · Steam Power Plant · Exergoeconomic Analysis

1 Introduction

Due to steadily increasing global energy demands, energy savings and cost reduction are paramount considerations driving the quest for more efficient thermal systems, especially thermal power plants. Consequently, thermodynamic analyses are frequently required for improving these systems. Traditional energy analysis methods of assessing thermal systems suffer limitations, including the treatment of work and heat interactions as equal forms of energy [1], and this has given rise to the exergy and exergoeconomic analysis

H. Altenbach et al. (Eds.): CAMPE 2023, pp. 3–12, LNME 2025.
https://doi.org/10.1007/978-3-031-82979-6_1

methods as preferred approaches. Conventional exergy and exergoeconomic analysis of thermal systems reveal rates of exergy destruction within system components, and the costs of exergy destruction and investments in parallel.

The effectiveness of conventional exergy and exergoeconomic analysis of thermal systems notwithstanding, limitations also exist in their assessments of thermal systems. Conventional exergy and exergoeconomic analysis cannot account for the exergy destruction in a component occurring due to inefficiencies of interconnected components and they do not reveal the influence of exergy destruction on the cost-effectiveness of thermal systems [2]. Furthermore, the achievable potential for improving thermal systems cannot be revealed in conventional exergy and exergoeconomic analysis [3]. Advanced exergy and exergoeconomic analysis methods have therefore been developed to handle the limitations of their conventional counterparts.

In advanced exergy and exergoeconomic analysis, the achievable potential for improving thermal systems is determined by splitting exergy destruction rates, exergy destruction costs and investment costs of components into unavoidable and avoidable portions [4, 5]. The avoidable portions represent the improvement potentials that can be practically realized for components and entire systems. The exergy destruction rates, exergy destruction costs and investment costs of processes that involve multiple interconnected components are revealed by splitting them into endogenous and exogenous portions [6], where the endogenous portions reveal the exergy destruction rates, exergy destruction costs and investment costs caused by the inefficiencies within components while exogenous portions reveal those caused by inefficient processes in interconnected components. Both the unavoidable/avoidable and endogenous/exogenous portions may be further combined to reveal additional information. The goal of the present study was to obtain these for the boiler of a gas-fired steam power plant.

2 Literature

Instances abound in the literature of the advanced exergy analysis of thermal systems to determine the endogenous and exogenous portions of their exergy destruction rates. For example, the advanced exergy analysis performed on a marine steam power plant by Koroglu and Sogut [3] revealed that the possible turbine and boiler improvement were 1.3% and 6%, respectively. Wang et al. [7] in a similar study on a supercritical coal-fired power plant improvement, found that most of the exergy destruction rate (90%) occurring in the turbine was endogenous, which indicated that processes in the interconnected components contributed only 10% of the total exergy destruction rate in the turbine. Exergy destruction rates in the feedwater were 70% endogenous, while other plant components similarly had higher endogenous exergy destruction rates.

The advanced exergy analysis of a combined cycle plant was performed by Wang et al. [8]. The endogenous exergy destruction rates in the compressor (19.7 MW) and combustion chamber (237 MW) were higher than the exogenous exergy destruction rates (i.e., the interconnection effects), which was only 1 MW for both components. The avoidable exergy destruction rates in the two components were insignificant (i.e., <1 MW).

Thermal plants have been similarly analyzed to determine the avoidable and unavoidable portions of their exergy destruction rates. Analyzing a co-generation plant using the advanced exergy analysis method, Vukić et al. [9] found that 7.66% of the exergy destruction rate in the boiler was avoidable, indicating a significant improvement potential. A higher proportion of avoidable exergy destruction rate (70%) was obtained for the pumps, which was followed by the compressed air station (24.44%). On a tri-generation plant for heat, cold and power production, Anvari et al. [10] performed advanced exergy and exergoeconomic analysis and found that 32%, 33% and 58% of the total exergy destruction rate, cost rate of exergy destruction and investment cost, respectively, for the plant, were avoidable.

Following an advanced exergy analysis of a refrigerating system, Tsatsaronis and Morosuk [11] concluded that, when compared to the outcomes of advanced exergy analysis, the results of conventional exergy analysis could be misleading. For instance, the conventional exergy analysis of the system prioritized the third throttling valve (TV3) for improvement, whereas the TV2 was prioritized by an advanced exergy method that assessed the total avoidable exergy destruction rates in the system components. Advanced exergy and exergoeconomic analyses have also been applied to a geothermal power plant [12], distillation system [13], gasoline absorption–stabilization process [14], integrated solar-assisted gasification cycle [15], etc.

Parametric studies have also been undertaken to assess the impact of diverse parameters on the advanced exergy performances of thermal systems. Boyaghchi and Molaie [16] varied the turbine inlet temperature and compressor pressure ratio of a combined cycle power plant to determine the extent to which exergy destruction rates could be minimized. Increasing the turbine inlet temperature and compressor pressure ratio led to increased avoidable endogenous exergy destruction rates in the combustion chamber. Mossi and Goni [17] investigated the humidification ratio's effect on the performances of a combined cycle power plant. Avoidable endogenous exergy destruction rates at the topping cycle turbine and compressor decreased when the humidification ratio and the combustion chamber exit temperature were increased. Another parametric study on the effect of the cold side temperature difference in the heat exchanger of a combined cycle power plant [18] showed that reducing the cold side temperature difference of the heat exchanger lowered its endogenous exergy destruction rate.

Existing parametric studies have dwelt largely on the influence of the investigated parameters on exergy destruction rates in components of thermal systems, especially combined cycle plants. The present paper, however, investigates parametric effects on the advanced exergy and exergoeconomic performance of a gas-fired steam turbine plant's boiler (superheater).

3 Research Methodology

The boiler unit of the 220 MW reheat-regenerative cycle Egbin thermal power plant, Lagos, Nigeria, was investigated. The boiler incorporates both superheater (SH) and reheater (RH) as a single component and is designed for firing with either natural gas as primary fuel or low/high pour fuel oil as secondary fuel. A full description of the plant (the process diagram and state points) is presented in Ref. [19].

First-tier splitting of exergy destruction ($\dot{\Xi}_{D,k}$), exergy destruction cost ($\dot{C}_{D,k}$) and investment cost ($\dot{Z}_k$) rates for the kth component were performed to obtain their avoidable portions ($\dot{\Xi}_{D,k}^{AV}$, $\dot{C}_{D,k}^{AV}$ and $\dot{Z}_{k}^{AV}$, , respectively) and unavoidable portions ($\dot{\Xi}_{D,k}^{UN}$, $\dot{C}_{D,k}^{UN}$ and $\dot{Z}_{k}^{UN}$). The first-tier splittings are respectively stated as:

$$\dot{\Xi}_{D,k} = \dot{\Xi}_{D,k}^{UN} + \dot{\Xi}_{D,k}^{AV} \tag{1}$$

$$\dot{C}_{D,k} = \dot{C}_{D,k}^{UN} + \dot{C}_{D,k}^{AV} \tag{2}$$

$$\dot{Z}_{k} = \dot{Z}_{k}^{UN} + \dot{Z}_{k}^{AV} \tag{3}$$

where the unavoidable portions are respectively obtained as:

$$\dot{\Xi}_{D,k}^{UN} = \dot{\Xi}_{P,k}\left(\frac{\dot{\Xi}_{D}}{\dot{\Xi}_{P}}\right)_{k}^{UN} \tag{4}$$

$$\dot{C}_{D,k}^{UN} = c_{F,k}\dot{\Xi}_{D,k}^{UN} \tag{5}$$

$$\dot{Z}_{D,k}^{UN} = \dot{\Xi}_{P,k}\left(\frac{Z}{\dot{\Xi}_{P}}\right)_{k}^{UN} \tag{6}$$

$\left(\frac{\dot{\Xi}_{D}}{\dot{\Xi}_{P}}\right)_{k}^{UN}$ and $\left(\frac{Z}{\dot{\Xi}_{P}}\right)_{k}^{UN}$ are the specific unavoidable exergy destruction and specific unavoidable investment cost which are obtained from a relationship between investment cost and exergy destruction in a thermal system component [20], $\dot{\Xi}_{P,k}$ is the product exergy rate, while $c_{F,k}$ is the unit cost of fuel exergy. The avoidable portions of the exergy destruction, cost and investment can be eliminated via technical, design, and operational improvements to the components and represent the potential for component improvements. The avoidable portions are by obtained subtracting the unavoidable portions from their total. While the unavoidable portions cannot be eliminated, even though the best available near-future technologies are applied [4].

The alternative first-tier splitting involves splitting the $\dot{\Xi}_{D,k}$, $\dot{C}_{D,k}$ and $\dot{Z}_{k}$ to obtain their endogenous and exogenous portions, i.e.,

$$\dot{\Xi}_{D,k} = \dot{\Xi}_{D,k}^{EN} + \dot{\Xi}_{D,k}^{EX} \tag{7}$$

$$\dot{C}_{D,k} = \dot{C}_{D,k}^{EN} + \dot{C}_{D,k}^{EX} \tag{8}$$

$$\dot{Z}_{k} = \dot{Z}_{k}^{EN} + \dot{Z}_{k}^{EX} \tag{9}$$

where the $\dot{\Xi}_{D,k}^{EN}$, $\dot{C}_{D,k}^{EN}$ and $\dot{Z}_{k}^{EN}$ are the endogenous portions of $\dot{\Xi}_{D,k}$, $\dot{C}_{D,k}$ and $\dot{Z}_{k}$, respectively, while $\dot{\Xi}_{D,k}^{EX}$, $\dot{C}_{D,k}^{EX}$ and $\dot{Z}_{k}^{EX}$ are the respective exogenous portions. The endogenous portions are caused by processes/operational inefficiencies within the kth component, while the exogenous portions are caused by the inefficiencies during interactions with other components besides the kth component. The $\dot{\Xi}_{D,k}^{EN}$ may be obtained using a number of proposed methods, however, this study adopts the decomposition method [21]. The $\dot{C}_{D,k}^{EX}$ and $\dot{Z}_{k}^{EX}$ in Eqs. (8) and (9) are obtained as:

$$\dot{C}_{D,k}^{EN} = c_{F,k}\dot{\Xi}_{D,k}^{EN} \tag{10}$$

$$\dot{Z}_{D,k}^{EN} = \dot{\Xi}_{P,k}^{EN}\left(\frac{Z}{\dot{\Xi}_P}\right)_k^{Real} \tag{11}$$

$\left(\frac{Z}{\dot{\Xi}_P}\right)_k^{Real}$ is the unavoidable investment cost at real state. The exogenous portions are obtained by subtracting the endogenous portions from their total.

The second-tier splitting involves merging the two first-tier splits, i.e.:

$$\dot{\Xi}_{D,k} = \dot{\Xi}_{D,k}^{UNEN} + \dot{\Xi}_{D,k}^{AVEN} + \dot{\Xi}_{D,k}^{UNEX} + \dot{\Xi}_{D,k}^{AVEX} \tag{12}$$

$$\dot{C}_{D,k} = \dot{C}_{D,k}^{UNEN} + \dot{C}_{D,k}^{AVEN} + \dot{C}_{D,k}^{UNEX} + \dot{C}_{D,k}^{AVEX} \tag{13}$$

$$\dot{Z}_k = \dot{Z}_k^{UNEN} + \dot{Z}_k^{AVEN} + \dot{Z}_k^{UNEX} + \dot{Z}_k^{AVEX} \tag{14}$$

where the superscripts *UNEN, AVEN, UNEX* and *AVEX* on the terms in the RHS of Eqs. (7) to (9) are the unavoidable endogenous portions, the avoidable endogenous, the unavoidable exogenous and the avoidable exogenous portions of the kth component's $\dot{\Xi}_{D,k}$, $\dot{C}_{D,k}$ and $\dot{Z}_k$.

The kth component's total cost rate, $\dot{C}_{ov,k}^{AVEN}$, which is the sum of its avoidable endogenous exergy destruction and investment cost rates (Eq. (10)), is a parameter used in the advanced exergoeconomic analysis for designating the priority of components for improvement.

$$\dot{C}_{ov,k}^{AVEN} = \dot{C}_{D,k}^{AVEN} + \dot{Z}_k^{AVEN} \tag{15}$$

4 Results and Discussion

4.1 Parametric Advanced Exergy Performances of the Boiler

The current plant boiler (superheater, SH) temperature and pressure were both varied in the ranges of 500–580°C and 120–160 bars, respectively, to determine their effects on the first- and second-tier splits of exergy destruction rates in the SH, obtained with Eqs. (1), (4) and (7). These are shown in Fig. 1. The SH temperature has more marked effects on the various splits of SH exergy destruction rates in comparison to the SH pressure. All the splits of SH exergy destruction rates experience declines with increasing SH temperature.

The SH avoidable exergy destruction rate ($\dot{\Xi}_{D,SH}^{AV}$) declined by 5.9% (3.3 MW) with increasing temperature, while increasing the SH temperature and pressure (i.e., from 542.14 to 580°C and 140 to 160 bar) decreased the $\dot{\Xi}_{D,SH}^{AV}$ by 2.6% (1.4 MW) as shown

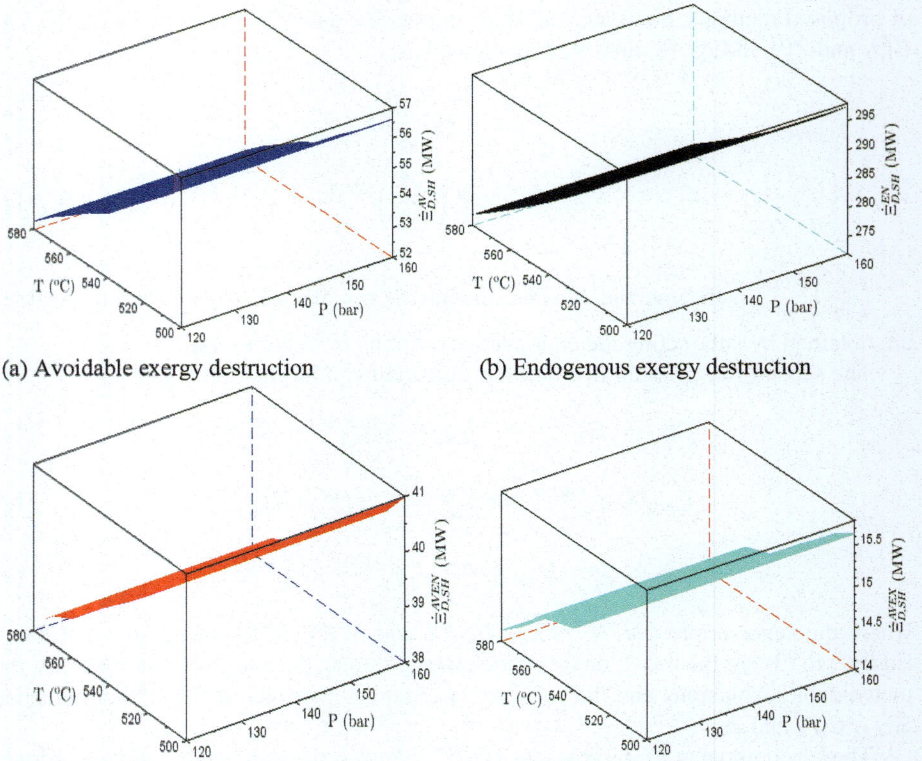

(a) Avoidable exergy destruction
(b) Endogenous exergy destruction
(c) AVEN exergy destruction
(d) AVEX exergy destruction

Fig. 1. Splits of exergy destruction rates in the SH for varying SH temperatures and pressures.

in Fig. 1(a). When the current plant temperature and pressure increased, the endogenous exergy destruction rates declined by 4% (i.e., 11.5 MW) as shown in Fig. 1(b). Similarly, increasing the current plant SH pressure and temperature reduced the avoidable endogenous and avoidable exogenous exergy destruction rates by 3% (1.2 MW) and 1.4% (0.2 MW), respectively (Figs. 1(c) and (d), respectively).

4.2 Advanced Exergoeconomic Performances of the Boiler

The effects of operating temperature and pressure of the SH on the advanced exergoeconomic performance of the plant were also investigated. Effects of SH pressure (at constant temperature) on its sum of cost rates of avoidable endogenous exergy destruction and avoidable endogenous investment ($\dot{C}_{D,SH}^{AVEN} + \dot{Z}_{SH}^{AVEN}$) are presented in Fig. 2. It is evident that the $\dot{C}_{D,SH}^{AVEN} + \dot{Z}_{SH}^{AVEN}$ of the SH can generally be reduced by decreasing the current plant SH pressure—it decreases by 3% (150 \$/h) when the SH operating pressure is reduced from 135 to 130 bar.

Conversely, the effects of varying the temperature of the SH at constant pressure on the $\dot{C}_{D,SH}^{AVEN} + \dot{Z}_{SH}^{AVEN}$ are shown in Fig. 3. The $\dot{C}_{D,SH}^{AVEN} + \dot{Z}_{SH}^{AVEN}$ slightly decreased with

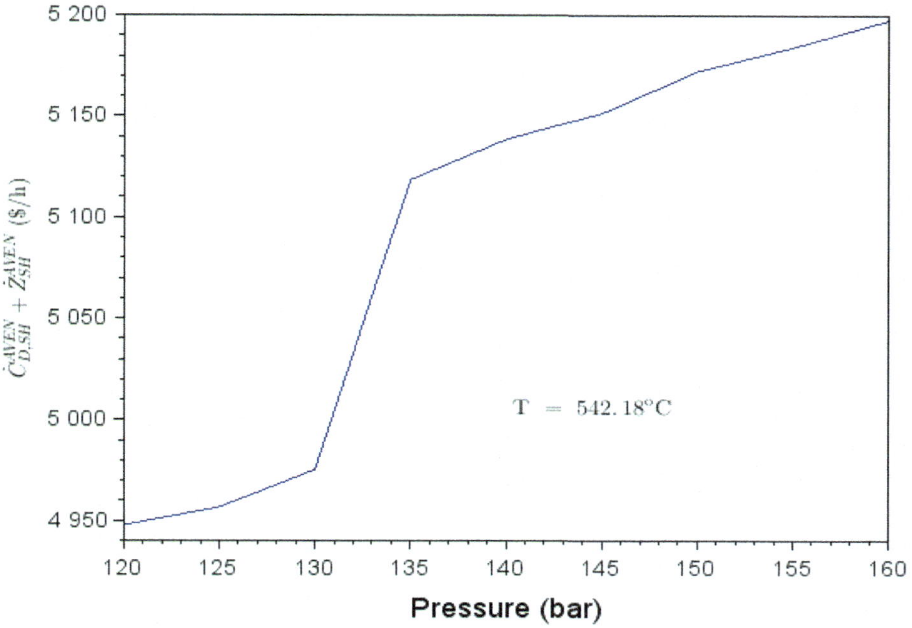

Fig. 2. Overall cost rates in the superheater at varying pressure and constant temperature.

deceasing operating SH temperature. Noticeably, a 56% increase in $\dot{C}^{AVEN}_{D,SH} + \dot{Z}^{AVEN}_{SH}$ (i.e., 2800$/h) was recorded when the operating temperature increased to 590 °C.

Finally, the effects of varying both operating temperature and pressure of the SH on the $\dot{C}^{AVEN}_{D,SH} + \dot{Z}^{AVEN}_{SH}$ are presented in Fig. 4. This clearly shows the predominant influence of SH temperature on the $\dot{C}^{AVEN}_{D,SH} + \dot{Z}^{AVEN}_{SH}$ - the $\dot{C}^{AVEN}_{D,SH} + \dot{Z}^{AVEN}_{SH}$ decreased with decreasing temperature while the change with pressure was imperceptible. Decreasing the current plant operating temperature to 500 °C would reduce the $\dot{C}^{AVEN}_{D,SH} + \dot{Z}^{AVEN}_{SH}$ by 7% i.e., 350 $/h, while increasing the current plant operating temperature to 580 °C would potentially increase the $\dot{C}^{AVEN}_{D,SH} + \dot{Z}^{AVEN}_{SH}$ by 30%, i.e., 1500 $/h.

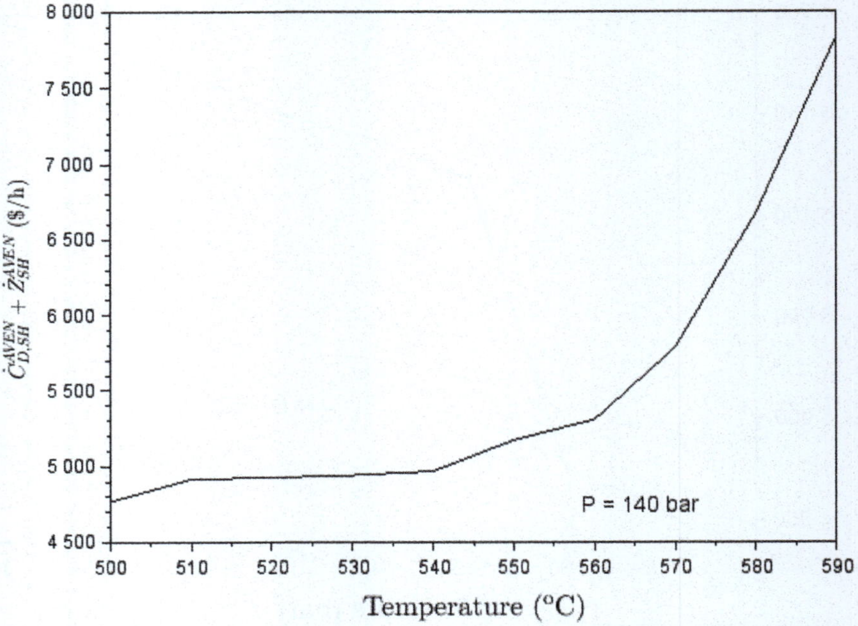

Fig. 3. Overall superheater cost rates at varying temperatures and constant pressure.

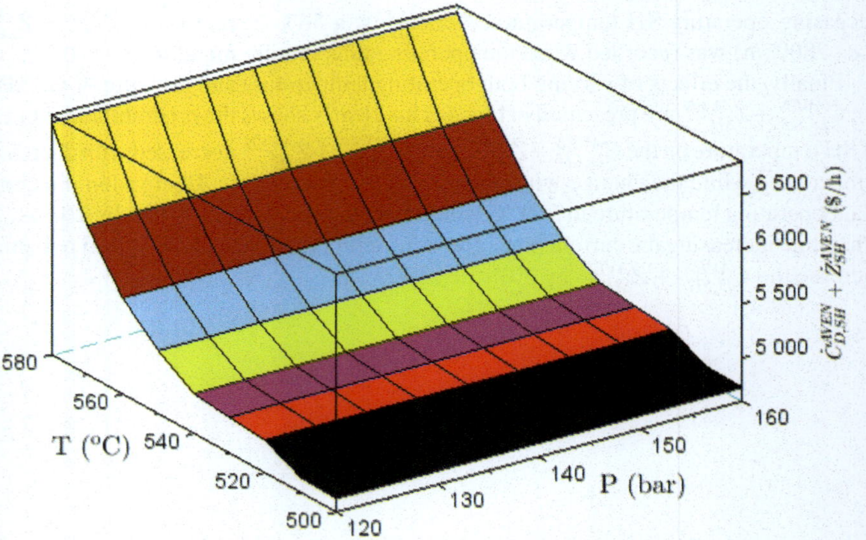

Fig. 4. Overall cost rates in the superheater at varying pressure and temperature.

5 Conclusion

A parametric advanced exergy and exergoeconomic analysis of a gas-fired steam power plant has been performed to ascertain the effects of operating temperatures and pressures on the advanced exergy and exergoeconomic performances of the boiler (superheater), which is the most significant component of the system. The study findings show that:

a) The temperature of the superheater during plant operation is a more important parameter than the operating pressure, having more significantly affects on the different portions of exergy destruction rates.
b) The avoidable-, endogenous-, avoidable endogenous- and avoidable exogenous-exergy destruction rates in the superheater can be effectively reduced during operation by increasing its operating temperature.
c) Decreasing the superheater operating temperatures will equally reduce its total investment costs and the costs incurred by its avoidable endogenous exergy destruction.

References

1. Kotas, T.J.: The exergy method of thermal plant analysis. Anchor Brendon Ltd, Tiptree, Essex (1996)
2. Petrakopoulou, F., Tsatsaronis, G., Morosuk, T.: Advanced exergoeconomic analysis applied to a complex energy conversion system. J. Eng. Gas Turbines Power **134**, 1–7 (2012)
3. Koroglu, T., Sogut, O.: Advanced exergy analysis of an organic Rankine cycle waste heat recovery system of a marine power plant. J. Therm. Eng. **3**, 1136–1148 (2017)
4. Tsatsaronis, G., Park, M.: On avoidable and unavoidable exergy destructions and investment costs in thermal systems. Energy Convers. Manage. **43**, 1259–1270 (2002)
5. Esmaeilion, F., Soltani, M., Nathwani, J., Al-Haq, A., Dusseault, M.: Conventional and advanced exergy and exergoeconomic analysis of a novel multi-fueled cogeneration system with desalination. Energy Rep. **9**, 5704–5720 (2023)
6. Vučković, G., Stojiljković, M., Vukić, M.: First and second level of exergy destruction splitting in advanced exergy analysis of an existing boiler. Energy Convers. Manage. **104**, 8–16 (2015)
7. Wang, L., Yang, Y., Morosuk, T., Tsatsaronis, G.: Advanced thermodynamic analysis and evaluation of a supercritical power plant. Energies **5**, 1850–1863 (2012)
8. Wang, S., Fu, Z., Zhang, G., Zhang, T.: Advanced thermodynamic analysis applied to an integrated solar combined cycle system. Energies **11**, 1–16 (2018)
9. Vukić, M., Vuckovic, D., Stojiljkovic, M.: Avoidable and unavoidable exergy destruction and exergoeconomic evaluation of the thermal processes in a real industrial plant. Therm. Sci. **16**, S433–S446 (2012)
10. Anvari, S., Saray, R., Bahlouli, K.: Conventional and advanced exergetic and exergoeconomic analyses applied to a tri-generation cycle for heat, cold and power production. Energy **91**, 925–939 (2015)
11. Tsatsaronis1, G., Morosuk, T.: Advanced exergetic analysis of a refrigeration system for liquefaction of natural gas. Int. J. Energy Environ. Eng. **1**, 1–18 (2008)
12. Güler, O.V., Gürbüz, E.Y., Georgiev, A.G., Keçebaş, A.: Advanced exergoeconomic assessment of CO_2 emissions, geo-fluid and electricity in dual loop geothermal power plant. Energies **16**(8), 3466 (2022)

13. Wei, Z., Zhang, B., Wu, S., Chen, Q., Tsatsaronis, G.: Energy-use analysis and evaluation of distillation systems through avoidable exergy destruction and investment costs. Energy **42**, 424–433 (2012)
14. Lei, Y., Chen, Y., Yang, Y., Liu, X., Luo, H., Yan, W.: Advanced exergy analysis for a novel gasoline absorption-stabilization process. ACS Omega **6**(23), 15332–15347 (2021). https://doi.org/10.1021/acsomega.1c01658
15. Jadidi, E., Khoshgoftar Manesh, M.H., Delpisheh, M., Onishi, V.C.: Advanced exergy, exergoeconomic, and exergoenvironmental analyses of integrated solar-assisted gasification cycle for producing power and steam from heavy refinery fuels. Energies **14**(24), 8409 (2020). https://doi.org/10.3390/en14248409
16. Boyaghchi, F., Molaie, H.: Sensitivity analysis of exergy destruction in a real combined cycle power plant based on advanced exergy method. Energy Convers. Manage. **99**, 374–386 (2015)
17. Mossi, I.A.K., Goni, B.K.: Advanced exergy analysis of a combined gas power cycle with humidification. Int. J. Green Energy **17**(15), 990–1004 (2020)
18. Soltani, S., Yari, M., Mahmoudi, S.M.S., Morosuk, T., Rosen, M.A.: Advanced exergy analysis applied to an externally-fired combined-cycle power plant integrated with a biomass gasification unit. Energy **59**, 775–780 (2013)
19. Azubuike, U.G., Njoku, H.O., Ekechukwu, O.V.: Advanced exergoeconomics of a steam power plant based on double-tier splitting of exergy destruction rates. Proc. Inst. Mech. Eng. Part A J. Power Energy (2023). https://doi.org/10.1177/09576509231170906
20. Cziesla, F., Tsatsaronis, G., Gao, Z.: Avoidable thermodynamic inefficiencies and costs in an externally fired combined cycle power plant. Energy **31**, 1472–1489 (2006)
21. Penkuhno, M., Tsatsaronis, G.: A decomposition method for the evaluation of component interactions in energy conversion systems for application to advanced exergy-based analyses. Energy **133**, 388–403 (2017)

Aeroelastic State of Turbine Rotor During Harmonic Blade Oscillations

Lyubov Kolodyazhnaya[1], Yuriy Bykov[1(✉)], and Romuald Rządkowski[2]

[1] Anatolii Pidhornyi Institute of Mechanical Engineering Problems of the National Academy of Sciences of Ukraine, Kharkiv, Ukraine
yubykoff@gmail.com

[2] Robert Szewalski Institute of Fluid-Flow Machinery of the Polish Academy of Sciences, Gdansk, Poland

Abstract. The rotor blades of the last stages of steam turbines have a considerable length and weight, and are subjected to significant unsteady loads, so the problem of the appearance of self-excited oscillations or auto-oscillations is relevant. Such oscillations are quite dangerous, reduce the blade system's service life, and affect the flow characteristics in the turbine. During the operation of a 370 MW steam turbine, damage was detected in the last stage rotor blades in the root region. Numerical studies did not reveal the conditions for the occurrence of oscillations with a dangerous amplitude. In this regard, a numerical analysis of the aeroelasticity of the last stage rotor blades was performed using an improved numerical method for modeling unsteady aerodynamics and harmonic oscillations of the blades. The results of the numerical study are presented in the form of a distribution of the work of aerodynamic forces on blade displacement for different values of interblade phase angles, as well as by blade height for the first five natural oscillation modes. The results of the calculations showed the presence of conditions for the excitation of blade oscillations for the first natural mode in one of the steam turbine operating modes.

Keywords: Aeroelasticity · Axial turbine · Aerodynamic work · Flutter · Energy method

1 Introduction

During the development of modern turbines and compressors, considerable attention is paid to improving efficiency, which is achieved by creating stages with increased pressure and temperature drops, with longer blades with a slender profile [1]. The combination of such factors contributes to the emergence of aeroelastic phenomena that can cause excessive oscillations of the blades, which will threaten the integrity of the entire structure. Therefore, the task of increasing the reliability of turbomachines is becoming increasingly important.

The cause of aerodynamic excitation of oscillations of the blade system of an axial turbine is the unsteady flow, the unsteady force interaction of aerodynamic, elastic, and inertial forces acting on the blades [2], which is the basis of the aeroelastic state

H. Altenbach et al. (Eds.): CAMPE 2023, pp. 13–22, LNME 2025.
https://doi.org/10.1007/978-3-031-82979-6_2

in turbomachines [3]. Aerodynamically excited vibrations, especially of turbine rotor blades [4], can lead to damage or even destruction of the structure and therefore represent the most important part of the task of providing operational reliability.

The phenomenon of blade oscillation excitation is dangerous and can cause damage, especially for long blades of the last stages of steam turbines [5]. However, similar aeroelasticity phenomena occur in other bladed machines, particularly wind turbines [6], fans [7], compressor blade systems, etc. The best method for determining the aeroelastic state is to conduct experimental studies [8], which can be reduced by using the analysis of acoustic measurements [9].

The method of modeling the aeroelastic phenomena of blade systems is to solve the problem of interaction between two physical media - liquid and elastic [10]. For an inviscid incompressible flow, simplified gridless panel methods can be used [11]. One of the low-cost and fastest methods is to solve the problem in the frequency domain. This approach uses linearization of the equations of unsteady flow and is effective only for small oscillations in non-complex flows [12], however, it is also used to determine the effect of seals [13], and even for coupled aeroelasticity problems [14]. For a more complete representation of the interaction between the flow and the motion of structural elements, it is necessary to simultaneously solve the equations of motion of a viscous turbulent flow and blade oscillations with the exchange of data at the contact boundaries [15]. To reduce the cost of computation, simplifications are used, such as two-dimensional calculation in sections [16] and multigrid methods [17].

To solve the problem of the aeroelastic state, two software packages were used: the first uses a numerical method for calculating the spatial transonic periodically unsteady flow of an ideal gas through an isolated row of oscillating blades. The method is based on the direct numerical integration of a complete nonlinear system of unsteady gas dynamics equations (Euler's equations) [15]. The second one uses a mathematical model of the coupled problem of unsteady gas dynamics and elastic vibrations of the blade row of a turbine stage in a three-dimensional viscous gas flow. The unsteady Reynolds-averaged unsteady Navier-Stokes equations were chosen as the aerodynamic model, and their numerical integration was performed using the commercial package ANSYS Fluent Academic Edition 19R3 [18]. This method has been well-proven in modeling unsteady flows in various types of turbomachinery [19].

The dynamic model of an oscillating blade is based on the use of the modal approach and the finite element method to calculate the natural modes and natural frequencies of blade oscillations [20].

This paper presents numerical studies of the aeroelastic state of a turbine blade row under harmonic oscillations of the blades for the calculation of three-dimensional flow of ideal and viscous gas, taking into account the uniform pressure distribution behind the stage in the circumferential direction and along the height of the blade.

The aim of this work is to study the aeroelastic characteristics of the turbine blade row by means of numerical analysis of aeroelastic (harmonic) vibrations in a three-dimensional flow of ideal and viscous gas.

2 Object of Study

During the operation of a powerful 370 MW steam turbine, damage was detected in the root section of the last stage blades [21]. We consider a three-dimensional flow of ideal and viscous gas around the rotor blades of the last stage, the oscillations of which are described using the modal approach. The turbine blade row consists of 53 blades. The meridional section of the flow part and the tangential sections that form the turbine blade are shown in Fig. 1. Each of the segments of the computational domain is discretized using a hybrid deformed H-O mesh. The parameters of the computational mesh are determined according to the recommendations [22].

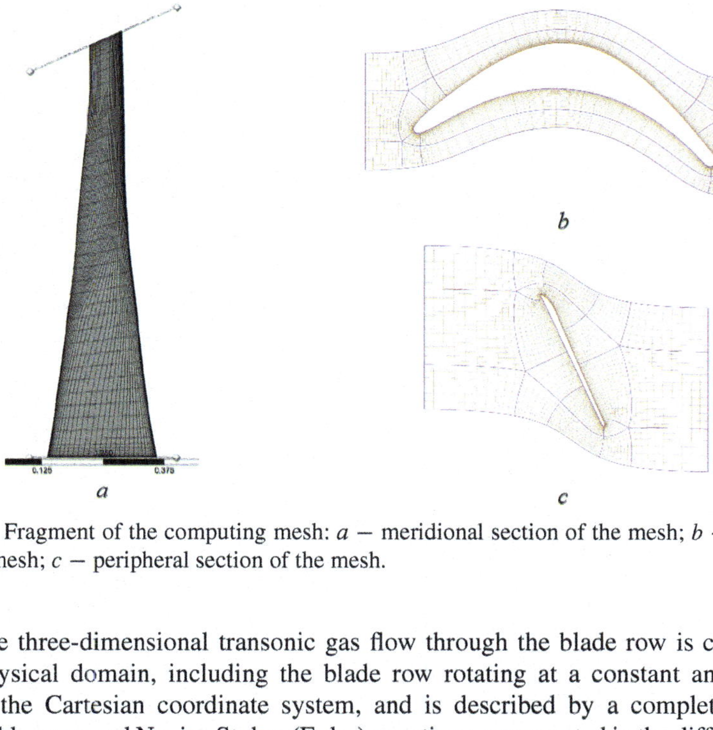

Fig. 1. Fragment of the computing mesh: *a* – meridional section of the mesh; *b* – root section of the mesh; *c* – peripheral section of the mesh.

The three-dimensional transonic gas flow through the blade row is considered in the physical domain, including the blade row rotating at a constant angular velocity in the Cartesian coordinate system, and is described by a complete system of Reynolds-averaged Navier-Stokes (Euler) equations, represented in the differential form as conservation laws [18, 23].

The operation mode of the turbine blade row is characterized by the following distribution of flow parameters:

- total pressure and temperature before the stage $P_0 = 39400$ Pa, $T_0 = 348$ K;
- flow angles in the circumferential (α) and radial (γ) directions;
- static pressure behind the stage $P_2 = 10400$ Pa;
- rotor speed $n = 3000$ rpm.

Vibrations of the blades were determined taking into account the first five natural oscillation modes. The values of the blades' natural frequencies are given in Table 1.

Table 1. Blade natural frequencies.

Natural mode	1	2	3	4	5
ν_i, Hz	119.43	159.24	238.85	358.3	477.7

3 Methods of Study

In this work, a numerical method is used [18, 24], which sequentially integrates the equations of viscous and ideal gas flow (Reynolds-averaged Navier-Stokes and Euler equations) with a specified law of blade oscillations. The Navier-Stokes equations are complemented by the equations of the SST turbulence model of Menter [18, 25].

All blades perform harmonic oscillations in each of their natural modes according to the identical law with a constant phase angle δ (interblade phase angle (IBPA)):

$$q_{ij} = q_{i0} \cdot sin[2\pi \nu_i t + (j - 1)\delta]$$

where q_{ij} – modal coefficient; i – number of the natural mode; j – number of the blade; q_{i0} – amplitude of oscillations; ν_i is the natural frequency of the i-th mode; δ is the interblade phase angle (IBPA) of oscillations of neighboring blades.

The aeroelastic state of the system "gas flow – blade row" without taking into account mechanical damping is determined by the aerodamping coefficient D, equal to the aerodynamic work W taken with a minus sign, performed by the aerodynamic force on the blade movement for one period of oscillations [15].

4 Results of Study and Discussion

The calculations were performed for harmonic vibration of the blades at IBPA = 0°, 180°, ±90°, ±45°, ±135°, taking into account each of the natural modes separately for an ideal gas and a viscous gas. The computational domain included the required number of blade channels for each value of IBPA (1 for 0°, 2 for 180°, etc.). The equation of state of an ideal gas corresponding to the full steam parameters at the rotor inlet was used for modeling. The initial blade vibration amplitude q_{i0} for determining the aerodynamic work was 0.5 mm, which corresponds to the average amplitude level of the experiment [21], and is often used when performing numerical modeling of aeroelastic vibration of blades [20]. The character of energy exchange between the gas flow and the oscillating blade for an ideal gas is shown in Fig. 2, which shows the dependence of the aerodamping coefficient D on the IBPA.

The aerodamping coefficient of the rotor blades at different IBPA for ideal gas for the 1st natural mode showed that the energy transfer to the blade (flutter) occurred at IBPA = −45° and +90° (Fig. 2a). For the other natural vibration modes, the aerodamping coefficient corresponds to the dissipation of the energy of the oscillating blade into the main flow (aerodamping).

Figure 3 shows the distribution by blade height of the aerodamping coefficient for IBPA = 0°, 180°, ±90° for the first 5 natural modes.

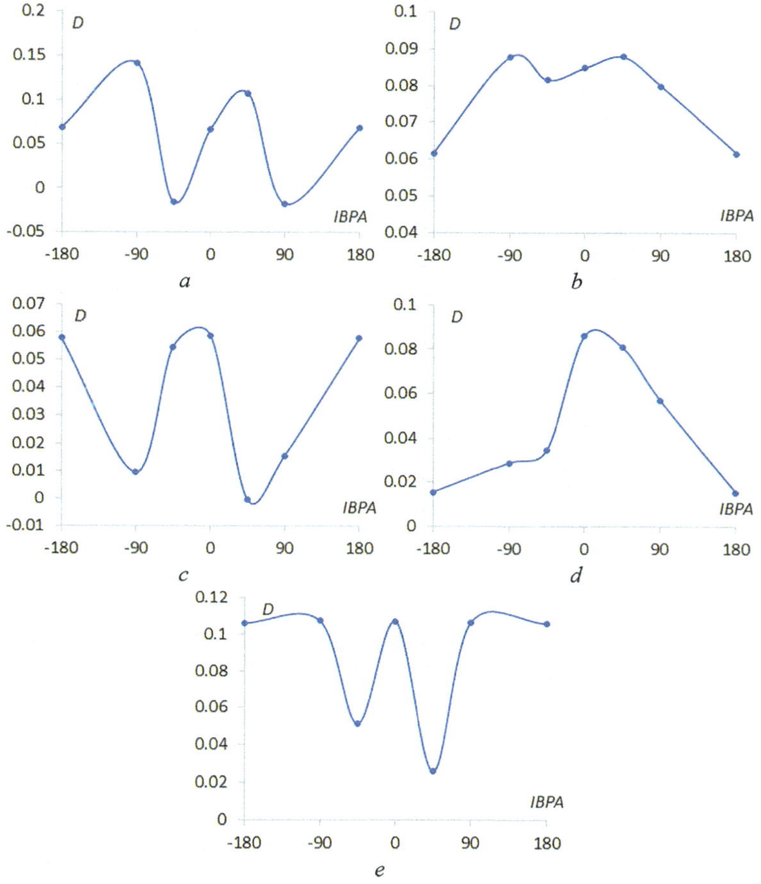

Fig. 2. Dependence of the aerodamping coefficient averaged over the blade height on the IBPA, ideal gas: *a* – 1st mode; *b* – 2nd mode; *c* – 3rd mode; *d* – 4th mode; *e* – 5th mode.

The graphs in Fig. 3 show that the majority of the aerodynamic work is concentrated in the peripheral part of the blade. The oscillation excitation work has positive values only for the first natural mode for the value of IBPA $= 90°$.

Under the same boundary and initial conditions, the aeroelastic state was calculated using a viscous turbulent flow model. In this case, the computational domain included all rotor blades. The Peng-Robinson equation of state of a real gas was used for modeling [18]. The dependence of the aerodamping coefficient on the IBPA for a viscous gas is shown in Fig. 4.

As for an ideal gas, a negative aerodamping coefficient (the possibility of flutter) is observed for the first natural mode, in the range of IBPA $= -135°\ldots -34°$. The general form of the dependence of the aerodamping coefficient on the IBPA for a viscous gas is different: local minima of aerodamping are observed only for the second and fourth natural modes, in contrast to the 1st, 2nd, 3rd, and 5th natural modes for an ideal gas.

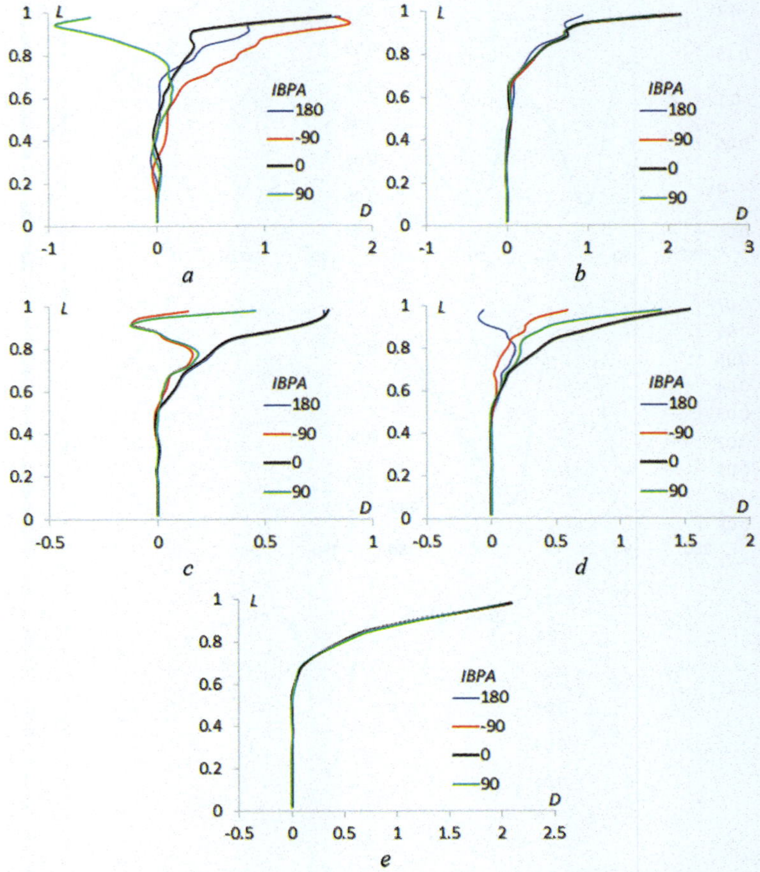

Fig. 3. Change of the aerodamping coefficient by blade height for different IBPA, ideal gas: *a* – 1st mode; *b* – 2nd mode; *c* – 3rd mode; *d* – 4th mode; *e* – 5th mode.

Figure 5 shows the distribution of the dimensionless aerodamping coefficient *D* by blade height for IBPA = 0°, 180°, and ±90° for a viscous gas.

As with the results for an ideal gas, the main work of aerodynamic forces is observed in the peripheral region of the blade, except for the fifth mode, where the level is approximately the same in the middle and on the periphery. The distribution of the coefficient *D* along the blade height is approximately the same as for an ideal gas only for the first two natural modes, while for the other modes, for a viscous gas, regions with both positive and negative aerodamping are observed.

Comparison of the aerodamping values for the first natural mode in the calculation with an ideal (Fig. 3a) and viscous (Fig. 5a) gas reveals a difference in the modeling results. The calculation using the viscous gas model demonstrates a much larger range with a higher probability of the possible occurrence of a flutter for the first natural mode of the blade oscillations for a given steam turbine operation mode.

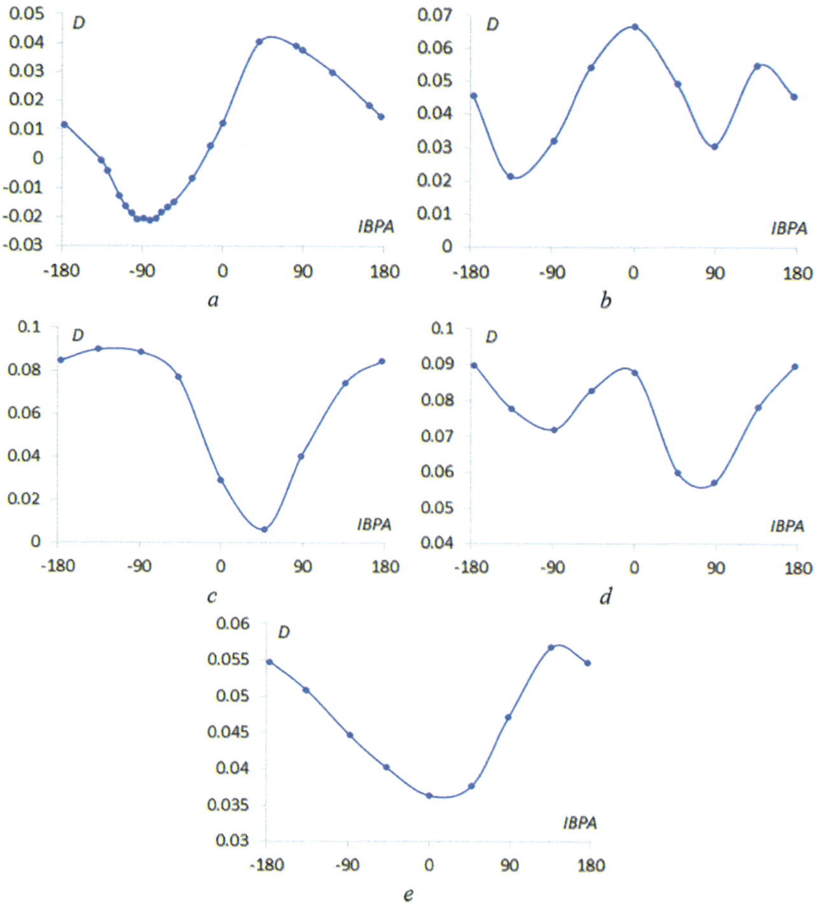

Fig. 4. Dependence of the aerodamping coefficient averaged over the blade height on the IBPA, viscous gas: *a* – 1st mode; *b* – 2nd mode; *c* – 3rd mode; *d* – 4th mode; *e* – 5th mode.

Other authors carried out a numerical study of the aeroelastic characteristics of the same turbine rotor, using their own code implementing the 3d RANS TBR model and the k-ε turbulence model built on the basis of the ANSYS CFX package [26]. As a result of the modeling, the dependence of the aerodamping coefficient on the IBPA is obtained, for each of the first 6 natural forms. The oscillations on the 2nd-6th forms were damped for all inter-blade phase angles. Negative damping values were observed on the first natural form, but the first form was damped for fully coupled oscillations. A comparative analysis of the results of studies by the proposed method and the work [26] demonstrates similarity in the distribution nature of the aerodamping coefficient for all natural forms, including negative values for the first form and IBPA = –90°. The proposed method offers almost twice as much aerodynamic work on excitation of vibrations for the first natural form, for other forms the difference is minimal. As a result of the comparative analysis the conclusion is made about qualitative and in most cases quantitative correspondence with the results obtained by other authors.

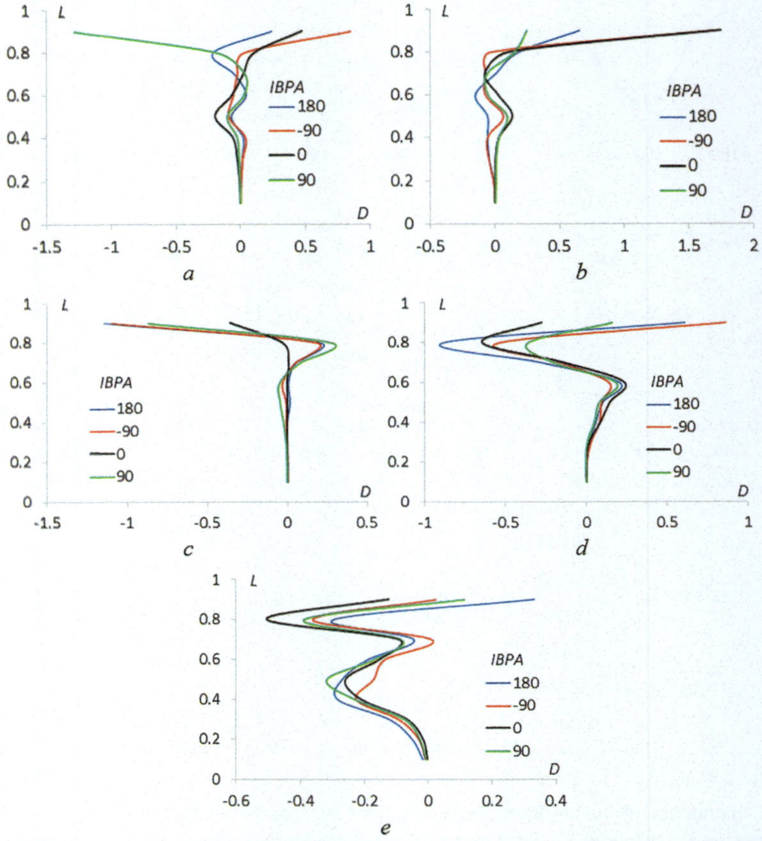

Fig. 5. Change of the aerodamping coefficient by blade height for different IBPA, viscous gas: *a* – 1st mode; *b* – 2nd mode; *c* – 3rd mode; *d* – 4th mode; *e* – 5th mode.

5 Conclusions

During the operation of a powerful 370 MW turbine, damage was detected in the area of the root of the last stage rotor blade in some operating modes. To solve the possible conditions for the occurrence of flutter, a numerical study of the aeroelastic state of the rotor was carried out using a mathematical model and a numerical method for solving the coupled problem of unsteady gas dynamics and elastic oscillations of blades in a three-dimensional flow of ideal gas (Euler's equation) and viscous gas (Navier-Stokes equation) of the turbine at given harmonic oscillations. As a result, the dependences of the aerodynamic forces on the excitation of blade oscillations for different interblade phase angles were obtained. The analysis of the results of ideal gas calculations confirmed the emergence of unstable modes of the turbine blade tip at IBPA = -45° and 90° for the 1st natural mode. The analysis of the results of the viscous gas calculation showed the occurrence of possible unstable modes of turbine blade vibration in a much larger range of IBPA = $-135^\circ \ldots -34^\circ$ for the 1st natural mode. As a result, the possible occurrence of self-excited oscillations (flutter) for a given turbine operation mode on the first natural mode of blade oscillations was confirmed. The next step in the correct accounting of the energy exchange between the flow and the blades is to solve the coupled problem of aerodynamics and elastic oscillations.

References

1. Sajedin, A., Allport, J., Amoozgar, M., Marandi, O.F.: Blade thickness effects on viscous flutter in a radial turbocharger turbine. Eng. Fail. Anal. **136**, 106139 (2022). https://doi.org/10.1016/j.engfailanal.2022.106139
2. Duan, X., Cong, J., Dai, Z., Cheng, J., Zhang, X., Jing, J.: Influence mechanism and quantificational evaluation of key factors affecting flutter stability of a transonic fan. Aerosp. Sci. Technol. **138**, 108312 (2023). https://doi.org/10.1016/j.ast.2023.108312
3. Terry, B., Wiebe, R., Motley, M.: Interaction between instabilities in vertical-axis turbine blades. J. Eng. Mech. **149**(6), 04023028 (2023). https://doi.org/10.1061/JENMDT.EMENG-6993
4. Win Naung, S., Nakhchi, M.E., Rahmati, M.: Prediction of flutter effects on transient flow structure and aeroelasticity of low-pressure turbine cascade using direct numerical simulations. Aerosp. Sci. Technol. **119**, 107151 (2021). https://doi.org/10.1016/j.ast.2021.107151
5. Sun, T., Petrie-Repar, P., Vogt, D.M., Hou, A.: Detached-eddy simulation applied to aeroelastic stability analysis in a last-stage steam turbine blade. ASME. J. Turbomach. **141**(9), 091002 (2019). https://doi.org/10.1115/1.4043407
6. Sabale, A.K., Gopal, N.K.V.: Nonlinear aeroelastic analysis of large wind turbines under turbulent wind conditions. AIAA J. **57**(10), 4416–4432 (2019). https://doi.org/10.2514/1.J057404
7. Stapelfeldt, S., Vahdati, M.: Improving the flutter margin of an unstable fan blade. ASME J. Turbomach. **141**(7), 071006 (2019). https://doi.org/10.1115/1.4042645
8. Liu, Y., Chen, Y., Cong, P., Yang, L., Yang, M., Mo, D.: Numerical simulation and experimental validation of the influences of stator angle on fan blade flutter. J. Eng. Thermophys. **43**(6), 1519–1527 (2022)
9. Li, Z., Wang, Y., Qiao, B., Wen, B., Chen, X.: Experimental investigation of aeroelastic instabilities in an aeroengine fan: using acoustic measurements. Aerosp. Sci. Technol. **130**, 107927 (2022). https://doi.org/10.1016/j.ast.2022.107927
10. Huang, H., Yang, M., Wang, D.: Efficient acoustic fan flutter analysis using the influence coefficient method and a sliding plane. J. Aerosp. Eng. **36**(1), 06022005 (2023). https://doi.org/10.1061/(ASCE)AS.1943-5525.0001513
11. Prasad, C.S., Kolman, R., Pešek, L.: Meshfree reduced order model for turbomachinery blade flutter analysis. Int. J. Mech. Sci. **222**, 107222 (2022). https://doi.org/10.1016/j.ijmecsci.2022.107222
12. Ojha, V., Fidkowski, K.J., Cesnik, C.E.S.: Adaptive high-order fluid-structure interaction simulations with reduced mesh-motion errors. AIAA J. **59**(6), 2084–2101 (2021). https://doi.org/10.2514/1.J059730

13. Corral, R., Greco, M., Vega, A.: Tip-shroud labyrinth seal effect on the flutter stability of turbine rotor blades. ASME J. Turbomach. **141**(10), 101006 (2019). https://doi.org/10.1115/1.4043962
14. Huang, H., Liu, W., Petrie-Repar, P., Wang, D.: An efficient aeroelastic eigenvalue method for analyzing coupled-mode flutter in turbomachinery. ASME J. Turbomach. **143**(2), 021010 (2021). https://doi.org/10.1115/1.4048294
15. Rzadkowski, R., Gnesin, V., Kolodyazhnaya, L., Szczepanik, R.: Unsteady rotor blade forces of 3D transonic flow through steam turbine last stage and exhaust hood with vibrating blades. In: Asset Intelligence through Integration and Interoperability and Contemporary Vibration Engineering Technologies, LNME, pp. 523–531. Springer, Cham (2019). https://doi.org/10.1007/978-3-319-95711-1_52
16. Sváček, P.: On numerical simulation of fluid—structure interaction problems using variational multiscale methods. J. Comput. Appl. Math. **427**, 115125 (2023). https://doi.org/10.1016/j.cam.2023.115125
17. Phan, H.M., He, L.: Efficient modeling of mistuned blade aeroelasticity using fully-coupled two-scale method. J. Fluids Struct. **115**, 103777 (2022). https://doi.org/10.1016/j.jfluidstructs.2022.103777
18. ANSYS Fluent Theory Guide. ANSYS, Inc., Canonsburg, PA (2019)
19. Bykov, Y., Khoryev, O., Korotaiev, P., Dedkov, V., Agibalov, Y.: Numerical Investigation of Unsteady Flow in Draft Tube with Ribs. In: 2022 IEEE 3rd KhPI Week on Advanced Technology, KhPI Week 2022—Conference Proceedings, pp. 589–594, IEEE (2022). https://doi.org/10.1109/KhPIWeek57572.2022.9916461
20. Gnesin, V., Bykov, Y., Kolodyazhnaya, L., Donchenko, V., Petrov, O.: Modeling of aeroelastic instability of fan blades. In: Advances in Mechanical and Power Engineering, CAMPE 2021, LNME, pp. 14–23. Springer, Cham (2023). https://doi.org/10.1007/978-3-031-18487-1_2
21. Rzadkowski, R., Troka, P., Manerowski, J., Kubitz, L., Kowalski, M.: Nonsynchronous rotor blade vibrations in last stage of 380 MW LP steam turbine at various condenser pressures. Applied Sciences (Switzerland) **12**(10), 4884 (2022). https://doi.org/10.3390/app12104884
22. Rusanov, A., Subotin, V., Shvetsov, V., Rusanov, R., Palkov, S., Palkov, I., Chuhai, M.: Development of the flow part of reactive type HPC of K-325-23.5 series steam turbine based on the use of modern computer technologies. J. Mech. Eng. **24**(4), 6–16 (2021). https://doi.org/10.15407/pmach2021.04.006
23. Rusanov, A., Shvetsov, V., Alyokhina, S., Pashchenko, N., Rusanov, R., Ishchenko, M., Slaston, L., Sherfedinov, R.: The efficiency increase of the steam turbine low pressure cylinder last stage by the blades spatial profiling. J. Mech. Eng. **23**(1), 6–14 (2020). https://doi.org/10.15407/pmach2020.01.006
24. Lampart, P., Gardzilewicz, A., Rusanov, A., Yershov, S.: The effect of stator blade compound lean and compound twist on flow characteristics of a turbine stage—numerical study based on 3D NS simulations. Am. Soc. Mech. Eng. Press. Vessels Pip. Div. PVP **397**(II), 195–204 (1999)
25. Rusanov, A., Chugay, M., Rusanov, R.: Advanced computer technologies in the new flow part development for reactive type HPC steam turbine of T-100 series. In: Advances in Mechanical and Power Engineering, CAMPE 2021, LNME, pp. 55–63. Springer, Cham (2023). https://doi.org/10.1007/978-3-031-18487-1_6
26. Rzadkowski, R., Kubitz, L., Gnesin, V., Kolodyazhnaya, L.: Flutter of long blades in a steam turbine. J. Vib. Eng. Technol. **6**(4), 289–296 (2018). https://doi.org/10.1007/s42417-018-0040-5

Experimental Studies of Diffuser Fins Effect in Pump-Turbine Draft Tube on Pressure Pulsations in Turbine Mode

Andrii Rusanov, Oleg Khoryev, Yevgen Ahibalov, Pavlo Korotaiev, and Yuriy Bykov(✉)

Anatolii Pidgorny Institute of Mechanical Engineering Problems of the National Academy of Sciences of Ukraine, 2/10 Pozharskogo St., Kharkiv 61046, Ukraine
yubykoff@gmail.com

Abstract. One of the methods of reducing the level of pressure pulsations in the flow part of a Francis pump-turbine for a heads up to 200 m by installing straightening fins in the suction pipe diffuser is experimentally investigated. The model of the Dniester PSP hydraulic machine with two types of runners was chosen as the object of study: RK5217, which is installed on units 1–4 of the plant, and its modification RK5217M2. The tests were conducted on the hydrodynamic test stand for a model with runner diameter of $D_1 = 350$ mm. The stand has the status of a national heritage, meets the requirements of IEC 60193 for model studies of hydraulic machines and has an efficiency measurement error of ±0.25%. The runner blades and fins were made using 3D printing technology from PLA plastic, which significantly reduced the cost and time of the work. The study was conducted both without fins and with two versions of fins of different lengths installed in the diffuser. The pressure sensor is located in the diffuser behind the fins at a distance of $1.5D_1$ from the outlet of the runner. A comparison of the pulsation characteristics of the models with fins 170 and 300 mm long and without them at reduced rotational speeds corresponding to the minimum, nominal, and maximum heads at the Dniester PSP is presented. The pulsation characteristics of models with different runners are compared. It is shown that the length of the fins practically does not affect the pulsation characteristics in the draft tube diffuser.

Keywords: Pump-Turbine · Runner · Draft Tube · Hydrodynamic Test Stand · Experimental Studies · Pressure Fluctuations

1 Introduction

At the end of 2022, global renewable energy capacity reached 3,372 GW, increasing by a record 295 GW, or 9.6%, over the year [1].

Almost half of all new capacity in 2022 was added in Asia. The largest contribution came from China, which added 141 GW of new capacity. Renewables in Europe and North America increased by 57.3 GW and 29.1 GW, respectively. South America increased its capacity by 18.2 GW, Africa by 2.7 GW, Oceania by 5.2 GW, and the Middle East had the highest ever increase of 3.2 GW.

H. Altenbach et al. (Eds.): CAMPE 2023, pp. 23–32, LNME 2025.
https://doi.org/10.1007/978-3-031-82979-6_3

Although hydropower accounts for the largest share of global renewable energy capacity – 1,250 GW, solar and wind power accounted for 90% of all new renewable energy capacity in 2022. In hydropower, the installed capacity in the world increased by 21 GW (+2%) over the year, which is in line with recent years.

Due to the outstripping development of wind and solar energy, the issue of regulating imbalances in the daily electricity consumption schedule is becoming extremely important. One of the main and most promising ways to accumulate energy and create a reliable flexible regulatory reserve is to build new pumped storage plants [2].

The integration of renewable energy sources into the energy sector requires hydraulic turbines to be more maneuverable, which leads to a drop in efficiency, cavitation, hydroacoustic resonance, and flow instability, which in turn lead to pressure fluctuations, vibration, noise, and premature component wear [3]. In modern conditions, hydroturbines often have to operate in suboptimal modes [4]. It is reported that some existing turbines experience significant vibrations when operating even at 70% of their rated capacity [5]. Depending on the mode of operation, various phenomena can occur in Francis-type turbines, including the formation of an interblade vortex and a precessing vortex in the draft tube at partial load, and an oscillating cavitation vortex at full load [6].

At partial loads, a rotating vortex rope can form, which affects the turbine efficiency [7]. This vortex rope has a precession frequency of 0.2 to 0.4 of the runner speed and causes significant pressure pulsations and vibrations [8]. Thus, reducing the vortex rope is one of the main goals to raise the efficiency of a hydro turbine, increase its service life, and improve maneuverability [9].

Another important form of unsteadiness is the upper partial load instability, which can occur when operating at 70 to 85% of the rated power and is associated with the transformation of the vortex rope cross-section from a circle to an ellipse. This instability causes strong vibration and affects the efficiency and service life of the machine [10].

There are several methods to reduce the vortex rope and its consequences [11]. The method of active flow control is the injection of water or air into the suction pipe. This method can be either active or passive, depending on the method of air injection [12], but it reduces turbine efficiency by 0–3%, and an increase in air injection can lead to cavitation problems, which affects turbine erosion.

The grooves on the runner cone have shown some improvements in the flow characteristics in the draft tube at partial load [13]. The fins and baffles in the draft tube restrain and regulate the flow, reducing the amplitude of the rope [14], but they do not completely reduce its intensity. The use of fins has also been reported to improve the pressure recovery coefficient in the draft tube [11]. New methods are emerging, for example, an adjustable system of guide vanes in the draft tube for a high-head model of a Francis turbine [15]. One of the methods to reduce the amplitude of the vortex rope is to modify the shape of the runner blades, for example, the trailing edge [16]. This approach makes it possible to significantly reduce pressure fluctuations with a small increase in efficiency at partial modes. The geometric characteristics of the draft tube also play a significant role in the formation of the vortex. Studies [17] demonstrate that the inclination of the cone axis of the diffuser can be essential for reducing the intensity of the vortex rope.

Obtaining reliable measurements of pressure fluctuations requires experimental studies on a full-scale turbine under real operating conditions [5], but such studies are limited by the capabilities of measuring equipment and design features. Therefore, research on turbine models is widely used [7, 15, 18], which also allows for velocity measurements and flow visualization [10]. To improve numerical methods, it is useful to conduct simultaneous experimental and numerical studies [7].

2 Object of Study

The object of study is the model of the ORO5217 pump-turbine of the Dniester PSP with the RK5217 runner, which is installed on 1–4 hydraulic units, and with the modified RK5217M2. The RK5217 blades have a traditional design: the trailing edge is located in the radial plane, the leading edge is vertical in the meridional projection, and the leading edge is straight in the frontal projection, inclined towards the rotor rotation by 10°. The RK5217M2 is a modification of the RK5217, with blades designed using circumferential lean. With this method, the blade profiles remain unchanged, only their relative position and the shape of the leading and trailing edges change. The lean obtained by shifting the hub profile in the direction of rotor rotation is assumed to be positive, and the opposite is assumed to be negative. This approach allows us to optimize the blade shape and improve the flow structure in the runner channel and at the inlet to the draft tube. The modified RK5217M2 has a negative circumferential lean φ = -45°. Its hub profile is shifted by 18.2° relative to the original RK5217 in the direction opposite to the rotation, while the other profiles are shifted by angles that depend linearly on the height of the runner at the leading edge (Fig. 1).

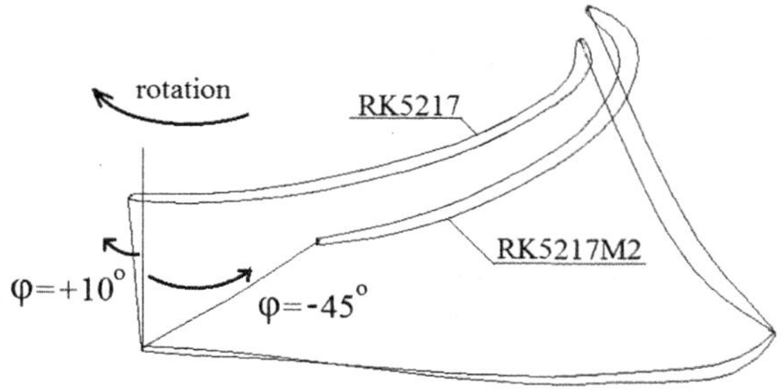

Fig. 1. Runner blades RK5217 and RK5217M2.

3 Description and Characteristics of the Hydrodynamic Test Stand

Experimental studies of pump-turbine models with runner diameter D_1 = 350 mm were carried out on the IMEP ECS-30 hydrodynamic test stand. The test stand has no analogues in Ukraine; in 2011, it received the status of a national heritage and meets all the requirements of the international standard IEC 60193 for model acceptance tests of hydraulic machines of various types [19]. The main parameters of the experimental equipment of the experimental stand ECS-30 are: diameter of the runner of the model 300–400 mm; head up to 25 m in turbine, up to 30 m in pump modes; flow rate up to 0.300 m^3/s, supply up to 0.500 m^3/s; power of the DC drive motors of circulating pumps up to 160 kW; power of the balancing motor-generator 200 kW. The error in determining the efficiency does not exceed ±0.25%.

The flow part of the pump-turbine models consists of a spiral casing, a stator, a guide vane, a runner and a draft tube. The spiral casing has circular cross-sections, 360° coverage angle; the stator has 20 columns; the guide vane has 20 blades of positive curvature with a height of $0.14D_1$, the diameter of the blade rotation axis is $1.2D_1$; runners with 7 blades; draft tube with KU-3RO elbow with a constant diameter outlet. The runner blades were made of PLA plastic by 3D printing, which significantly reduced the cost and time of the research.

The tests were conducted in turbine mode at a head of $H = 6$ m, with the opening of the guide vane from 10 to 42 mm in 2 mm increments. Pressure fluctuations in the draft tube diffuser were studied for three variants: without fins and with short and long fins. The profiles of both variants of the fins have the same shape of an equilateral right triangle, whose catenary is 25 mm and is located in the radial plane. The vertical length of the long fins is 300 mm, and the short fins are 170 mm. The number of fins is 4. To reduce hydraulic losses, the top and bottom of the fins are chamfered at angle of 45° (Fig. 2). The fins are also made of plastic by 3D printing.

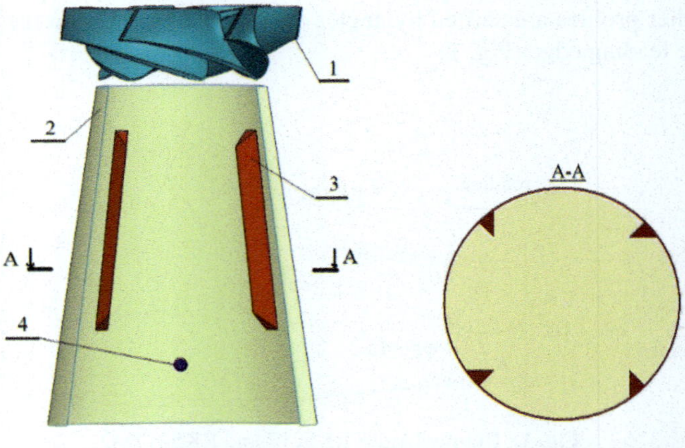

1 – runner; 2 – diffuser; 3 – 300 mm long fin; 4 – pressure sensor DMK331

Fig. 2. Placement of fins and sensor in the draft tube diffuser.

4 Results of Experimental Studies

The energy characteristics of the models were obtained and analyzed. Table 1 shows the parameters of the most efficient modes of all the considered variants. The relative value of the efficiency η^* in this work is the ratio of its current value to the maximum value of the efficiency of the original version of the RK5217, in percent.

Table 1. Parameters of optimal turbine modes for models with and without fins installed in the diffuser for two runner options.

Runner option	RK5217			RK5217M2		
Parameter	Q_I', m^3/s	n_I', min^{-1}	η^*, %	Q_I', m^3/s	n_I', min^{-1}	η^*, %
No fins	0.312	78.0	100.00	0.313	79.0	100.74
170 mm fins	0.324	77.0	99.55	0.311	77.5	100.66
300 mm fins	0.317	77.0	99.45	0.317	79.0	100.34

The installation of fins of both lengths had little effect on the values of the optimal reduced flow rates Q_I' and rotational speeds n_I'. The presence of fins and an increase in their length is expected to lead to a decrease in the maximum efficiency, and for RK5217M2 it is somewhat less than for the original version.

Previously, numerical studies of pressure fluctuations in the suction pipe in the absence of fins and for fins with a length of 170 and 300 mm were conducted [20]. The results of the studies revealed a 13% reduction in the overall level of pulsations when using 300 mm fins compared to a tube without fins at 60% load. Numerical studies were performed using a software package for modeling unsteady flows in the flow parts of turbomachinery of various types [21, 22].

Pulsation studies were performed to determine the amplitudes and frequencies of pressure pulsations in the entire range of operating modes of the Dniester PSP. They were conducted, as well as energy studies, at a head of $H = 6$ m.

To measure the unsteady pressure, the sensors of the DMK331 type manufactured by BD SENSORS were used, the membranes of which were installed flush with the surface of the flow part. Pulsations were measured at the point of the draft tube diffuser located at a distance of $1.5D_1$ below the runner (Fig. 2).

The signal from the sensors was processed using the "Spider 8" dynamic process measurement system from HBM, followed by processing with the "Catman" program and presentation of the results in the form of an amplitude-frequency spectrum.

Graduation dependencies between the control pressure and instrument readings were determined by static calibration using a load-piston manometer of the MP-6 type. The graduation dependence was entered into the program to determine the value of the amplitude of pressure pulsations.

The measurement results are presented in the form of spectra, where the dimensionless amplitude is the standard deviation of the pressure from the mean value *rms* divided by the head H in percent, and the frequency is normalized to the runner rotation frequency.

The study of pressure fluctuations was carried out in turbine mode at reduced rotation speeds $n'_I = 80, 85, 91, 95$ min^{-1}, which correspond to the optimal, maximum, rated, and minimum heads of the Dniester PSP, in the range of reduced flow rates Q'_I from 0.100 to 0.450 m^3/s at guide vane openings $a_0 = 8; 10; 12; 16; 20; 24; 28; 34$ mm. Based on the results of the study, the pulsation characteristics are shown below.

Figure 3 shows the dependence of the dimensionless total pressure pulsations *rms/H* in the draft tube diffuser on the flow rate for four rotation speeds of the RK5217 runner $n'_I = 80, 85, 91$, and 95 min^{-1} with and without fins.

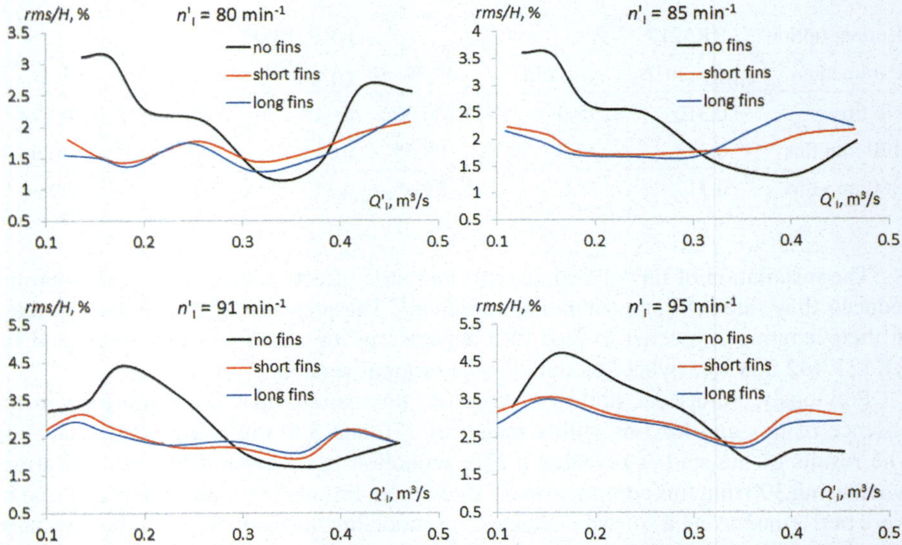

Fig. 3. Pressure pulsation amplitudes in the diffuser with different fins and rotation frequencies n'_I for the runner RK5217.

From the graphs comparing the amplitudes of pulsations with different fins installed in the diffuser of the draft tube for the flow part with RK5217 (Fig. 3), it is clear that at flow rates less than $Q_I' = 0.300$ m^3/s, the installation of fins reduces the amplitude of pressure pulsations by up to 52% compared to the tube without fins, for flow rates greater than 0.300 m^3/s in the tube with fins, the pulsations are higher, but their amplitude remains within the allowed limits of the Dniester PSP operation. There is no significant difference in the pulsation amplitudes for long and short fins.

Figure 4 shows similar to Fig. 3 graphs of pressure pulsation amplitude dependence for the runner RK5217M2.

Similar results were obtained for the flow part with RK5217M2 (Fig. 4). In the partial load mode (less than $Q_I' = 0.260$ m^3/s), the amplitude of pressure pulsations with the fins installed is reduced by up to 54% compared to the diffuser without fins. For flow rates close to optimal and higher, the installation of fins increases the amplitude of pulsations, but leaves it within the allowed operating limits, except for the rotation frequency $n'_I = 80$ min^{-1}, where for a flow rate of $Q_I' = 0.360$ m^3/s the amplitude increase is 77% for

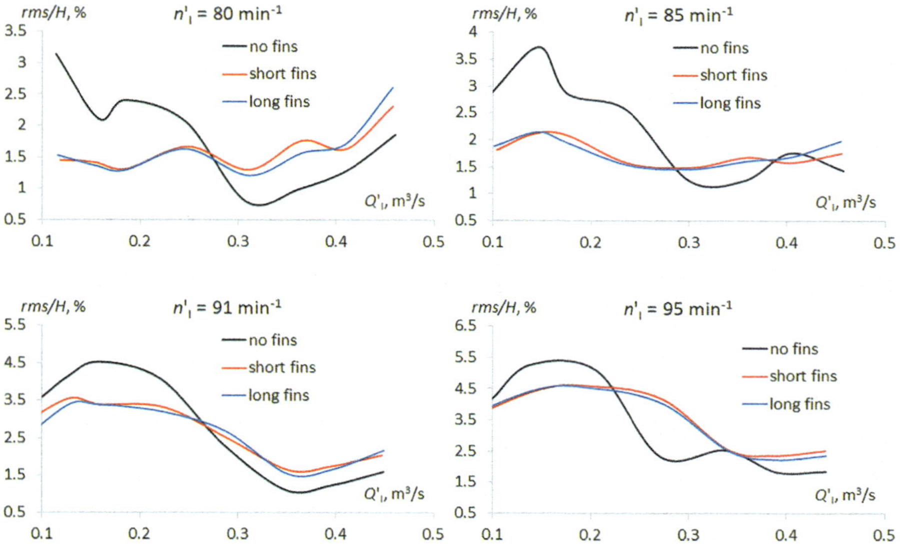

Fig. 4. Amplitudes of pressure pulsations in the diffuser with different fins and rotation frequencies n'_{I} for the runner RK5217M2.

long fins. The spectra of pressure fluctuations in the diffuser with the runner RK5217 for the partial load ($Q_{\mathrm{I}}' = 0.237\ \mathrm{m^3/s}$) and close to the rated flow ($Q_{\mathrm{I}}' = 0.408\ \mathrm{m^3/s}$) and the rotation frequency $n'_{\mathrm{I}} = 91\ \mathrm{min}^{-1}$ are shown in Fig. 5.

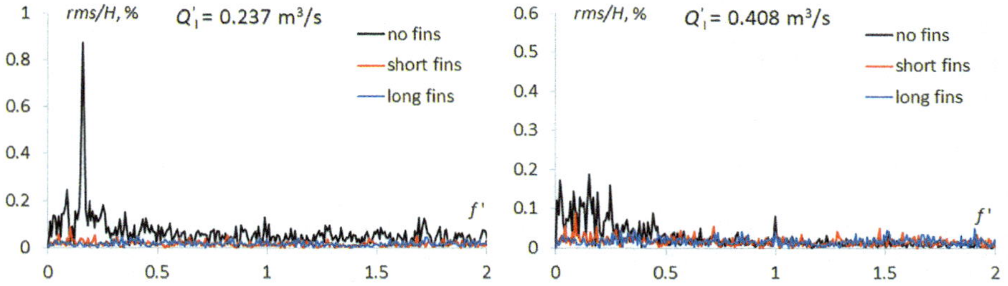

Fig. 5. Spectra of pressure pulsations in the diffuser for the runner RK5217 ($n'_{\mathrm{I}} = 91\ \mathrm{min}^{-1}$).

When examining the pressure pulsation spectra for the flow part of the RK5217, it is clear that the installation of fins significantly reduces the amplitude of pulsations at frequencies below the rotational (Fig. 5). Pulsations at a frequency of 0.2 of the rotational, which is responsible for the precession of the vortex rope, almost disappear when the fins are installed. At other frequencies, a decrease in amplitude is also observed. Since the amplitude of pulsations at frequencies below the rotation frequency decreases with increased turbine flow rates (Fig. 5, $Q_{\mathrm{I}}' = 0.408\ \mathrm{m^3/s}$), the installation of fins in this mode does not have a significant effect on reducing pulsations. Figure 6 shows similar spectra of pressure fluctuations in the diffuser with the runner RK5217M2.

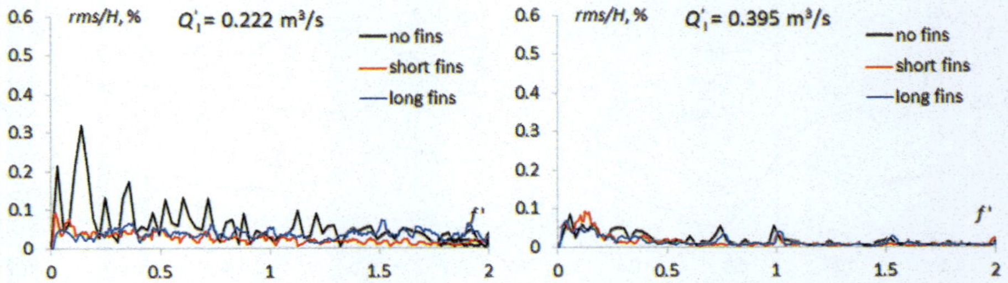

Fig. 6. Spectra of pressure pulsations in the diffuser for the runner RK5217M2 ($n'_{\mathrm{I}} = 91\ \mathrm{min}^{-1}$).

When operating with the runner RK5217M2 in partial load mode $Q_{\mathrm{I}}' = 0.222\ \mathrm{m}^3/\mathrm{s}$, the amplitude of pulsations at the vortex rope precession frequency (0.2 rotation frequency) is significantly lower than with RK5217 (Fig. 5), while when using fins, these pulsations nearly disappear. Also, when the fins are installed, the pulsation amplitudes decrease in almost the entire frequency range; the difference in amplitudes for long and short fins is insignificant.

A similar method of reducing pulsations in the draft tube was investigated in [14]. In this paper, numerical and experimental studies of the effect of the number of vertical baffles of rectangular cross section installed in the draft tube cone were carried out. From the simulation results, it was found that in the 66% rated flow rate mode, the installation of four baffles reduced the pulsations by up to 0.44 times compared to a cone without baffles. The reduction in pulsation was due to the decrease in vortex rope intensity, at a frequency of 0.3 of the runner speed. This result is in good agreement with the obtained results, where the pulsation amplitude for the mode 66% of the rated flow rate decreases from 0.8 to 0.5 times for different runner speeds, with the amplitude decrease observed for a frequency that also corresponds to the precession of the vortex rope, as in the paper [14]. The slight difference in the results may be due to the different relative size and shape of the baffle cross-section.

5 Conclusion

Experimental studies of the effect of fins in the diffuser of the draft tube of a Francis turbine model on the head of 200 m were carried out. Fins of different lengths and runners with different circumferential lean were used in the study. It was determined that the installation of fins in the draft tube diffuser can significantly reduce the amplitude of low-frequency pressure pulsations (frequency less than the rotation frequency) at partial load turbine operating modes. Increasing the length of the fins does not have a significant impact on the level of pulsations over the entire turbine operating range. The circumferential lean of the runner blades affects the maximum level of pulsations; no significant effect was observed on the ability of the fins to reduce the pulsation amplitude. The installation of the fins in the diffuser has a slight effect on the values of the optimal reduced flow rates and rotational speeds, but leads to a decrease in the optimal efficiency.

References

1. IRENA Renewable capacity statistics 2023, https://www.irena.org/Publications/2023/Mar/Renewable-capacity-statistics-2023, last accessed 2023/07/11
2. Hydropower capacity should more than double by 2050 to meet climate goals, Irena says. The National News, https://www.thenationalnews.com/business/energy/2023/02/14/hydropower-capacity-should-more-than-double-by-2050-to-meet-climate-goals-irena-says/, last accessed 2023/07/11
3. Pereira, J.G., Vagnoni, E., Favrel, A., Landry, C., Alligné, S., Nicolet, C., Avellan, F.: Prediction of unstable full load conditions in a Francis turbine prototype. Mech. Syst. Signal Process. **169** 108666 (2022). https://doi.org/10.1016/j.ymssp.2021.108666
4. Kumar, S., Cervantes, M.J., Gandhi, B.K.: Rotating vortex rope formation and mitigation in draft tube of hydro turbines—a review from experimental perspective. Renew. Sustain. Energy Rev. **136**, 110354 (2021). https://doi.org/10.1016/j.rser.2020.110354
5. Jamali, R., Sohani, A., Hemmatpour, K., Behrang, M., Ghobeity, A.: Experimental study of pressure pulsation in a large-scale hydropower plant with Francis turbine units and a common penstock. Energy Conversion and Management **X**(16), 100308 (2022). https://doi.org/10.1016/j.ecmx.2022.100308
6. Zhao, W., Presas, A., Egusquiza, M., Valentín, D., Egusquiza, E., Valero, C.: Increasing the operating range and energy production in Francis turbines by an early detection of the overload instability. Measurement **181**, 109580 (2021). https://doi.org/10.1016/j.measurement.2021.109580
7. Lai, X., et al.: Experimental and numerical investigation of vortex flows and pressure fluctuations in a high-head pump-turbine. Renew. Energy **211**, 236–247 (2023). https://doi.org/10.1016/j.renene.2023.04.092
8. Pasche, S., Gallaire, F., Avellan, F.: Origin of the synchronous pressure fluctuations in the draft tube of Francis turbines operating at part load conditions. J. Fluids Struct. **86**, 13–33 (2019). https://doi.org/10.1016/j.jfluidstructs.2019.01.013
9. Shahzer, M.A., Cho, Y., Shamsuddeen, M.M., Kim, J.-H.: Investigation of cavitating vortex rope instabilities and its suppression inside a Francis turbine model with Thoma number variation. Phys. Fluids **35**(3), 033310 (2023). https://doi.org/10.1063/5.0140973
10. Amini, A., Vagnoni, E., Favrel, A., Yamaishi, K., Müller, A., Avellan, F.: Upper part-load instability in a reduced-scale Francis turbine: an experimental study. Exp. Fluids **64**(6), 110 (2023). https://doi.org/10.1007/s00348-023-03649-0
11. Wang, L., et al.: Research on the vortex rope control techniques in draft tube of Francis turbines. Energies **15**(24), 9280 (2022). https://doi.org/10.3390/en15249280
12. Bucur, D.M., Dunca, G., Bunea, F., Ciocan, G.D.: Experimental analysis of the operation of a small Francis turbine equipped with an innovative aeration device. IOP Conf. Ser.: Earth Environ. Sci. **240**(4), 042010 (2019). https://doi.org/10.1088/1755-1315/240/4/042010
13. Chen, Z., Baek, S.-H., Cho, H., Choi, Y.-D.: Optimal design of J-groove shape on the suppression of unsteady flow in the Francis turbine draft tube. J. Mech. Sci. Technol. **33**(5), 2211–2218 (2019). https://doi.org/10.1007/s12206-019-0423-x
14. Zhou, X., Wu, H., Shi, C.: Numerical and experimental investigation of the effect of baffles on flow instabilities in a Francis turbine draft tube under partial load conditions. Adv. Mech. Eng. **11**(1), 1687814018824468 (2019). https://doi.org/10.1177/1687814018824468
15. Joy, J., Raisee, M., Cervantes, M.J.: Experimental investigation of an adjustable guide vane system in a Francis turbine draft tube at part load operation. Renew. Energy **210**, 737–750 (2023). https://doi.org/10.1016/j.renene.2023.04.096
16. Zhang, F., Xiao, R., Zhu, D., Liu, W., Tao, R.: Pressure pulsation reduction in the draft tube of pump turbine in turbine mode based on optimization design of runner blade trailing edge profile. J. Energy Storage **59**, 106541 (2023). https://doi.org/10.1016/j.est.2022.106541

17. Zhou, X., Wu, H., Cheng, L., Huang, Q., Shi, C.: A new draft tube shape optimisation methodology of introducing inclined conical diffuser in hydraulic turbine. Energy **265**, 126374 (2023). https://doi.org/10.1016/j.energy.2022.126374
18. Jhankal, N.K., Kumar, A., Mangla, M.: Establishment of correlation for the pressure fluctuations on the low-head Francis turbine in the draft tube cone from the model test. Flow Meas. Instrum. **91**, 102360 (2023). https://doi.org/10.1016/j.flowmeasinst.2023.102360
19. Khoryev, O., Korotaiev, P., Agibalov, Y., Bykov, Y., Maksymenko-Sheiko, K.: Experimental studies of pump-turbine flow part models at heads of 80–120 m. In Advances in Mechanical and Power Engineering, CAMPE 2021, Lecture Notes in Mechanical Engineering, pp. 24–33. Springer, Cham (2023). https://doi.org/10.1007/978-3-031-18487-1_3
20. Bykov, Y., Khoryev, O., Korotaiev, P., Dedkov V., Agibalov Y.: Numerical investigation of unsteady flow in draft tube with ribs. In: 2022 IEEE 3rd KhPI Week on Advanced Technology, KhPIWeek, Kharkiv, Ukraine, 2022, pp. 1–6. (2022). https://doi.org/10.1109/KhPIWeek57572.2022.9916461
21. Rusanov, A., et al.: Development of the flow part of reactive type HPC of K-325-23.5 series steam turbine based on the use of modern computer technologies. J. Mech. Eng. **24**(4), 6–16 (2021). https://doi.org/10.15407/pmach2021.04.006
22. Rusanov, A., et al.: The efficiency increase of the steam turbine low pressure cylinder last stage by the blades spatial profiling. J. Mech. Eng. **23**(1), 6–14 (2020). https://doi.org/10.15407/pmach2020.01.006

Liquid Vibrations Analysis of Baffled Reservoirs with Fuzzy Concepts Implementation

Neelam Choudhary[1], Kyryl Degtyariov[2(✉)], Vasyl Gnitko[2], Denis Kriutchenko[2], and Elena Strelnikova[2,3]

[1] Bennet University, Plot No. 9-11, Techzone II, Greater Noida, Uttar Pradesh 201310, India
[2] Anatolii Pidhornyi Institute of Mechanical Engineering Problems NAS of Ukraine, Kharkiv, Ukraine
Kdegt89@gmail.com
[3] V.N. Karazin, Kharkiv National University, Svobody Sq., 4, Kharkov 61022, Ukraine

Abstract. The main objective of this paper is to elaborate effective numerical methods for analyzing liquid vibrations in partially filled rigid reservoirs with baffles under coupled action of vertical and horizontal excitations. The novelty of the proposed approach consists in involving the fuzzy logic concepts for estimating uncertainties in loading parameters. It is supposed that the load frequencies and amplitudes are characterized by triangular fuzzy numbers. First, the crisp boundary value problem is formulated for receiving own modes and frequencies of the liquid vibrations in the reservoirs without baffles, with circular horizontal, and vertical baffles. After receiving the solutions of these spectral problems by boundary element methods, the forced vibrations of liquid are considered. Modes, obtained at solving the spectral problems are used onwards as basic functions in series for an unknown potential and a free surface elevation. The forced vibrations are considered in a weakly nonlinear formulation, that leads to second order differential equations relative to generalized coordinates. Using the triangular fuzzy numbers for describing uncertainties in initial data, the fuzzy initial value problem is reduced to a system of fuzzified differential equations. The numerical solution of this system has been received and analyzed.

Keywords: Sloshing · Rigid Shells of Revolution · Boundary Element Method

1 Introduction

Shells and shell structures with compartments partially filled with liquids are widely used in various industrial areas, such as aerospace, oil and gas industry, power engineering and transport. These reservoirs often contain aggressive or flammable aggregates, and operate under intensive loads. At suddenly applied load, the sloshing phenomenon is observed in tanks containing liquids, such as rocket fuel tanks, tanks in marine and space transportation, oil storage facilities, dams, as well as in reactors and nuclear vessels. The intensive liquid movement can create a high localized pressure on the tank walls, which in turn can lead to the structure failure or stability losing. Therefore, the design of such tanks

H. Altenbach et al. (Eds.): CAMPE 2023, pp. 33–43, LNME 2025.
https://doi.org/10.1007/978-3-031-82979-6_4

requires a careful preliminary study of the liquid behaviour in the containers at operation. To reduce the sloshing amplitude, baffles of various types are used. Their sizes, locations, properties of materials, can be determined quite accurate by experimental research. But such experiments are too expensive, and often can lead to the destruction of samples. So, the computer simulation of dynamic processes occurring in reservoirs with liquid is an effective and powerful technology of virtual testing in order to obtain optimal design options. Nowadays, new significant challenges are associated with modelling the liquid storage tanks that operate under uncertain aeolian or intensive suddenly applied loadings. So, involving the concepts of fuzzy mathematics for adequate design of the reservoirs under such loads is a topical and urgent problem.

2 Literature Review

A lot of numerical and analytical methods has been developed in recent decades to study both sloshing phenomenon in storage reservoirs and stress-strain states of industrial equipment. The analytical methods are mostly applied for estimating strength and dynamic characteristics of simple geometry reservoirs [1]. Advanced numerical methods based on immersion approach and concern with launch vehicle strength and stability were elaborated in [2]. The finite volumes method was further developed in [3]. New variant of coupled finite (FEM) and boundary element methods (BEM) was proposed in [4] for studying the fluid-structure interaction problems.

Due to developing modern numerical methods and advanced computer technologies, it has become possible to study the dynamic behaviour of liquids in reservoirs of complicated shapes, with baffles [5] and floating covers [6]. Different kinds of baffles were proposed as damper devices. In [7] the effect of porous baffles was simulated using BEM. The baffled shell was presented as the multiply connected domain [8], and multi-domain iso-geometric BEM was applied for frequencies estimation.

Especially topical are problems of reservoirs stability during filling process [9], when uncertain Aeolian loads [10] become predominant. The similar problems are arisen at carrying out strength analysis of structures subjected to seismic excitations [11]. Theoretical [5] and experimental [12] sloshing analysis indicated changing in the frequency spectrum due to baffles installation. This complicates frequency detuning, especially for uncertain loading parameters. So, it is topical to elaborate methods of estimating the dynamic characteristics of liquid-filled reservoirs involving fuzzy mathematics concepts [13].

3 Research Methodology

3.1 Basic Equations of Crisp Boundary Value Problem

Consider liquid vibrations in a rigid partially-filled shell of revolution subjected to coupled longitudinal-lateral excitation, Fig. 1a).

Note that $S_1 = S_1(t)$ and $S_1 = S_1(t)$ are the time-dependent shell wetted surface, and the liquid free surface, respectively.

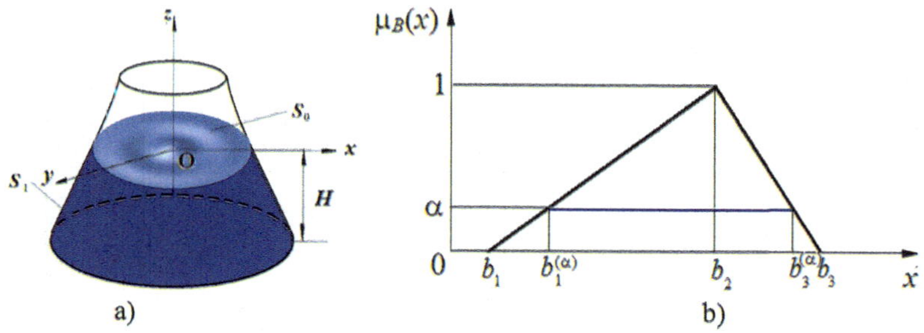

Fig. 1. Partially filled shell of revolution, a) and membership function, b).

Denote as $S_0(0) = \Sigma_0$ and $S_1(0) = \Sigma_1$. The free surface Σ_0 coincides with the plane z = 0 at the state of rest. For the shell of revolution, the surface Σ_0 is a circle of radius R. Consider the area $Q(t)$, occupied by the liquid, in cylindrical coordinate system $Q(t) = \{0 \le \theta \le 2\pi, \quad 0 \le r \le r(z), \quad -H \le z \le \zeta(r, \theta, t)\}$, where $r = r(z)$ is the shell meridian, the function $\zeta = \zeta(\theta, r, t)$ describes the time-dependent free surface elevation. Supposing the liquid is an ideal and incompressible one, neglecting the surface tension, and applying the mass conversation law, we obtain the continuity equation $\mathrm{div}\boldsymbol{V} = 0$, where $\mathbf{V}$ is the liquid velocity. Let the external force $\mathbf{F}_Q$ with an acceleration $\mathbf{a}$ is applied to the liquid-filled shell. Using the momentum conservation law, we receive the equation of liquid movement as

$$\rho\left(\frac{\partial \boldsymbol{V}}{\partial t} + (\mathbf{V} \cdot \nabla)\boldsymbol{V}\right) = -\nabla p + \boldsymbol{a}, \tag{1}$$

where ρ is the liquid density, and p is the liquid pressure. If the fluid flow is vortex-free, then there exists a scalar velocity potential $\Phi = \Phi\,(x, y, z, t)$, and the continuity equation is reduced to the Laplace equation $\nabla^2\Phi = 0$. So, it follows from Eq. (1), that

$$p - p_0 = -\rho\left[\frac{\partial \Phi}{\partial t} + (g + a_z(t))z + a_x(t)x + \frac{1}{2}(\nabla\Phi, \nabla\Phi)\right]. \tag{2}$$

Here $a_x(t)$, $a_z(t)$ are driving force accelerations in the lateral and longitudinal directions, g is the gravity acceleration, p_0 is an atmospheric pressure. The boundary value problem for the Laplace equation in the time-dependent fluid domain $Q(t)$ is formulated. The non-penetration condition is applied at the wetted surface S_1 as

$$\partial\Phi/\partial\mathbf{n}|_{S_1} = 0, \tag{3}$$

where $\mathbf{n}$ is the external unit normal to the wetted surface S_1. The kinematic and dynamic boundary conditions on the free surface S_0 are following:

$$\frac{\partial \Phi}{\partial \mathbf{n}} = \left.\frac{\partial\zeta/\partial t}{\sqrt{1 + |\nabla\zeta|^2}}\right|_{S_0}, \; \frac{\partial \Phi}{\partial t} + (g + a_z(t))z + a_x(t)x + \frac{1}{2}(\nabla\Phi, \nabla\Phi) = 0. \tag{4}$$

The solvability condition for the Neumann problem (1–4) was obtained in [14] as

$$\iiint_{Q(t)} dQ(t) = 0. \tag{5}$$

The initial data for the boundary value problem (3–5) are given in the form

$$\zeta(x, y, 0) = \zeta_0(x, y) = 0, \quad \left|\frac{\partial \Phi}{\partial \mathbf{n}}\right|_{\Sigma_0} = \varphi(x, y, 0). \tag{6}$$

So, the problem consists in determining the unknown functions Φ and ζ from the Laplace equation with boundary conditions (3–4) and initial data (6).

It was proved in [14] that the unknown functions can be presented as series

$$\begin{aligned}\Phi(r, z, \theta, t) = {} & R_1(t)\Phi_{11}(r, z)\cos\theta + R_2(t)\Phi_{11}(r, z)\sin\theta + R_3(t)\Phi_{01}(r, z) \\ & + R_4(t)\Phi_{21}(r, z)\cos 2\theta + R_5(t)\Phi_{21}(r, z)\sin 2\theta,\end{aligned} \tag{7}$$

$$\begin{aligned}\zeta(r, \theta, t) = {} & \beta_1(t)\zeta_{11}(r, z)\cos\theta + \beta_2(t)\zeta_{11}(r, z)\sin\theta + \beta_3(t)\zeta_{01}(r, z) \\ & + \beta_4(t)\zeta_{21}(r, z)\cos 2\theta + \beta_5(t)\zeta_{21}(r, z)\sin 2\theta,\end{aligned} \tag{8}$$

where $\beta_j(t)$, $R_j(t)$ are generalized coordinates. In [15], the relationship between $\beta_j(t)$ and $R_j(t)$ is obtained that leads to the following system of ordinary differential equations with respect to $\beta_j(t)$:

$$\begin{aligned}&\ddot{\beta}_k + \omega_{11}^2(g + a_z)\beta_k \\ &\quad + f_k\big(\beta_1, \beta_2, \beta_3, \beta_4, \beta_5, \dot{\beta}_1, \dot{\beta}_2, \dot{\beta}_3, \dot{\beta}_4, \dot{\beta}_5, \ddot{\beta}_1, \ddot{\beta}_2, \ddot{\beta}_3, \ddot{\beta}_4, \ddot{\beta}_5, a_x\big) = 0, k = 1, 2 \\ &\ddot{\beta}_3 + \omega_{01}^2(g + a_z)\beta_3 + f_3\big(\beta_1, \beta_2, \dot{\beta}_1, \dot{\beta}_2, \ddot{\beta}_1, \ddot{\beta}_2\big) = 0, \\ &\ddot{\beta}_k + \omega_{21}^2(g + a_z)\beta_k + f_k\big(\beta_1, \beta_2, \dot{\beta}_1, \dot{\beta}_2, \ddot{\beta}_1, \ddot{\beta}_2\big) = 0, \ k = 4, 5.\end{aligned} \tag{9}$$

System (9) is solved under zero boundary conditions.

3.2 Spectral Boundary Value Problems Solutions

The basic functions Φ_{lk} and ζ_{lk} are solutions of the spectral linear boundary value problems [5], when the boundary surfaces Σ_0, Σ_1 are time-independent. The basic functions Φ_{lk} are solutions of the following spectral problems:

$$\nabla^2\Phi_{lk}, \quad \left.\frac{\partial \Phi_{lk}}{\partial \mathbf{n}}\right\|_{\Sigma_1} = 0, \quad \frac{\partial \Phi_{lk}}{\partial \mathbf{n}} = \left.\frac{\partial \zeta_{lk}}{\partial t}\right|_{\Sigma_0}. \tag{10}$$

The next relation is fulfilled on the free surface Σ_0:

$$\zeta_{kJ}(r) = \left.\frac{\partial \varphi_{lk}(r, z)}{\partial z}\right|_{S_0} = \frac{\chi_{lk}^2}{g}\varphi_{lk}(r, 0), \tag{11}$$

where χ_{lk} are fundamental frequencies [14]. Dropping indices lk, and using Green's third identity, the following presentation for partial potentials Φ_{lk} is received:

$$2\pi\,\Phi(P_0) = \int\int_S \frac{\partial\Phi}{\partial\mathbf{n}}\frac{1}{|P-P_0|}dS - \int\int_S \Phi\frac{\partial}{\partial\mathbf{n}}\frac{1}{|P-P_0|}dS$$
$$, S = \Sigma_1 \cup \Sigma_0. \tag{12}$$

Here $P - P_0$ is the Cartesian distance between points P and P_0 of the surface S. Using (11–12) and satisfying boundary conditions (3), (4), obtain the singular integral equations system [5]

$$2\pi\varphi_1 + \int\int_{\Sigma_1} \Phi_1\frac{\partial}{\partial\mathbf{n}}\left(\frac{1}{|P-P_0|}\right)d\Sigma_1 - \frac{\chi^2}{g}$$
$$\int\int_{\Sigma_0} \Phi_0\frac{1}{|P-P_0|}d\Sigma_0$$
$$+\int\int_{\Sigma_0} \Phi_0\frac{\partial}{\partial z}\left(\frac{1}{|P-P_0|}\right)d\Sigma_0 = 0,$$
$$-\int\int_{\Sigma_1} \Phi_1\frac{\partial}{\partial\mathbf{n}}\left(\frac{1}{|P-P_0|}\right)d\Sigma_1 - 2\pi\Phi_0 + \frac{\chi^2}{g}$$
$$\int\int_{\Sigma_0} \Phi_0\frac{1}{|P-P_0|}d\Sigma_0 = 0, \tag{13}$$

where the potential values of the free surface and shell walls are denoted by Φ_0 and Φ_1. For shells of revolution, Eqs. (13) are reduced to a one-dimensional system of singular integral equations, and for its solution the effective numerical BEM-based method is applied [16]. The following systems are received for the partial potentials:

$$2\pi\Phi(z_0)\int_\Gamma \Phi(z)\Theta(z,z_0)r(z)d\Gamma - \int_0^R q(\rho)\Xi(P,P_0)\rho d\rho = 0, P_0 \in \Sigma_1, \tag{14}$$

$$\int_\Gamma \Phi(z)\Theta(z,z_0)r(z)d\Gamma - \int_0^R q(\rho)\Xi(P,P_0)\rho d\rho = 0, P_0 \in \Sigma_0, q = \partial\Phi/\partial\mathbf{n},$$
$$\Theta(z,z_0) = 4/\sqrt{a+b}\left\{\frac{1}{2r}\left[\frac{r^2-r_0^2+(z_0-z)^2}{a-b}E_l(k) - F_l(k)\right]n_r + \frac{z_0-z}{a-b}E_l(k)n_z\right\},$$
$$\Xi(P,P_0) = 4/\sqrt{a+b}\,F_l(k), a = r^2+r_0^2+(z-z_0)^2, b = 2rr_0.$$

Here the generalized elliptical integrals are introduced as

$$E_l(k) = (-1)^l\left(1-4l^2\right)\int_0^{\pi/2} \cos 2b_1 l\theta\sqrt{1-k^2\sin^2\theta}d\theta, \tag{15}$$

$$F_l(k) = (-1)^l\int_0^{\pi/2}\frac{\cos\, 2b_1 l\theta d\theta}{1-k^2\sin^2\theta}d\theta,$$
$$k^2 = 2b/(a+b). \tag{16}$$

In (14)-(16) parameter l corresponds to the first index in the partial potentials Φ_{lk}. It defined the wave number; the value of parameter b_1 characterizes the presence of vertical baffles [5]. For the unbaffled shells $b_1 = 1$, and for shells with vertical baffles $b_1 = 2$. When functions Φ_{lk} are received together with frequencies $_{lk}$, functions ζ_{lk} are calculated using Eq. (10). It allows us to compose system (9), and estimate the free surface level changing via time for both unbaffled shells and shells with vertical baffles. For shells with horizontal circular baffles, the multi-domain method is used [5].

3.3 Simulation of Liquid Vibrations in Reservoirs with Fuzzy Methods

Suppose that external force accelerations correspond to harmonic excitation as follows.

$$a_x = a_0 \cos f_0 t, \; a_z = a_1 \cos f_1 t,$$

where f_0, f_1 are the loading frequencies, and a_0, a_1 are the loading amplitudes.

The crisp boundary value problem is reduced to system (9) of differential equations with zero initial conditions, depending on such parameters: a_0, a_1, f_0, f_1. But the deterministic analysis did not provide necessary information in studying uncertainties in the frequencies and amplitudes of the load oscillations. So, the crisp parameters are fuzzified as triangular fuzzy numbers, using non-axisymmetric membership functions $\mu_B(x)$ presented in Fig. 1b). The membership functions $\mu_B(x)$ is accepted as follows:

$$\mu_B(x) = \begin{cases} (x - b_1)/(b_2 - b_1), & b_1 \le x \le b_2, \\ (b_3 - x)/(b_3 - b_2), & b_2 \le x \le b_3, \\ 0, & x < b_1, x > b_3. \end{cases}$$

Crisp intervals B_α, defined by α-cut operation, are received as.

$$B_\alpha = \left[b_1^{(\alpha)}, \, b_3^{(\alpha)}\right], b_1^{(\alpha)} = (b_2 - b_1)\alpha + b_1, \;\; b_3^{(\alpha)} = (b_2 - b_3)\alpha + b_3.$$

The α-cut intervals are shown in Fig. 1b). So, all crisp parameters are fuzzified in the following way:

$$\begin{aligned} & f_0 \leftrightarrow F_0 = (f_{01}, f_{02}, f_{03}), \quad f_1 \leftrightarrow F_1 = (f_{11}, f_{12}, f_{13}), \\ & a_0 \leftrightarrow A_0 = (a_{01}, a_{02}, a_{03}), \quad a_1 \leftrightarrow A_1 = (a_{11}, a_{12}, a_{13}), \end{aligned} \tag{17}$$

Using fuzzified parameters (17), the numerical solution of system (9) is obtained by method described in [13]. The α-cuts intervals for the free surface elevation ζ are estimated for different α. This allows us to analyse the influence of loading uncertainties.

4 Results and Discussion

4.1 Benchmark Testing

To represent unknown functions as series (7) and (8), it is necessary to find basis functions that are solutions of the spectral problem. First, the required number of boundary elements for calculations of eigenmodes and frequencies with suitable accuracy is determined. Test calculations are carried out for the rigid cylindrical shell partially filled with liquid, with $R = 1$m, $H = 1$m, Fig. 2a). It follows from formulae (7), (8), that the only first modes for the three wavenumbers (0, 1,2) should be considered.

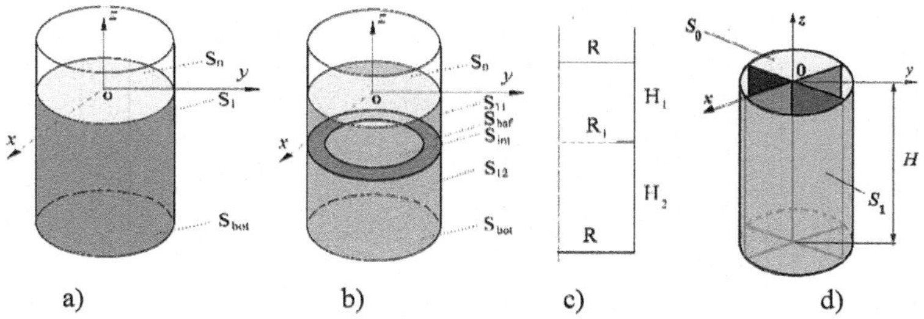

Fig. 2. Cylindrical tanks with baffles

Table 1 below shows the numerical and analytical values of the frequency parameter χ_{lk}^2/g for different wave numbers.

Convergence is achieved when the numbers of boundary elements along the wall N_R, bottom N_b of the shell, and along the radius of the free surface N_R are equal to 150. Therefore, the total number of elements is 450.

Table 1. Comparison of analytical and numerical results.

Boundary elements number, *NR*	Wave number, *l*		
	0	1	2
	Frequency parameter /g		
25	3.8310	1.7516	3.0431
75	3.8289	1.7512	3.0424
150	3.8283	1.7509	3.0409
Analytical method, [14]	3.8281	1.7508	3.0407
Frequency,	6.1257	4.1413	5.4582

For validation of the proposed method, the natural axisymmetric frequencies of baffled cylindrical shells are calculated at $H_1 = 0.5$m, $H_1 = 0.9$m, with $R_1 = 0.7$m, and $H = 1.0$m, Figs. 2b, c). The comparison of results obtained by proposed BEM and the analytically oriented method of [14] is demonstrated in Table 2.

These results are in a good agreement, so the proposed approach has been validated.

4.2 Simulation of Forced Liquid Vibrations in Reservoirs with Fuzzy Methods

According to statistical data of [17], the frequency ranges of structures excited by wind f_w and earthquakes f_e are following: $f_w \in (0.03, 10.0)$ and $f_s \in (0.3, 12.0)$ with expected values 0.5 Hz, 4,5Hz. Then for load frequencies the following presentations as fuzzy triangular numbers are implemented $F_0 = (0.03, 4.6, 12.0)$, $F_1 = (0.03, 4.6, 12.0)$. In doing so, we consider the at first stage the combined spectrum for load frequencies with expected value, typical for seismic loads. In [17] the typical load amplitudes ranges are

Table 2. Comparison of numerical results for at $l = 0$.

baffle position	method	$k = 1$	$k = 2$	$k = 3$	$k = 4$
$H_1 = 0.5$m	BEM	3.7593	7.0127	10.1763	13.3281
	[14]	3.7592	7.0105	10.1735	13.3244
$H_1 = 0.9$m	BEM	2.2882	6.2001	9.6094	12.8101
	[14]	2.2866	6.1972	9.6087	12.8088

presented as $a_w \in (0.01, 0.3)$ and $a_e \in (0.05, 0.6)$ for the wind and seismic loads without indicating the expected values. So, for fuzzifying the load amplitudes the next triangular fuzzy number are involved $a_0 = (0.1, 0.3, 0.6)$, $a_1 = (0.03, 0.15, 0.3)$.

Next, the liquid-filled cylindrical shells with and without baffles are considered, Fig. 2. Suppose that tank radius is $R = 1$m and filling level $H = 5$m. Liquid vibrations in the shell with circular baffle are analysed, supposing that $R_1 = 0.7$m, and $H_2 = 4.5$m, Fig. 2b),2c). The fundamental modes and frequencies of the cylindrical quarter tank are also estimated, Fig. 2d). The first four frequencies for all abovementioned cylindrical shell are presented in Table 3 for wave umbers 0,1,2. Namely these wave numbers are the subject of interest, as follows from series (7), (8). From the results given in Table 3, the following frequency ranges are received: (4.2424, 11.4295) for the unbaffled shell, (2.6340,11.2043) for the shell with the circular baffle, and (5.4582, 11.4295) for the quarter tank. It should be noted that all these intervals are inside the load frequency interval. Following [18], the α-cut intervals are calculated for the free surface elevation at different α for each tank, that makes it possible to estimate the uncertainties introduced by the load parameters.

Figures 3 and 4 below demonstrate time-histories of the $\alpha-$ cut bounds for function ζ for unbaffled and baffled shells. Here $\alpha = 0.5$ is considered, because this $\alpha-$ cut interval for the load frequency turned out to be equal $F_\alpha = (2.265, 8.25)$. As the results from Table 3 show, this interval contains the first lowest frequencies for all wave numbers, and for all reservoirs under consideration, so all these modes should be involved in series (8) for estimating the liquid level at solving system (9). Figure 3a) shows the lower and upper bounds of $\alpha-$ cut for the unbaffled shell. The comparison of the upper bound simulations in linear (black line) and weakly non-linear formulations is demonstrated in Fig. 3b). It should be noted that using linear formulation leads to the appearance of an unphysical growth of the sloshing amplitude by time, whereas weakly non-linear formulation allows to obtain correct results.

Number 2 in all figures below corresponds to the crisp solutions (black lines), by numbers 1 (blue lines) and 3 (green lines) the lower and upper bounds of $\alpha-$ cuts are marked. The time-histories of the level changings ζ are shown at point with coordinates $(r, \theta, z) = (1.0, 0.0, 5.0)$ in the free surface. Figures 4a) and 4b) show the lower and upper bounds of $\alpha-$ cuts of the shell with the circular baffle, and the quarter shell, respectively.

From the results obtained, one can conclude that the baffle installation leads to essentially mitigation of the liquid vibrations. In the tank without baffles, the highest

Table 3. Frequencies of liquid vibrations in different cylindrical tanks, Hz.

l	k	unbaffled	horizontal baffle, $H_2 = 4.5$m	vertical baffle
0	1	6.1248	4.7242	6.1248
	2	8.2919	7.7948	8.2919
	3	9.9847	9.7035	9.9847
	4	11.4295	11.2043	11.4295
1	1	4.2424	2.6340	5.4582
	2	7.2286	6.6452	8.1067
	3	9.1472	8.9621	9.8843
	4	10.7120	10.6436	12.6574
2	1	5.7709	3.8755	5.7709
	2	8.1068	7.0949	8.1068
	3	9.8843	9.7762	9.8843
	4	11.3609	11.1443	11.3609

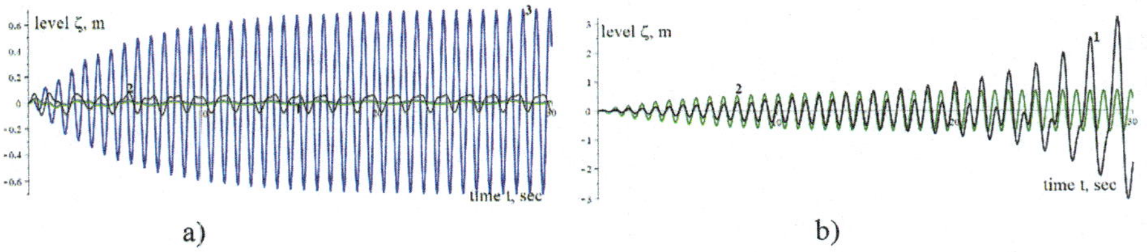

Fig. 3. Comparison of linear and nonlinear solutions, a), α – cut bounds for function ζ via time for unbaffled shell, b).

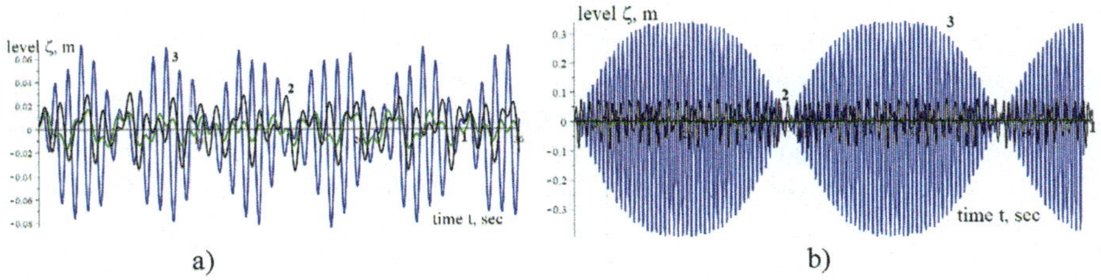

Fig. 4. Time histories of α – cut bounds of ζ for shell with circular baffle, a), for quarter tank, b).

amplitudes are observed, reaching 0.6 m, while in the tank with the horizontal baffle, they are close to 0.1 m. In the quarter tank, the beating mode is observed, but with the half amplitude compared to the unbaffled tank. The disadvantages of the approach are harmonic loading constraints, neglections of damping and viscosity.

5 Conclusion and Future Research

The fluid vibrations in fluid-filled reservoirs under external loads is a process with many uncertainties, that require involving fuzzy mathematics concepts in mathematical modelling. The boundary element method with fuzzy concepts implementation is applied to analyzing the liquid vibrations in the partially filled rigid reservoirs with baffles under coupled vertical and horizontal excitations. The problems are considered in linear and weakly non-linear formulations, and the adequacy of using the non-linear formulation is proven. Three kinds of cylindrical shells are analysed, and the estimation of the filling level is carried out. The sloshing amplitudes are significantly reduced when baffles are installed. Changing in the frequency spectrum due to baffles installation does not lead to the amplitude increasing, at least for the considered load parameters diapasons. Moreover, in the tank without baffles, the highest amplitude has reached 0.6 m, while in the baffled tanks, is reduced by at least half. In the future, it is planned to study liquid behaviour in reservoirs at impulsive and seismic loads.

Acknowledgments. The authors are very grateful to professor Stavros Syngellakis, Wessex Institute of Technology, UK, for his constant supporting.

References

1. Balas, O.-M., Doicin, C.V., Cipu, E.C.: Analytical and numerical model of sloshing in a rectangular tank subjected to a braking. Mathematics **11**, 949–955 (2023)
2. Gontarovskyi, P., Smetankina, N., Garmash, N., Melezhyk, I.: Numerical analysis of stress-strain state of fuel tanks of launch vehicles in 3d formulation. integrated computer technologies in mechanical engineering—2020. ICTM 2020. Lecture Notes in Networks and Systems, pp. 609–619, Springer, Cham (2021), https://doi.org/10.1007/978-3-030-66717-7_52
3. Rusanov, A., Khorev, O., Agibalov, Y., Bykov, Y., Korotaiev, P.: Numerical and Experimental Research of Radial-Axial Pump-Turbine Models with Spliters in Turbine Mode, In: ICTM 2020, Lecture Notes in Networks and Systems, 188, (2021), DOI:/https://doi.org/10.1007/978-3-030-66717-7_36
4. Gimperlein, H., Oezdemir, C., Stephan, E.P.: A time-dependent FEM-BEM coupling method for fluid-structure interaction in 3d. Appl. Numerical Math. **152**, 49–65 (2020). https://doi.org/10.1016/j.apnum.20
5. Strelnikova, E., Choudhary, N., Kriutchenko, D., Gnitko, V., Tonkonozhenko, A.: Liquid vibrations in circular cylindrical tanks with and without baffles under horizontal and vertical excitations. Eng. Anal. Boundary Elem. **120**, 13–27 (2020). https://doi.org/10.1016/j.enganabound.2020.07.02m
6. Taverniers, S., Svyatoslav Korneev, S., Somarakis, C., Lew A.J. A finite element method to compute the damping rate of oscillating fluids inside microfluidic nozzles. Physics / Computational Physics, arXiv preprint arXiv:2307.00094 (2023)
7. Zang, Q., Liu, J., Yu, L., Lin, G.: Boundary element analysis of liquid sloshing characteristics in axisymmetric tanks with various porous baffles. Appl. Ocean Res. **93**, 101963 (2019)
8. Liu, J., Zang, Q., Ye, W., Lin, G.: High performance of sloshing problem in cylindrical tank with various barrels by isogeometric boundary element method. Eng. Anal. Boundary Elem. **114**, 148–165 (2020)

9. Zhang, C., et al.: Numerical simulation of thermal–structural behaviour of liquid helium tank during filling process. Fusion Eng. Des. **173**, 112798 (2021)
10. Liu, J., Yan, B., Mou, Z., Gao, Y., Niu G., Li, X.: Numerical study of aeolian vibration characteristics and fatigue life estimation of transmission conductors. PLoS One. **17**(1), e0263163 (2022). https://doi.org/10.1371/journal.pone.0263163
11. Misura, S., Smetankina, N., Misiura, I.: Optimal design of the cyclically symmetrical structure under static load. In: Nechyporuk, M., Pavlikov, V., Kritskiy, D. (eds.) Integrated Computer Technologies in Mechanical Engineering—2020. ICTM 2020. Lecture Notes in Networks and Systems, pp. 256–266. Springer, Cham (2021), https://doi.org/10.1007/978-3-030-66717-7_21
12. Dong-lin, W., Li, A., Hao, H., Song, M., Ming, L., Liang-bin, C.: Numerical simulation and full-scale experimental verification for sloshing and vibration of aircraft fuel tank. J. Engineering Mechanics **39**(4), 219–229 (2022). https://doi.org/10.6052/j.issn.1000-4750.2021.02.0130
13. Mazandarani, M., Xiu, L.: A review on fuzzy differential equations. Eng. Mech. **39**(4), 219–229 (2022)
14. Raynovskyy, I.A., Timokha, A.N.: Sloshing in Upright Circular Containers: Theory, Analytical Solutions, and Applications. CRC Press/Taylor and Francis Group (2020)
15. Zhou, D., Wang, J.D., Liu, W.Q.: Nonlinear sloshing of liquid in rigid cylindrical container with a rigid annular baffle under lateral excitation. Shock. Vib. **2019**(5398038), 1–18 (2019). https://doi.org/10.1155/2019/5398038
16. Gnitko,V., Karaiev, A., Degtyariov, K., Strelnikova, E.: Singular boundary method in a free vibration analysis of compound liquid-filled shells. WIT Trans. Eng. Sci. **126**, 189–200 (2019). https://doi.org/10.2495/BE420171
17. Iancovici, M., Ionică, G., Pavel, F., Moţa, F., Nica, B.: Nonlinear dynamic response analysis of buildings for wind loads A new frontier in the structural wind engineering. J. Build. Eng. **47**, 103708 (2022). https://doi.org/10.1016/j.jobe.2021
18. Wang, Z.W., Ge, N., Li, C.W.: Structural vibration mode fuzzy control based on bpneural network algorithm. J. Shandong Univ. (Eng. Science) **50**(243), 17–23 (2020)

Influence of Cylinder Diameter and Position on Thermal and Hydrodynamic Interaction of Shedding Vortices with the Heat Exchanger Wall

Basile Perly[1], Izabela Wardach-Święcicka[2], and Dariusz Kardaś[2](✉)

[1] Polytech Orléans, 8 Rue Léonard de Vinci, 45072 Orléans, France
[2] Robert Szewalski Institute of Fluid-Flow Machinery of the Polish Academy of Science, Fiszera 14, 80-231 Gdańsk, Poland
dk@imp.gda.pl

Abstract. The work is focused on optimizing a counter-current heat exchanger under laminar conditions using von Kármán vortices. Cylindrical obstacles placed in the hot channel generate vortices to mix the fluid and enhance the heat transfer between the hot and cold fluids. Numerical analysis is aimed at examining the influence of cylinders position and their diameter on the value of local and global heat transfer coefficient. Simulations are performed using one cylinder at the center of the hot channel, then two cylinders placed diagonally. The results show that increasing the diameter of the cylinder placed at the center of the cavity leads to an enhancement in heat transfer. However, it also results in a proportional increase in pressure drop. The results also demonstrate that it is preferable to position two small cylinders diagonally and closer to the walls to promote the mixing of the cold fluid within the thermal boundary layer with the hot fluid at the center of the cavity. Additionally, the horizontal gap between the two cylinders has a significant role in terms of local heat transfer. The geometry, therefore, should depend on the desired objective: whether one aims to increase heat transfer at a specific location or to enhance overall heat transfer along the heat exchanger.

Keywords: vortex shedding · cylinders · heat transfer coefficient · pressure drop

1 Introduction

Currently, the selection of heat exchangers is a typical engineering problem, often of great technical and financial importance, but still using classic models and commercial programs. Depending on the problem, the improvement of heat exchange between two fluids is achieved by changing the geometry, e.g. using partitions and generating disturbances, or by increasing the flow rate. Increasing the flow velocity and thus the Reynolds number leads to a higher heat transfer coefficient, which translates into more intensive heat exchange. The more turbulent the flow, the better the heat exchange.

In technology, however, there are situations when there is a need to intensify heat exchange in laminar flow. When it is additionally necessary to maintain the fluid volume,

H. Altenbach et al. (Eds.): CAMPE 2023, pp. 44–52, LNME 2025.
https://doi.org/10.1007/978-3-031-82979-6_5

the flow rate cannot be increased and thus turbulence is caused. The research presented here concerns the intensification of heat exchange in laminar flow by using baffles inside the channel. To improve the heat exchange in the laminar regime, the phenomenon of the von Karman vortex path is used. The very phenomenon of a vortex path behind an obstacle in an unlimited space was studied by Karman [1], Karman & Rubach [2]. The works of Kovasznay [3], and the more recent articles by Strykowski and Sreenivasan [4] by Nodberg [5] should be also mentioned. More recently, the vortex path has been numerically modeled by Iman and Mohsen [6] Gaza et al. [7], and Rajani et al. [8]. Some papers present the DNS results of the generation of vortices sourced by different obstacles, but still, they are considering only the cold laminar flow [9–13]. In the literature, there are only a few works devoted to heat exchange in laminar flows with obstacles, but they do not include studies on the impact of walls and the size of obstacles on the generation of vortices and on the heat transfer intensification simultaneously [14]. In the work presented here, first of all, we go beyond the subject of vortices generated behind obstacles in laminar flow but in limited space, and, secondly, we analyze the impact of the generated vortices on heat transfer intensification. Therefore, we aimed to fill the research gap and analyze these problems simultaneously, i.e. we consider the intensification of heat exchange by the generation of the vortices through the obstacles with different sizes in wall-limited space. This issue has a practical meaning. In the technology of heat exchangers, there are some situations when there is a need to enhance heat exchange in laminar flow. Especially, when it is additionally necessary to maintain the fluid volume. In such cases, the flow rate cannot be increased to obtain more turbulent flow which causes better mixing and intensify the heat exchange between cold and hot fluid. We have analyzed the geometry of the heat exchanger in the frame of the NASA project. The technical requirements for the heat exchanger placed in the international space station concerned the size and the weight of the unit, which causes the low Reynolds number condition.

Due to the heat exchange designing problem, the subject of the analysis will be the two-dimensional low Reynolds flow in the area limited by the edges. This interaction between the vortices generated by cylinders and the boundary area is of particular interest due to heat transfer and pressure drop.

The first part aims to present the geometry of the simple counter-current heat exchanger and the reference case of this study. The second part of the conducted numerical research is aimed at determining how the diameter of the cylinder affects the thermal and hydrodynamic parameters of the flow. Then, the third part concerns heat transfer with two cylinders positioned relative to each other and relative to the flow in different configurations.

2 Methodology and Reference Geometry

In the present work, a 2D laminar and incompressible flow passing through a typical counter-flow heat exchanger is considered. The laminar flow simulations allow to obtain the direct numerical solution for a complete two-dimensional and time-dependent solution of the Navier–Stokes equations. The advantages of such kind of flow modeling (low Reynolds numbers) give us a direct numerical solution without considering any additional models. Thus, we can have the exact solution for the instantaneous fluid velocity as a function of position and time, like in the case of Direct Numerical Simulations (DNS) and the modeling of turbulent flows. Therefore, we do not need any experimental data to validate the numerical results. The geometry of the considered device is presented in Fig. 1. The 200mm length heat exchanger consists of a hot channel with a height of D = 10mm. A cold fluid flows through two channels 5mm high each in the opposite direction. The two fluids are separated by a 1mm thick aluminum wall. The time-dependent simulations are carried out until the volume average temperature in the whole domain does not change anymore.

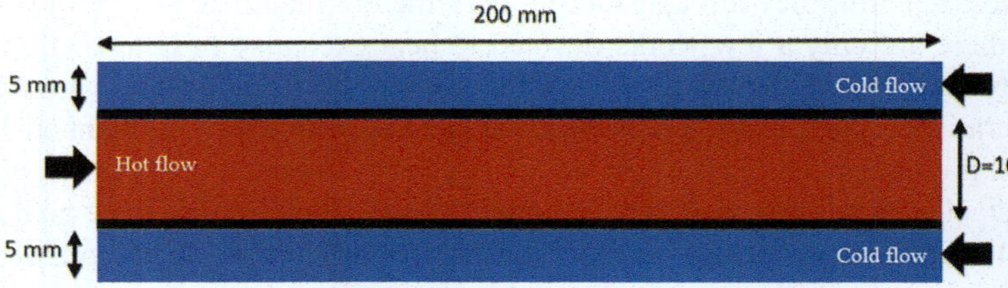

Fig. 1. Reference case of the heat exchanger

Boundary conditions and fluid properties.

At the inlet of the hot channel, the fluid velocity is uniform at 0.1m/s, and the Reynolds number is Re = 1000. The fluid temperature is $T_{hot} = 400$K. At the inlet of the cold fluid, the velocity is also 0.1m/s, the Reynolds number is Re = 500, and the temperature is $T_{cold} = 300$K. Fluid properties are described in Table 1.

Table 1. Fluid properties

Properties	Value
Density	$998.2\left[kg/m^3\right]$
Dynamic viscosity μ	$1 \times 10^{-3}[kg/(ms)]$
Thermal conductivity	$0.6[W/(mK)]$

The local heat transfer coefficient α is calculated from the heat flux q and the temperature difference between hot T_{hot} and cold T_{cold} fluids.

$$\alpha = \frac{q}{T_{hot} - T_{cold}}\left[W/\left(m^2K\right)\right] \quad (1)$$

The 2D Ansys Fluent program is used to calculate heat and flow in the laminar regime. The distribution of the heat transfer coefficient along the reference (pure without obstacles) heat exchanger is shown in Fig. 2. The coordinate x = −0.1m denotes the hot fluid inlet, and x = 0.1m represents the hot fluid outlet. The average heat transfer coefficient (HTC) for the reference case is therefore 398.2 $W/\left(m^2K\right)$. The local HTC is very high at the ends of the heat exchanger because the temperature difference is significant: the thermal boundary layer is not developed yet, which enhances the heat transfer.

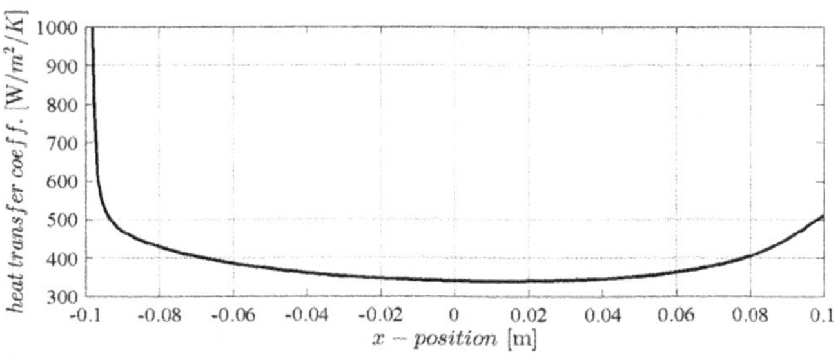

Fig. 2. Heat transfer coefficient distribution along the heat exchanger

3 Study of the Heat Exchanger with One Cylinder

In this part, a single cylinder is added at the center of the heat exchanger in the hot channel (at position x = 0). The simulations are performed with a cylinder diameter D_c ranging from 0.1D to 0.3D, with D the height of the hot channel, resulting in Reynolds numbers Re_{Dc} based on the cylinder diameter

$$Re_{Dc} = \frac{uD_c}{\nu} \tag{2}$$

between 100 and 300. The Ansys Fluent program with a two-dimensional solver is used for the calculations. The mesh around the cylinder allows to have a smooth transition from the circular mesh of the cylinder to the rectangular cells of the heat exchanger. Figure 3 shows an example of the mesh downstream of the cylinder (from $x = 0$ to $x = 0.05$m). The whole numerical domain consists of 100,000 cells. This means that in the case of a 200mm long heat exchanger, the average size of one cell is about 0.2mm x 0.2mm. In the boundary layers, the mesh is appropriately dense. The value of y+ is 1. Nevertheless, this value for laminar flow is not important because we do not model the flow in the boundary layer zone. We solve a full set of time-dependent mass, momentum, and energy equations, so our solutions are the DNS results. We analyzed the influence of the mesh size on the solution and for the considered number of cells, the more dense grid did not affect the obtained values.

The cylinder in the center of the channel sheds the vortices, which continue to propagate in the channel. Figure 4 shows the instantaneous distribution of the temperature field with visible traces of vortices. Since the averaged parameters, such as the heat transfer coefficient and the heat flux, are important in heat transfer, the averaged temperature field has been calculated and shown in Fig. 5.

The averaged local HTC distribution (Fig. 6) clearly shows that as the cylinder diameter increases, the local maximum of heat transfer coefficient downstream of the

Fig. 3. Mesh of the heat exchanger with one cylinder

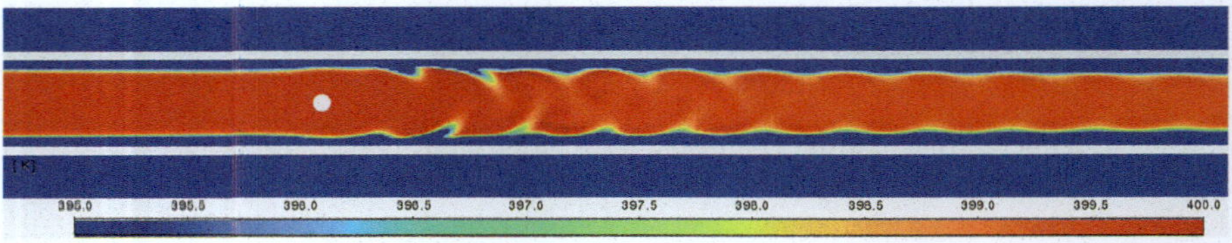

Fig. 4. Temperature in the heat exchanger at a specific time

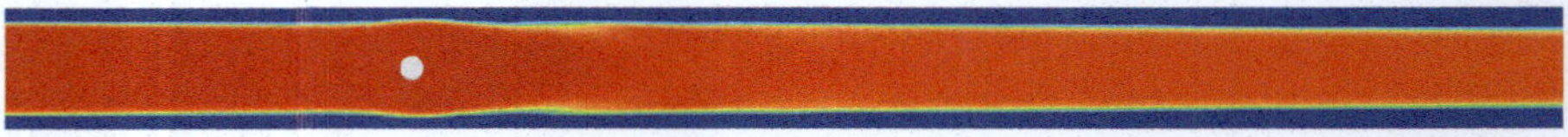

Fig. 5. Averaged temperature in the heat exchanger (hot channel) with data sampling.

cylinder also increases. More specifically, the HTC increases significantly at the cylinder's location ($x = 0$) due to the increased flow velocity, which reduces the thickness of the hydraulic and thermal boundary layers, leading to an enhancement in heat flux. It appears that the value of HTC is almost 1.2, 1.4, and 1.6 times higher in comparison to the base case (without obstacles). The larger the diameter of the cylinder, the more intensification of the heat transfer. The same takes place behind the cylinder. On the other hand, the presence of vortices downstream of the cylinder allows for mixing the fluid, leading to a significant increase in the HTC. The local HTC increases as the cylinder diameter increases because the vortices have more energy. To compare the different cases globally, the average HTC and the pressure drop in the hot channel were calculated and the results are presented in Fig. 7. As observed earlier, the average heat transfer coefficient increases with the cylinder diameter. For the larger cylinder, the HTC value is about 10% higher than in the case without any obstacle (top graph). This increase is correlated with the pressure drop in the cavity due to the presence of the cylinder.

For larger cylinder, the pressure drop is about 35% higher than in the base case (bottom graph). Indeed, the cylinder enhances fluid mixing by increasing friction with the heat exchanger walls, resulting in a more significant pressure drop. The configuration with a single cylinder at the center of the channel shows good results, but the work aims to have a greater impact on the thermal boundary layer while minimizing the pressure drop. Therefore, in the following section, the case with two cylinders positioned closer to the walls is considered.

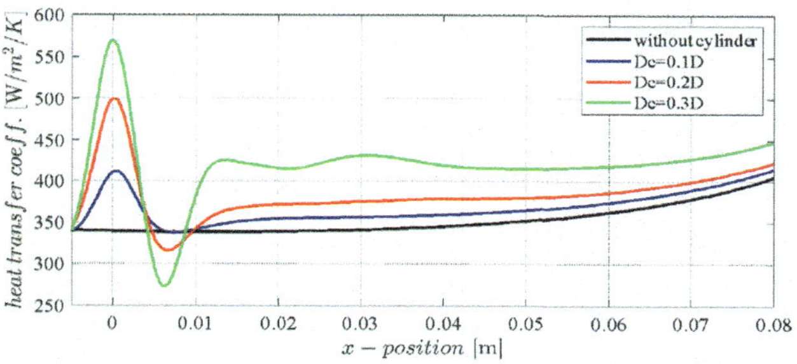

Fig. 6. Local heat transfer coefficient depending on the diameter of the cylinder

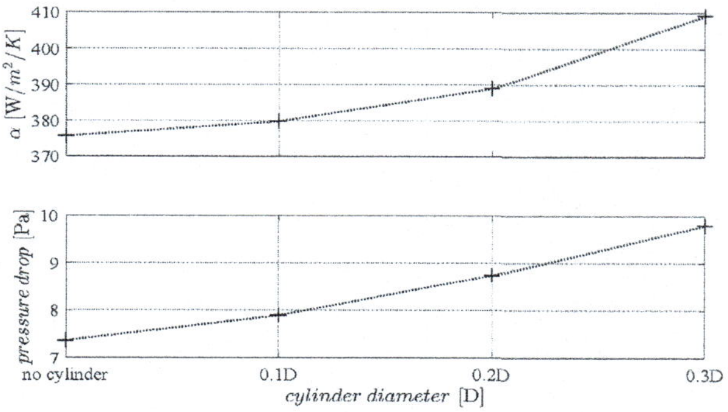

Fig. 7. Average heat transfer coefficient (top) and pressure drop (bottom) in the channel with one cylinder

4 Heat Exchanger with Two Cylinders

Preliminary studies have revealed that the heat flux increases more significantly when the cylinder is closer to the wall (its diameter is higher). However, it is essential to maintain a minimum distance to avoid disturbances to the vortices by the wall and to account for the hydraulic boundary layer along the wall, which can reduce the vortex intensity. A vertical gap of 0.3D between the center of the cylinder and the wall is a suitable compromise. In this case, the vortices facilitate the mixing of the cold fluid within the thermal boundary layer with the hot fluid at the center of the channel. In this part, two diagonally positioned cylinders with a diameter of 0.15D are considered. Now the horizontal distance between the two cylinders varies to observe its influence on the local and average heat transfer coefficient. It is important to avoid placing the cylinders too close to each other to prevent a significant increase in the flow velocity around the cylinders and to keep away from disturbing the vortices created by the two cylinders. Three cases are studied: a horizontal gap of 0.5D, 1D, and 2D. The last case is shown in Fig. 8 with the fluid temperature contour around the cylinders. As it is seen, the perturbation of the thermal boundary layers behind the cylinders occurs. To compare these three cases, the variation of the average heat transfer coefficient along the heat exchanger is calculated (Fig. 9). The location of the obstacles can be identified by the peaks.

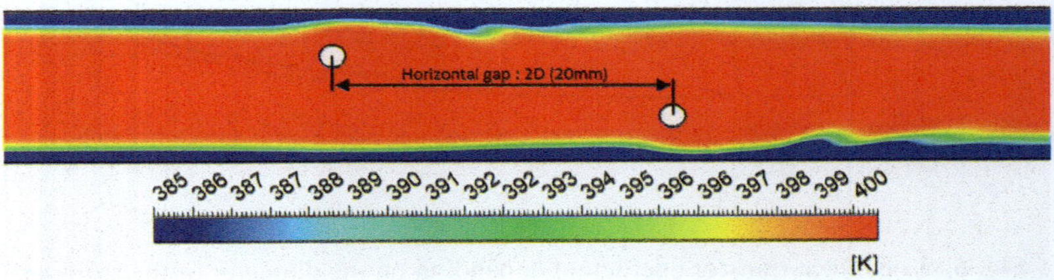

Fig. 8. Temperature contour in the hot channel (x = [−0.02m;0.4m]) for 2 cylinders with a horizontal gap of 2D, at a specific time

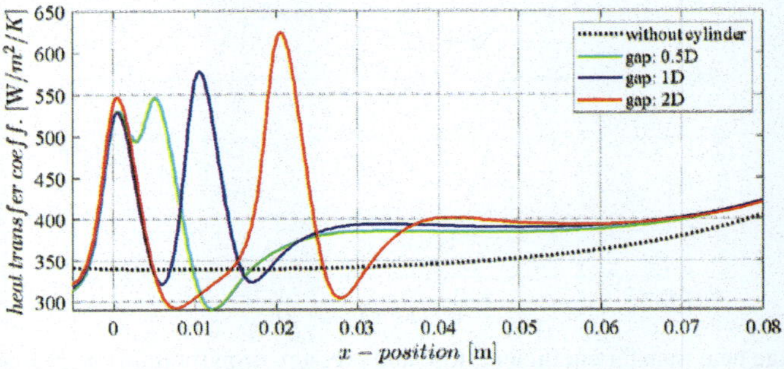

Fig. 9. Local heat transfer coefficient with 2 cylinders for different horizontal gap

If the horizontal gap is small, the increase in the HTC is concentrated around the cylinders (see Fig. 9 for 0.5D). On the contrary, if it's larger (e.g. gap 2D), the second peak is higher because the vortex power is greater due to the perturbation caused by the first obstacle. However, the local HTC stabilizes and becomes similar for the different cases. Nevertheless, the local HTC can be 50–80% higher in comparison to the reference case, depending on the gap. Thus, the choice of geometry depends on the objective. If the goal is to maximize local heat transfer around the cylinders, it is relevant to position them closer to enhance local mixing. However, if the aim is to maximize overall heat transfer, it is preferable to space them slightly apart to take advantage of vortex interferences and distribute the increased heat transfer at multiple locations.

The dependence of the average heat transfer coefficient on the pressure drop in the hot duct for two cylinders is presented in Fig. 10. The previous geometries with a single cylinder, as well as the reference case, are also represented for comparison.

All three cases with two cylinders exhibit a similar overall coefficient (ca. 405 W/(m^{2}K)), but case 1D offers the best performance. However, it can be noticed that increasing the horizontal gap allows to reduce the pressure drop a little. On the other hand, comparing the results from these geometries with the blue curve of a single cylinder, it can be seen that the pressure drop can significantly be reduced (from 9.75Pa to 8.75Pa) while maintaining the same overall heat transfer coefficient. Therefore, it is

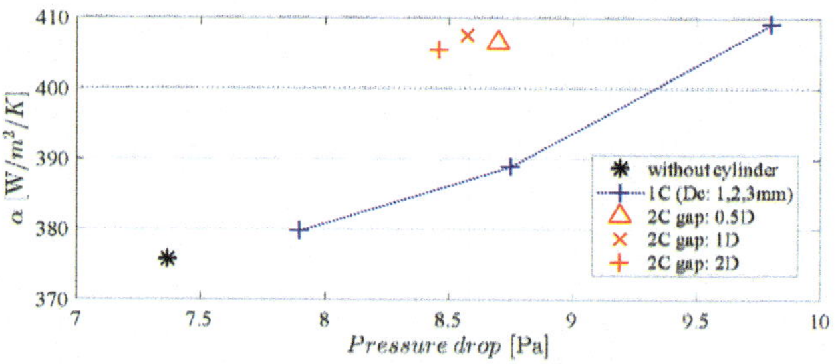

Fig. 10. Average heat transfer coefficient vs pressure drop

more interesting to use two smaller cylinders placed close to the walls and in a diagonal configuration, rather than a single large cylinder at the center of the channel, in order to maximize heat transfer while minimizing the pressure drop.

The case with two diagonal cylinders with a horizontal gap of 1D between them results in a 10% increase in the HTC compared to the reference case. This is achieved simply by adding two small cylindrical obstacles in the hot cavity. However, it should be noted that this also increases the pressure drop.

5 Conclusion

In this paper, the optimization of a counter-flow heat exchanger using cylindrical obstacles to generate von Kármán vortices, enabling fluid mixing within the hot cavity was investigated. This mixing consequently increases the heat transfer process, thereby enhancing the efficiency of the heat exchanger. This study was conducted by implementing one or two cylindrical obstacles within the hot cavity, varying the cylinder diameter, and subsequently adjusting the horizontal distance between the two obstacles. This approach enables comparison of different scenarios in terms of intensification of the heat exchange utilizing improvement of the value of local and average heat transfer coefficient, as well as the minimization of the pressure drop incurred due to the addition of obstacles.

The numerical simulations revealed that increasing the cylinder diameter leads to an augmentation in the heat transfer coefficient, even by almost 65% at the cylinder location and by almost 17% at the half of channel length in comparison to the case without an obstacle. The average HTC for the cases with one cylinder with the biggest considered diameter is 10% higher than in the base case. However, it was also observed that the pressure drop increases proportionally to it and its value increases even by 34% compared to the case without obstacle. The strategic arrangement of two cylinders in a diagonal configuration showcased the potential to achieve the best performance in terms of increased heat transfer and minimized pressure drop. Indeed, cylinders placed near the walls facilitated the mixing of cold fluid from the thermal boundary layer with the hot fluid at the center of the cavity. Furthermore, widening the horizontal gap between the two cylinders resulted in a slight reduction in pressure drop. Regarding the heat transfer coefficient, the optimal performance was achieved with a horizontal gap equal to the diameter of the cavity (D), with a 10% increase compared to the reference case (409 vs. 375 W/[m^2K]). Simultaneously, the pressure drop increased by 16% (8.6 vs. 7.4 Pa).

Finally, the use of cylindrical obstacles to generate vortices holds promise for optimizing heat exchangers while preserving a low Reynolds flow regime. The incorporation of an increased number of diagonal obstacles stands as a potential avenue for substantially amplifying mixing and, thereby, improving heat transfer efficiency.

In future work, we plan to make some experiments using PIV equipment for visualization of the mixing phenomena. The next step is to make a full 3D computation to get detailed information about the position and number of obstacles in the real heat exchanger geometry. Thus, the optimization of the mass and size of the real unit could be done properly.

Acknowledgments. This research was supported by computer centre CI TASK.

References

1. Von Karmán, T.: Über den Mechanismus des Wiederstandes, den ein bewegter Korper in einer Flüssigkeit erfahrt. Nachrichten von der Gesellschaft der Wissenschaften zu Göttingen, Mathematisch-Physikalische Klasse, 547–556 (1912)
2. Von Kármán, T., Rubach, H.: Über den mechanismus des flüssigkeits—und luftwiderstandes, Physikalische Zeitschrift, **13**, 49–59 (1912)
3. Kovasznay, L.S.G.: Hot-wire investigation of the wake behind cylinders at low Reynolds numbers. In: Proceedings of the Royal Society of London. Series A. Mathematical and Physical Sciences (1949), vol. 198, pp. 174–190
4. Strykowski, P.J., Sreenivasan, K.R.: On the formation and suppression of vortex 'shedding' at low Reynolds numbers. J. Fluid Mech. **218**, 71–107 (1990)
5. Nordberg, C.: An experimental investigation of the flow around a circular cylinder: influence of aspect ratio. J. Fluid Mech. **258**, 287–316 (1994)
6. Iman, H., Mohsen, S.: Numerical simulation of fluid flow and forced convection heat transfer from tandem circular cylinders using overset grid method. J. Fluids Struct. **28**, 309–327 (2012)
7. Al-Sumaily, G.F., Hussen, H.M., Chaichan, M.T., Dhahad, H.A., Thompson, M.C.: Numerical analysis of flow shedding over an obstacle at low Reynolds number. Int. J. Heat Fluid Flow **99**, 109098 (2023)
8. Rajani, B.N., Majumdar, S., Kandasamy, A.: Numerical simulation of laminar flow past a circular cylinder. Appl. Math. Model. **33**, 1228–2124 (2009)
9. Kumar, A., Ray, R.K.: A numerical simulation of shear flow past two equal sized square cylinders arranged in parallel at Re = 500. AIP Conf. Proc. **1863**, 490003 (2017)
10. Furquan, M., Mittal, S.: Flow-induced vibration of filaments attached to two side-by-side cylinders. Phys. Fluids **33**, 063609 (2021)
11. Yuhao, Y., Chunning, J., Narakorn, S.: On wake modulation and interaction features of a pair of dual-step circular cylinders in side-by-side arrangements. Phys. Fluids **33**, 093604 (2021)
12. Guo, F., Wu, G., Du, X., Mason, M.S.: Numerical investigation of flow around a square cylinder in accelerated flow. Phys. Fluids **33**, 104105 (2021)
13. Ortega-Jiménez V.M., Sanford, C.P.: Beyond the Kármán gait: knifefish swimming in periodic and irregular vortex streets J. Exp. Biol., 224 (2021)
14. Huang, Z., Xi, G., Zhang, W., Wen, S.: Mixed convection heat transfer from confined tandem square cylinders in a horizontal channel. Int. J. Heat Mass Transf. **66**, 625–631 (2013)

Energy Saving Technologies and Environmental Safety

Assessment of the Second-Order Phase Transition Effect on the Thermal Efficiency of a Theoretical Rankine Cycle

Anatolii Tarelin(✉), Iryna Annopolska, Yevhen Lukianov, and Andriy Tarelin

Department of Integrated Energy Technologies, Anatolii Pidhornyi Institute of Mechanical Engineering Problems of the National Academy of Sciences of Ukraine, Vul. Komunalnykiv, 2/10, Kharkiv 61046, Ukraine

tarel1364@gmail.com

Abstract. Increasing the thermal efficiency of the Rankine cycle by changing the thermodynamic characteristics of the working fluid of steam turbine plants by using the effect of a second-order phase transition under the targeted action of a magnetic field is considered. An analysis of experimental studies on the change in the physicochemical and thermodynamic characteristics of water under the influence of force fields is presented. The possible ranges of change in the specific heat capacity of water, volatility and hidden heat of vaporization are determined. Since an abrupt change in specific heat capacity occurs without heat supply or removal, it has been suggested that a second-order phase transition occurs at the first stage of the technological cycle after treatment with a force field, and a first-order phase transition occurs at the stage of vaporization. Numerical studies are performed for a specific simple reversible Rankine cycle with a change in heat capacity and hidden heat of vaporization. It is shown that even a small increase in the specific heat capacity only in the isobaric-isothermal process in front of the steam generator by 5% can lead to an increase in the thermal efficiency of the Rankine cycle by 1.3%. Areas for further directions of the research were selected. Unlike traditional methods for increasing the thermal efficiency of the cycle of steam turbine plants, the availability to use the proposed approach will not require significant costs and changes in the design of power plants.

Keywords: Efficiency of turbine plants · Thermodynamic characteristics · Influence of magnetic fields on the working fluid

1 Introduction

Improving the efficiency of modern steam turbine installations has been and remains one of the most urgent tasks in the economy sustainable development. Most of these plants operate on the Rankine cycle with intermediate superheating of steam and regenerative feed water heating.

Traditional methods of increasing the efficiency of the Rankine cycle, such as increasing the initial parameters of the steam, lowering the final pressure are almost over exhausted.

H. Altenbach et al. (Eds.): CAMPE 2023, pp. 55–63, LNME 2025.
https://doi.org/10.1007/978-3-031-82979-6_6

Some new ways for increasing unit capacities and efficiency of power plants have been started quite long ago and developed: steam-gas cycles [1]; the use of multicomponent low-boiling mixtures with a variable temperature of vaporization as a working fluid; the choice of optimal, in the thermodynamic sense, new working fluids [2, 3].

However, among a few substances with known thermodynamic behavior, water is still the most preferred working fluid for steam power plants.

The Rankine cycle used at thermal power plants is based on phase transitions. In the steam generator, water heats up and evaporates, i.e., a first-order phase transition occurs, the steam does work in the turbine, and then condenses (again a first-order phase transition) and the condensate enters the steam generator.

A characteristic feature of the presence of phase transitions is a sudden change at the transition point in the values of the functions of the thermodynamic state of the working fluid: heat capacity, entropy, enthalpy, isobaric-isothermal potential, and others.

In the case of the second-order phase transition, which can occur under the influence of force fields and precedes a phase transition of the 1st order, there is no change in entropy, i.e., no heat release or absorption. However, other thermodynamic parameters may change significantly.

Currently, studies of the impact on water of fields of various natures, which can significantly change the thermodynamic characteristics of water and water vapor, are being intensively developed.

Therefore, it seems appropriate to evaluate how such changes that occur during a second-order phase transition will affect the thermal efficiency of the Rankine cycle.

2 Literature Review

The influence of physical (in particular, magnetic fields) on water and water systems has been studied for a long time, but until now, the works of different researchers often contradict each other due to the lack of a theoretical basis that would allow to explain and to apply the results obtained.

All experimental work in this area can be divided into two large groups:

- research aimed at changing any physical and chemical properties of water in order to develop the theoretical basis for the influence of external physical fields;
- research aimed at obtaining specific changes in the properties of water systems in order to use the results in real technological processes.

In our opinion, most of the studies related to the first group over the past decade can be represented by the following few works. Valery Shalatonin et al. [4] in the study of deionized water revealed significant polar-dependent changes in water ab-sorption in the UV range (180–350 nm) and water evaporation. Ran Cai et al. [5] showed a decrease in surface tension and an increase in the viscosity of water. Yongfu Wang et al. [6] found that the coefficient of friction in magnetized water at different magnetic field strengths is less than in untreated water. S. Jawad et al. in a study [7] reported the effect of the magnetization process on the chemical and electrical properties of tap water (pH, electrical conductivity). Yanmin Li et al. in their studies [8] showed that a magnetic field can increase the freezing point of water and aqueous solution by about 3.0–5.0 K.

Kai-Tai Chang et al. in [9] found that the number of hydrogen bonds increases with increasing magnetic field strength. This means that this field can control the size of the water cluster.

In the second group of studies we are focused on works related to the thermophysical properties of water. Youkai Wang et al. [10] found an increase in evaporation by 38.98%, a decrease in specific heat capacity by 3.3%, and a decrease in boiling point by 2%. J.A. Duenas et al. [11] state that the treatment of deionized water with a magnetic field increases its evaporation by 20%. Yun-Zhu Guo et al. [12] showed an increase in water evaporation. A. Szcześ, et al. [13] showed that the conductivity of water decreases, and this decrease is inversely proportional to the flow rate. E.S. Malkin et al. [14] showed that when treating water with the help of the Ilios-M electromagnetic pulse apparatus, a decrease in the specific heat of vaporization of water by 6...10% and specific heat capacity by 9% was obtained. The maximum effect is obtained at the maximum pulse frequency.

All of the above works have, in our opinion, several significant shortages:

- it is stated that as a result of the action of the field, some parameter quantitatively changes only in one direction, and the magnitude of this change is proportional either to the absolute value (in the case of a constant field), or to the repetition rate of per-turbing pulses (in the case of an alternating field);
- there is no evidence of achieved repeatability of results;
- received, practically, the same result of the impact of the field in the case of using distilled or tap water as an object of study;
- little attention is paid to the parameters of the water flow and the parameters of the pipelines through which water flows both in the zone of influence of the field and in other zones.

However, the results of all experimental works clearly indicate the possibility of a significant change in the properties of water under the influence of fields of various physical natures.

The purpose of our research is to prove that these properties can be changed both upwards and downwards, which will make it possible to control the process in various technologies.

3 Research Methodology

3.1 Analytical Relations for Determining the Thermal Efficiency of the Rankine Cycle

Analytical relations for determining the thermal efficiency of the Rankine cycle when changing the specific heat of water and the hidden heat of vaporization under the influence of a force field are based on the classical thermodynamics' laws and on the theory of turbomachines.

Let us consider the simplest reversible Rankine cycle without intermediate overheating of steam and regenerative water heating (Fig. 1).

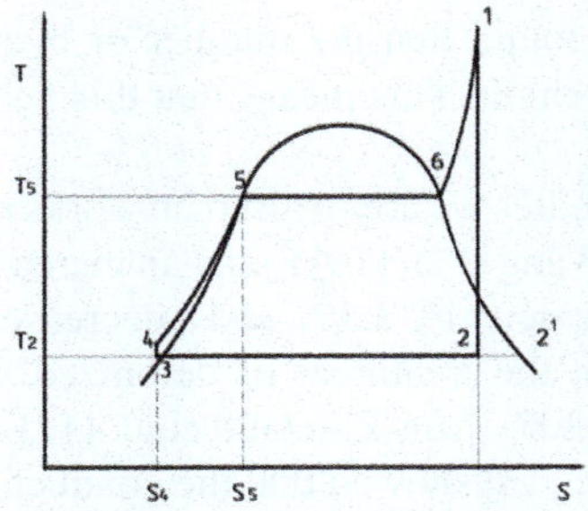

Fig. 1. Reversible Rankine cycle in T-S coordinates

The classical dependence for determining the thermal efficiency of a cycle is written as follows:

$$\eta_t = 1 - q_2/q_1 \tag{1}$$

where: q_1– heat supplied in the cycle; q_2- heat dissipation.

The supplied heat can be represented as the sum of the heat supplied during water heating (process 4–5), its evaporation (latent heat of vaporization r_Π, , process 5–6) and steam overheating (process 6–1).

For the simplest reversible Rankine cycle, if we neglect the value of the pump operation due to its smallness in comparison with the available enthalpy difference that occurs in the turbine, the dependence for determining the thermal efficiency of the cycle can be written as follows:

$$\eta t = 1 - T_3(S_1 - S_3)/\left(\int_{T_4}^{T_5} C_{P_B}(T)dT + T_5(S_6 - S_5) + \int_{T6}^{T_1} C_{P_\Pi}(T)dT\right), \tag{2}$$

where $C_{p\Pi}(T)$ is the specific heat capacity of steam, which varies in the temperature range T_6–T_1; $C_{pB}(T)$ - specific heat capacity of water in the range T_4–T_5; S_1, S_3, S_5, S_6 - entropy at the corresponding points of the cycle; $T_5(S_6–S_5) = r_\Pi$ - latent heat of vaporization; $T_3(S_1–S_3) = q_2$.

Accordingly,

$$\eta t = 1 - q_2/\left(\int_{T_4}^{T_5} (C_{pB}dT) + r_\Pi + \int_{T_6}^{T_1} (C_{p\Pi}dT)\right) \tag{3}$$

Under the influence of an electromagnetic or magnetic field on the working fluid in an isobaric-isothermal process in front of the steam generator (point 4), the specific heat of water C_{pB} will change by the value ΔC_{P_B}.

Thus, considering this effect, the expression for determining the efficiency of the thermodynamic cycle will change accordingly and take the form:

$$\eta t = 1 - q_2/\left(\int_{T_4}^{T_5} ((C_{pB} + \Delta C_{pB})dT) + r_\Pi + \int_{T_6}^{T_1} (C_{p\Pi}dT)\right). \tag{4}$$

3.2 Experimental Methods for Changing the Thermophysical Parameters of Water

For a long time, IPMash NAS of Ukraine has been carrying out work aimed at purposefully changing the thermophysical properties of the working fluid of heat machines [15]. In experimental work, both the input parameters of water and the climatic conditions of the work are controlled. Repeatability of results is up to 60%, because of the lack of accuracy of the measuring equipment and the probable presence of unaccounted factors affect.

The data obtained to this time allow us to assert that the rate of evaporation of distilled water can vary between 11% and +16%. The parameters of water flow and magnetic field induction such as physical and chemical properties, the optical density of water in the infrared range of wavelengths (+4.1% $\div$ −1.7%) and the solubility of oxygen in water change were identified. This fact can be used when calibrating magnetic systems for maximum impact efficiency. If we take water with different degrees of purification as an object of study or do not consider the trajectory and flow velocity, then the results differ radically. Thus, we anticipate that the thermophysical properties of the working fluid can be changed due to a second-order phase transition. In this case, under the action of a force field, hydrogen bonds in a water cluster are replaced by bonds of ion-ion, ion-dipole, and dipole-dipole interactions. As a result, new clusters are formed, but with different properties. The numerical values and the direction of change in properties are determined by the ratio between the number of new and old clusters, their sizes, and the number of "free" water molecules.

Thus, if the goal of research is to change the properties of water without introducing additional chemical reagents into the water, it is necessary to partially destroy and partially deform water clusters. The destruction or deformation of hydrogen bonds is not particularly difficult due to their low energy. Increasing the return time to the initial state is a significantly more difficult task. Solving this problem will bring this scientific direction to a new qualitative level.

Up to now, there is no information about the reagent-free control of the thermophysical properties of the working body of TPPs and CHPs and the change in the efficiency of the thermodynamic cycle associated with this effect.

This paper presents the results of experimental studies on the influence of a transverse magnetic field on the heat capacity of distilled water.

The studies were conducted on a thermogasdynamic stand developed and manufactured at the Institute of Mechanical Engineering of the National Academy of Sciences of Ukraine to study changes in the properties of water under the influence of force field, The functional diagram is shown in Fig. 2.

A constant amount of heat was transferred to a constant volume of treated water. The heating rate was chosen as a criterion parameter. Distilled water enters the stand input, which has undergone magnetic treatment in a transverse magnetic field generated by permanent magnets. The magnetic field strength can be changed.

The stand works as follows: water enters the heating chamber through the tap K2; when tap K4, structurally located in the upper part of the heating chamber is open; after filling the heating chamber (this is evidenced by the outflow of water through the tap K4), the tap K2 is close; in this way, a certain amount of still water is heated. Water can enter the heating chamber at different flow rates, the value of which is determined by the K2 tap. Flow control is carried out by a rotameter. Measurement of magnetic induction is carried out by teslameter EM4305.

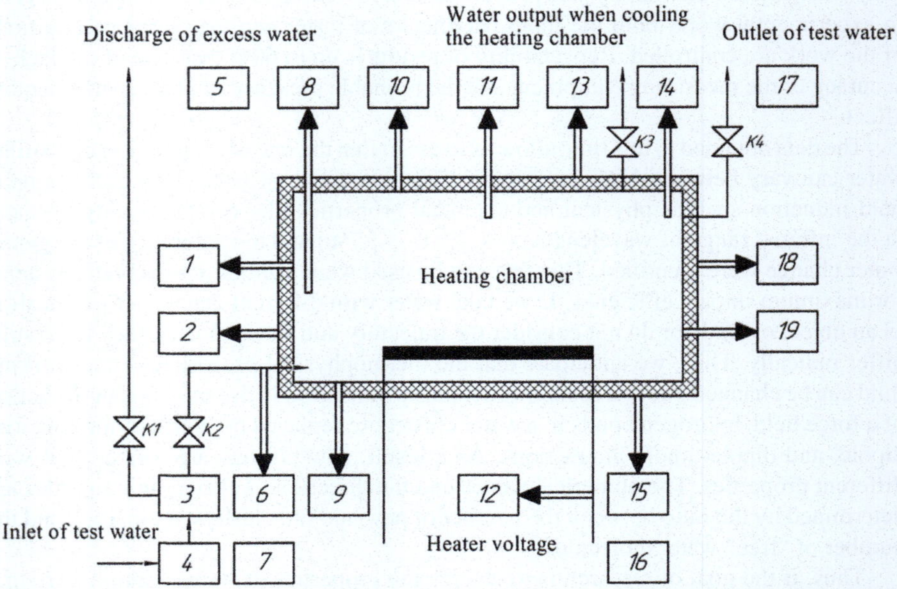

Fig. 2. Research stand for measuring the heating rate of liquid media (functional diagram). 1,9,10,19 - temperature sensors of the surface of the heating chamber; 2,13,15,18 - casing surface temperature sensors; 4-magnetic device; 5,7,16,17 - air temperature sensors; 6-inlet water temperature sensor; 8.14-water temperature sensors inside the heating chamber; 11-conductometric sensor; 12 - voltmeter.

The work on the stand was carried out according to the following method. The water was heated from 25 to 70°C. As an informative temperature value, the arithmetic mean of the readings of the sensors located inside the heating chamber was chosen. All temperature sensors mounted on the stand were selected using an exemplary mercury thermometer with an error of 0.1 °C.

To provide measurements in real time, the iButton TMEX™ SDK software kit by Dallas Semiconductor, US was installed on the PC. To exclude the influence of the supply voltage of heating elements on the heating rate, this parameter was calculated using the formula:

$$V = k/U * A * \tau, \qquad (5)$$

where:

$k = 45^0C/(B*A*c)$ – coefficient taking into account the heating temperature range;
V - the reduced heating rate;
U - electric voltage supplied to the heating elements;
A - electric current consumed by heating elements.

4 Results and Discussion

The results of experimental studies are shown in the form of graphs (Figs. 3 and 4). The final graphs included only those results that were obtained at the same ambient temperature of 20ºC and at a supply voltage of heating elements of 50V.

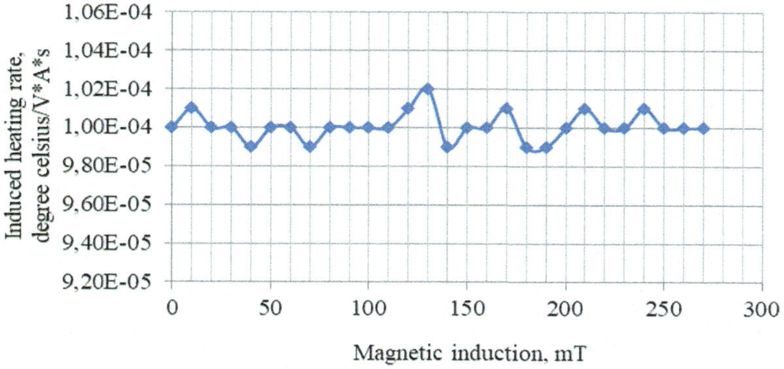

Fig. 3. Typical change in the reduced water heating rate for flow consumption of 100, 300, 400, 500, 600, 700, 800 dm^3/h.

The graph (Fig. 3) shows that at the indicated water flow rates the reduced heating rate changes by no more than 2%. This change is caused by the variability of the supply voltage of the heating elements.

The graph (Fig. 4) shows that the reduced heating rate during water treatment with a transverse magnetic field, considering the values of magnetic induction and water flow consumption, established during the research through the working section of the magnetic apparatus, can vary within -6% and + 5%. We believe that this corresponds to a change in the heat capacity of water within the same limits.

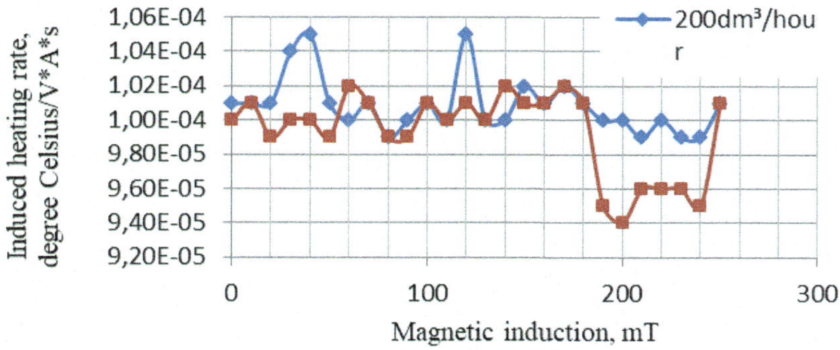

Fig. 4. Change in the reduced water heating rate at a flow consumption of 200 dm^3/hour and 250 dm^3/hour.

Thus, as follows from the literature review and our own experimental studies, it is possible to significantly change the thermodynamic characteristics of the working fluid. At the same time, there is a tendency for a simultaneous increase or decrease in specific heat capacity and evaporation, or latent heat of vaporization.

Using the obtained dependence 5, we estimate the influence of the electromagnetic field on water for the theoretical Rankine cycle with the following parameters [16]:

$P_1 = P_4 = P_5 = P_6 = 11$MPa; $P_2 = P_3 = 0.004$MPa; $t_1 = 550^0$C; $t_2 = t_3 = t_4 = 28.98^0$C; $t_5 = t_6 = 318.04$ °C; $q_1 = 3368.6$ kJ/kg; q2 = 1903.6 kJ/kg; Cycle efficiency = 0.435; $r_{\Pi} = 1256.16$ kJ/kg-K (P = 11MPa, $t_4 = 318.04$ °C); $C_{p\Pi} = 2.54$ (t1 = 550 °C).

Estimated efficiency is 0.435 with $C_{pB} = 4.19$ and $C_{p\Pi} = 3.89$ for raw water.

Such values can be taken only for evaluation calculations, for further research it is necessary to approximate the change in the heat capacity of water and steam from temperature at a given pressure and introduce the obtained analytical dependences under the integral sign.

Numerical studies have shown that with an increase in C_{pB} by 5%, the thermal efficiency of the Rankine cycle will be 0.447 (an increase of 1.2%).

The results obtained might be regarded as indicative (qualitative) only. An additional increase in cycle efficiency is possible by increasing the latent heat of vaporization and decreasing the latent heat of condensation, because at the same time, the heat supplied to the cycle will increase and the heat removed will decrease. Although a significant effect of changes in the thermodynamic parameters of water on the efficiency of the Rankine cycle has been theoretically proven, in fact, in the process of heating we have a new working fluid, the properties of which have not been fully studied, which indicates the need to continue research.

5 Conclusion

The conducted studies have shown that the use of a second-order phase transition, which occurs when a force field is purposefully applied to water, makes it possible to increase the thermal efficiency of the Rankine cycle. The direction of this action is determined, the results of which should be an increase in the specific isobaric heat capacity of water and a corresponding increase in the heat of vaporization.

Unlike traditional methods of increasing the efficiency of the cycle of steam turbine plants (Rankine cycle), the implementation of this method does not require a change in design, replacement of materials used in the creation of equipment and significant material costs. To be able to use a second-order phase transition, it is necessary to continue experimental studies of the effect of force fields on water. The goal is to obtain a positive effect at real flow consumption of the working fluid and temperatures of the thermo-dynamic cycle of steam turbine plants. The duration of change in properties should be long enough for use at any stage of the cycle.

Acknowledgements. This work is made by the support and within usage of lab and equipment of the IPMash NAS of Ukraine.

References

1. Dudnik, O.M.: Novi parogazovi ta gibrydni ustanovki na palyvnykh elementakh. Zbirka naukovykh prats' XV Mizhnarodnoi konferencii "Vugil'na teploeneregetyka: shlyakh reconstrucii ta rozvytku", pp. 40–43, Kyiv (2019)
2. Kim, K., Kim, J., Kim, C., Lee, Y., Lee, W.B.: Robust design of multicomponent working fluid for organic Rankine cycle. Ind. Eng. Chem. Res. **58**(10), 4154–4167 (2019), Publication Date: February 20, 2019. https://doi.org/10.1021/acs.iecr.8b04825, Copyright © 2019 American Chemical Society
3. International Journal of Engineering Applied Sciences and Technology, 2020 Vol. 5, Issue 6, ISSN No. 2455–2143, Pages 111–115 Published Online October 2020 in IJEAST (http://www.ijeast.com) Fundamentals of Thermal Power Generation Mohammed Elamin Department of Mechanical and Aerospace Engineering, North Carolina State University, Raleigh, NC 27695, United States https://www.researchgate.net/publication/346495460_FUNDAMENTALS_OF_THERMAL_POWER_GENERATION
4. Shalatonin, V., Pollack, G.H.: Effect of unipolar magnetic fields on UV absorption and evaporation of water. Chem. Phys. Impact **4**, 18 (2022). https://doi.org/10.1016/j.chphi.2022.100077
5. Cai, R., Yang, H., He, J., et al.: The effects of magnetic fields on water molecular hydrogen bonds. J. Molecu. Struct. **938**(1–3), 15–19 (2009). https://doi.org/10.1016/j.molstruc.2009.08.037
6. Wang, Y., Zhang, B., Gong, Z., et al.: The effect of a static magnetic field on thehydrogen bonding in water using frictional experiments. J. Molecul. Struct. **1052**(1), 102–104 (2013). https://doi.org/10.1016/j.molstruc.2013.08.021
7. Jawad, S., Karkush, M., Kaliakin, V.: Alteration of physicochemical properties of tap water passing through different intensities of magnetic field. J. Mech. Behav. Mat. **32**(1), id.246, 9pp (2023). https://doi.org/10.1515/jmbm-2022-0246
8. Li, Y., Lin, H., Yang, Z.: The effect of magnetic field on freezing point of water. Eng. J. Phys. Conf. Ser. **2194**(1) (2022). https://doi.org/10.1088/1742-6596/2194/1/012034
9. Chang, K.-T., Weng, C.-I.: The effect of an external magnetic field on the structure of liquid water using molecular dynamics simulation. J. Appl. Phys. **100**(4), 043917 (2006). https://doi.org/10.1063/1.2335971
10. Wang, Y., Wei, H., Li, Z.: Effect of magnetic field on the physical properties of water. Res. Phys. **8**, 262–267 (2018). https://doi.org/10.1016/j.rinp.2017.12.022
11. Duenas, J.A., Weiland, C., García-Selfa, I., Ruíz-Rodríguez, F.J.: Magnetic influence on water evaporation rate: An empirical triadic model. J. Magnet. Magnet. Mat. **53**, 168377 (2021). https://doi.org/10.1016/j.jmmm.2021.168377
12. Guo, Y.-Z., et al.: Evaporation rate of water as a function of a magnetic field and field gradient. Int. J. Molecul. Sci. **13**(12), 16916–16928 (2012). https://doi.org/10.3390/ijms131216916
13. Szcześ, A., Chibowski, E., Hołysz, L., et al.: Effects of static magnetic field on water at kinetic condition. Chem. Eng. Process. Process Intensif. **50**(1), 124–127 (2011). https://doi.org/10.1016/j.cep.2010.12.005
14. Malkin, E.S., Furat, I.E., Kovalenko, N.O., Sepik, A.V.: Vplyv magnitnoi obrobky na pytomy teplotu paroutvorennya vody. Ventylyatsia, osvitlennya ta teplogazopostachannya. Vyp. **17**, 77–83 (2014)
15. Energo i resusrsosberigayushie tekhnologii v energetike I energomashinostroenii/ Pod. Red. A.A. Tarelina. 2017. Kyev. Naukova Dumka. 272 p.
16. Teorytychni osnovy teplotekhniky. Vyznachennya efectyvnosti terrmodynamichnykh cycliv teplovykh dvyguniv: rozrakhunkova robota [Electronnyi resurs] : navch. posib. dlya stud. spetsialnostei 131 «Prykladna mekhanika» / V. V. Dubrovska, V.I. Shklyar: Electronni tekstovi dani Kyiv: KPI ім. Igora Sykorskogo, 31 p (2021)

Experimental Adjustment of a Double-Circuit Closed Loop System of Active Silencing of the Magnetic Field Generated by a Double-Circuit Power Line

Borys Kuznetsov[1] (✉), Ihor Bovdui[1], Olena Voloshko[1], Tetiana Nikitina[2], and Valerii Kolomiiets[2]

[1] Anatolii Pidhornyi Institute of Power Machines and Systems of the National Academy of Sciences of Ukraine, Komunalnykiv Str. 2/10, Kharkiv 61046, Ukraine
kuznetsov.boris.i@gmail.com

[2] Bakhmut Education Research and Professional Pedagogical Institute V.N. Karazin Kharkiv National University, 9a, Nosakov Str., Donetsk Region, Bakhmut 84511, Ukraine
tatjana55555@gmail.com, nnppiuipa@ukr.net

Abstract. The method of adjustment of a double-circuit closed loop robust system of active silencing of the magnetic field generated by double-circuit overhead transmission lines with a "Barrel" type arrangement of wires in multy-storey residential building based on experimental measurements magnetic field space-time characteristics developed at first time. The adjustment of the channel regulators carried out in such a way that their magnetic field space-time characteristics are orthogonal to each other with separate operation of only the first and only the second channels. The problem of design of two degree of freedom robust two circuit system of active shielding of initial magnetic field comes down to a decision of the multi-criteria game. The payoff vector game calculated based on Ampere-Laplace law. The game solution calculated based on the stochastic particles multiswarm optimization algorithms. Experimental adjustment of the model of double circuit system of active silencing of magnetic field, generated by double-circuit overhead transmission lines with a "Barrel" type arrangement of wires in multy-storey residential building based on experimental measurements space-time characteristics are given. The possibility of initial magnetic flux density level reducing to the living safe level has been experimentally shown.

Keywords: Overhead Power Lines · Magnetic Field · System of Active Silencing · Adjustment Field Experimental Study

1 Introduction

Overhead power lines often pass near residential buildings. These transmission lines generate a magnetic field inside residential buildings, the level of which usually exceeds the safe level of the magnetic field for residential buildings. Exposure to a power frequency magnetic field of even low induction for a long time leads to cancer in people living

H. Altenbach et al. (Eds.): CAMPE 2023, pp. 64–72, LNME 2025.
https://doi.org/10.1007/978-3-031-82979-6_7

nearby overhead power lines [1]. One of the most economically justified approaches for the safe operation of such residential buildings is to reduce the magnetic field to a level that is safe for living by means of active silencing [2–4].

2 Literature Review

Systems of active silencing of the power frequency magnetic field inside residential premises are widely used in world practice [4–6]. In Ukraine near residential buildings often pass double-circuit power lines with a "Barrel" type arrangement of wires. For active silencing of such magnetic field generated by double-circuit power lines in a multi-storey building at least two compensation windings are usually required. Its implementation with the necessary efficiency, about 5 [4], is provided only by methods of active silencing of the magnetic field [7–11].

However, in the works of [12–15], when synthesizing the system, uncertainties in the model of the initial magnetic field were not taken into account and there is no method for adjustment of a double-circuit closed loop system of active silencing of the magnetic field generated by double-circuit overhead transmission lines with a "Barrel" type arrangement of wires in multy-storey residential building based on experimental measurements magnetic field space-time characteristics.

This research contributes the development of method of experimental adjustment of a double-circuit closed loop system of active silencing of the magnetic field generated by a double-circuit power line. This improvement consists possibility to increase the value of the silencing factor of the initial magnetic field to the population safe level.

The paper consists of fifth sections. The first section is introduction. The second section concentrates on analysis of related works and problem statement. Third section describes research methodology. Fourth section describes the results. Fifth section is conclusion.

3 Research Methodology

For the geometric dimensions of double-circuit power transmission line with a "Barrel" type arrangement of wires, it is necessary to design an active silencing system for the magnetic field in the space of a multi-storey building model. Let us introduce the uncertainty vector δ of the parameters of the initial magnetic field model. In particular, the components of the uncertainty vector δ of the model parameters are the inaccurately specified geometric parameters of the power transmission line wires that change during operation, as well as the magnitude of the currents in these wires.

The currents in the wires have periodic hourly, daily, weekly, seasonal and annual changes. Note that a change in currents in a double-circuit transmission line leads not only to a change in the induction of the magnetic field generated by the transmission line, but also to a change in the spatiotemporal characteristics of this magnetic field. For the given instantaneous values of the current $I_k(\delta, t)$ in the k-th wire of the power transmission line at time t and their spatial location, we calculate the vector $\boldsymbol{B}_p(Q_j, \delta, t)$ of the instantaneous value of the induction of the initial magnetic field generated by all K wires of the power transmission line at points Q_j of the space of a multi-storey residential building, in which it is necessary to actively suppress the initial magnetic field. The vector $\boldsymbol{B}_p(Q_j, \delta, t)$ of the instantaneous value of the induction of the initial magnetic field calculated by applying a discrete approximation of the Ampere-Laplace law [16, 17] in the differential form of k-th conductors of a sufficiently small length based on the expression

$$\begin{aligned} \boldsymbol{B}_p(Q_j, \delta, t) = \tfrac{\mu_0}{4\pi} \sum_{k=1}^{K} I_{pk}(\delta, t) * \ldots \\ \ldots * \sum_{i=1}^{n} \tfrac{\boldsymbol{UL}_{ki} \times \boldsymbol{UR}_{ki}}{R_{ki}^2} \Delta L_{ki} \end{aligned} \tag{1}$$

where: μ_0 - permeability of the medium: ΔL_{ki} - length of i-th source element current of k power line wires: $\boldsymbol{UL}_{ki}$ - unit vector tangent to the i-th source element current ΔL_{ki} of k power line wires: R_{ki} - distance between point Q_j and i-th source element current of k power line wires: $\boldsymbol{UR}_{ki}$- unit vector tangent to R_{ki}.

Let us set the vector X of the desired design parameters of the active silencing system, the components of which are the geometric parameters of the compensating windings and the desired parameters of the regulators for open and closed current control of the compensating windings of the active jamming system with two degrees of freedom [15]. Let us also introduce the vector δ of parameters of uncertainties of the designed system [18–22], the components in the form of deviations of the real parameters of the designed system from their values adopted in the design of the system.

Then, for the given geometric dimensions of the compensating windings, which are components of the desired vector X of the designed system of active silencing of the magnetic field in the space of the model of a multi-storey building, and the back instantaneous values $I_{wk}(X, \delta, t)$ of the current in the wires of the compensating windings and their spatial arrangement, we calculate the vector $\boldsymbol{B}_w(Q_j, X, \delta, t)$ instantaneous value of the magnetic induction field generated by all compensating windings in space points Q_j of a multi-storey residential building, in which it is necessary to actively keep silent the initial magnetic field.

By analogy with (1), the vector $\boldsymbol{B}_w(Q_j, X, \delta, t)$ of the instantaneous value of the magnetic field induction generated by all compensating windings at the points Q_j is calculated by applying a discrete approximation of the Ampere-Laplace law in the differential form of k-th conductors of a sufficiently small length ΔL_{wki} based on the expression

$$\begin{aligned} \boldsymbol{B}_w(Q_j, X, \delta, t) = \tfrac{\mu_0}{4\pi} \sum_{k=1}^{K} I_{wk}(X, \delta, t) * \ldots \\ \ldots * \sum_{i=1}^{n} \tfrac{\boldsymbol{UL}_{wki} \times \boldsymbol{UR}_{wki}}{R_{wki}^2} \Delta L_{wki} \end{aligned} \tag{2}$$

where: ΔL_{wki} – length of i-th source element current of k silencing winding: $\boldsymbol{UL}_{wki}$ – unit vector tangent to the i-th source element current ΔL_{wki} of k silencing winding: R_{wki} – distance between points Q_j and i-th source element current of k silencing winding: $\boldsymbol{UR}_{wki}$ - unit vector tangent to R_{wki}.

Let us calculate the vector $\boldsymbol{B}_R(Q_j, X, \delta, t)$ of the instantaneous value of the induction of the resulting magnetic field, as the sum of the vector $\boldsymbol{B}_P(Q_j, X, \delta, t)$ of the instantaneous value of the magnetic field induction, generated by all wires of the power transmission line, and the vector $\boldsymbol{B}_w(Q_j, X, \delta, t)$ of the instantaneous value of the magnetic field induction generated by all compensating windings at the points Q_j in the space of a multi-storey residential building, in which it is necessary to actively suppress the initial magnetic field

$$\boldsymbol{B}_R(Q_j, X, \delta, t) = \boldsymbol{B}_p(Q_j, \delta, t) + \boldsymbol{B}_w(Q_j, X, \delta, t) \tag{3}$$

To implement the control algorithm of a closed loop system of active silencing of the initial magnetic field, feedback is used on the vector $\boldsymbol{B}_R(Q_j, X, \delta, t)$ of the instantaneous value of the induction of the resulting magnetic field in the space of a multi-storey residential building, in which it is necessary to actively keep silent the initial magnetic field. However, sanitary standards set safe level for the population with an induction of 0.5 μT limits the effective value of the induction vector in the residential area. Let us calculate the effective value $\boldsymbol{B}(Q_j, X, \delta)$ of the induction by integrating the square of the modulus of the vector $\boldsymbol{B}_R(Q_j, X, \delta, t)$ of instantaneous value over the period T of its change

$$B(Q_j, X, \delta) = \sqrt{\frac{1}{T}\int_0^T \left|\boldsymbol{B}_R(Q_j, X, \delta, t)\right|^2 dt}. \tag{4}$$

Then the problem of designing a robust system is reduced to computing the solution of the vector game

$$\begin{aligned} B(X, \delta) = \big[B(Q_1, X, \delta), B(Q_2, X, \delta)\ldots \\ \ldots B(Q_m, X, \delta)\big]^T \end{aligned} \tag{5}$$

The components of payoff vector $B(X, \delta)$ game (5) are the effective values $B(Q_i, X, \delta)$ of the induction vector at the m points Q_j in the silencing space.

At the same time, naturally, it is necessary to take into account constraints on the vector X of the desired design parameters of the active silencing system and on the vector δ of parameters of uncertainties of the designed system in the form of vector inequality and, possibly, vector equality

$$G(X, \delta) \le G_{\max}, H(X, \delta) = 0. \tag{6}$$

Note that the components of the payoff vector (5) and constraints (6) are nonlinear functions of the vector X of the desired design parameters of the active silencing system and on the vector δ of parameters of uncertainties of the designed system and their calculation is performed on the basis of the Ampere-Laplace law [16, 17].

Let consider the method of solving the formulated problem. In order to correctly solve the problem of multi-criteria optimization [23, 24], in addition to the vector optimization criterion and constraints, it is also necessary to have information about the binary relationship of preference of local solutions to each other [22]. The basis of this formal approach is the construction of areas of Pareto-optimal solutions [24]. This approach makes it possible to significantly narrow the range of possible optimal solutions to the original multi-criteria optimization problem and, therefore, to reduce the labor intensity of the decision-maker regarding the selection of a single variant of the optimal solution. The problem of finding a local minimum at one point of the considered space is, as a rule, multi-extreme, containing local minima and maxima, therefore, it is advisable to use stochastic multi-agent optimization algorithms for its solution. Let us consider the algorithm for finding a set of Pareto-optimal solutions of multi-criteria nonlinear programming problems based on stochastic multi-agent optimization.

To date, a large number of particle swarm optimization algorithms have been developed – PSO algorithms based on the idea of the collective intelligence of a particle swarm, such as the gbest PSO and lbest PSO algorithms. The application of stochastic multi-agent optimization methods for solving multi-criteria problems today causes certain difficulties and this direction continues to develop intensively. To solve the original multi-criteria problem of nonlinear programming with constraints, we will build a stochastic multi-agent optimization algorithm based on a set of particle swarms, the number of which is equal to the number of components of the vector optimization criterion [24].

4 Results

Consider design of system of active silencing of magnetic field generate by double-circuit power lines with a "Barrel" type arrangement of wires in multy-storey residential buildings, as shown in Fig. 1a.

a) b)

Fig. 1. Double-circuit power lines with a "Barrel" type arrangement of wires a) and its model b).

To conduct experimental studies, a model of a double-circuit power transmission line with a "Barrel" type arrangement of wires was developed, the photo of which shown in Fig. 1b.

A model of a two-circuit system of active silencing of such a magnetic field, containing two silencing windings, has also been developed. Figure 2a shows two windings of such a model.

These magnetic field sensors installed in the space of active silencing of the initial magnetic field to implement closed-loop magnetic field control algorithms for both silencing windings. Figure 2b also shows the third three-component magnetic field sensor, with the help of which the spatio-temporal characteristic of the resulting magnetic field is measured.

a) b)

Fig. 2. Two windings a) and magnetic field sensors b) of system of active silencing.

The control part of the active silencing system is shown in Fig. 3, including power amplifiers for supplying silencing windings, amplifiers for the measuring system of the space-time characteristic, amplifiers for magnetic field sensors and the current regulators of the silencing windings.

Fig. 3. Control part of the active silencing system.

To implement the closed-loop control of these silencing windings, two magnetic field sensors are used, photographs of which are shown in Fig. 4.

Consider the results of setting up the system. Figure 4a shows the spatio-temporal characteristic of the magnetic field generated by the power transmission line and only one winding of the first channel, and Fig. 4b shows the spatio-temporal characteristic of the magnetic field generated by the power transmission line and only one winding of the second channel.

The adjustment of the regulator of each channel was carried out in such a way that the spatio-temporal characteristic magnetic field during the operation of one channel was orthogonal to the spatio-temporal characteristic magnetic field during the operation of the

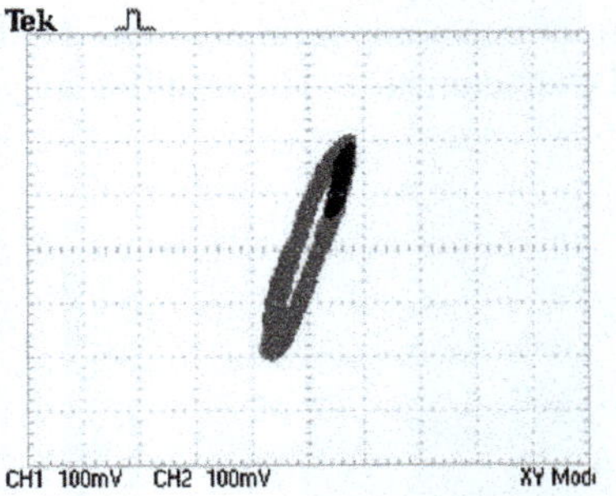

a)

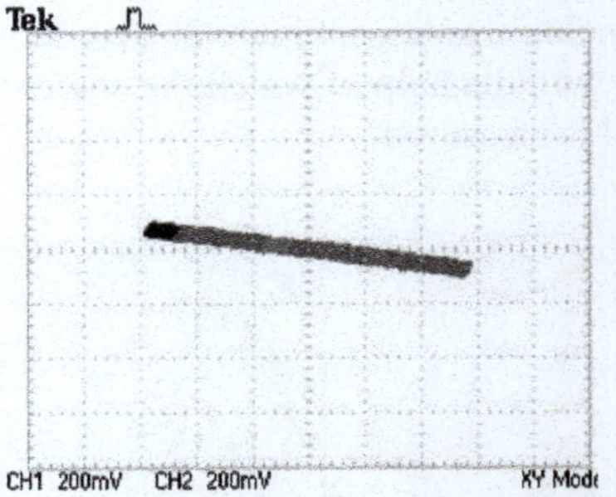

b)

Fig. 4. Spatio-temporal characteristic of the magnetic field generated by the power transmission line and only one winding of the first channel a), and generated by the power transmission line and only one winding of the second channel b).

second channel. With such an adjustment of the regulators of individual channels, with the joint operation of both channels in the system, the greatest efficiency of suppression of the initial magnetic field is realized. The experimentally confirmed level of silencing of the initial magnetic field in the entire silencing space is more than 5 units.

5 Conclusions

1. For the first time the method of adjustment of a double-circuit robust system of the active silencing of the magnetic field generated by double-circuit overhead transmission lines with a "Barrel" type arrangement of wires in multy-storey residential building developed. The method based on experimental measurements magnetic field space-time characteristics.
2. The adjustment of the regulator of each channel carried out in such a way that the magnetic field space-time characteristics during the operation of only first channel was orthogonal to the magnetic field space-time characteristics during the operation of the only second channel.
3. The problem of design of two degree of freedom robust two circuit system of active silencing of initial magnetic field comes down to a solution of the multi-criteria game. The vector payoff of this game calculated based on Ampere-Laplace law. The solution of this game calculated based on of multi-swarm multi-agent optimization algorithms from of Pareto-optimal set solutions.
4. Spatial location coordinates of the two silencing windings, the currents in the both shielding windings and the vector of parameters of regulators of open loop and closed loop control were determined during the design of two degree of freedom robust system of active silencing of magnetic field.
5. The field model of two circuit system of the active silencing of the magnetic field was developed. Calculated and field experimentally measured space-time characteristics of initial magnetic field of overhead power line and silencing coils of field model of system of the active silencing during their autonomous and simultaneous work are given.

6. With the help of the developed method of adjustment of a two circuit robust system of the active silencing based on experimental measurements space-time characteristics the level of the magnetic field generated by a overhead transmission lines with a "Barrel" type arrangement of wires in multy-storey residential building is reduced by more than 5 times.

Acknowledgment. The authors express their gratitude to the engineers Sokol A.V. and Shevchenko A.P. of the department of magnetism of technical object of Anatolii Pidhornyi Institute of Power Machines and Systems of the National Academy of Sciences of Ukraine for the creative approach and courage shown during the creation under fire, under martial law, of an experimental installation and successful testing of a laboratory model of the system of active silencing.

References

1. Global Cancer Statistics: GLOBOCAN Estimates of Incidence and Mortality Worldwide for 36 Cancers in 185 Countries. Enhanced Reader (2020). Accessed 28 Feb 2023
2. Bravo-Rodríguez, J., Del-Pino-López, J., Cruz-Romero, P.: A survey on optimization techniques applied to magnetic field mitigation in power systems. Energies **12**(7), 1332 (2019). https://doi.org/10.3390/en12071332
3. Salceanu, A., Paulet, M., Alistar, B.D., Asiminicesei, O.: Upon the contribution of image currents on the magnetic fields generated by overhead power lines. 2019 International Conference on Electromechanical and Energy Systems (SIELMEN) (2019). https://doi.org/10.1109/sielmen.2019.8905880
4. Thiab, H.G., JaduAli, K., Mutlaq, A.H.: The influence of the mixed electric line poles on the distribution of magnetic field. Indonesian J. Electr. Eng. Inform. (IJEEI) **10**(2), 292–301 (June 2022). ISSN: 2089-3272. https://doi.org/10.52549/ijeei.v10i2.3572
5. Victoria Mary, S., Pugazhendhi Sugumaran, C.: Investigation on magneto-thermal-structural coupled field effect of nano coated 230 kV busbar. Phys. Scr. **95**(4) (2020). https://doi.org/10.1088/1402-4896/ab6524
6. Del Pino Lopez, J.C., Cruz, R.P.: Influence of different types of magnetic shields on the thermal behavior and ampacity of underground power cables. IEEE Trans. Power Delivery **26**(4), 2659–2667 (2011). https://doi.org/10.1109/tpwrd.2011.2158593
7. Barsali, S., Giglioli, R., Poli, D.: Active shielding of overhead line magnetic field: design and applications. Electr. Power Syst. Res. **110**, 55–63 (2014). https://doi.org/10.1016/j.epsr.2014.01.005
8. Bavastro, D., Canova, A., Freschi, F., Giaccone, L., Manca, M.: Magnetic field mitigation at power frequency: design principles and case studies. IEEE Trans. Ind. Appl. **51**, 2009–2016
9. Beltran, H., Fuster, V., García, M.: Magnetic field reduction screening system for a magnetic field source used in industrial applications. 9 Congreso Hispano Luso de Ingeniería Eléctrica (9 CHLIE), Marbella (Málaga, Spain), pp. 84–99 (2005)
10. Canova, A., del-Pino-Lopez, J.C., Giaccone, L., Manca, M.: Active shielding system for ELF magnetic fields. IEEE Trans. Magnet. **51**(3), 1–4 (2015). https://doi.org/10.1109/tmag.2014.2354515
11. Canova, A., Giaccone, L.: Real-time optimization of active loops for the magnetic field minimization. Int. J. Appl. Electromagnet Mech **56**, 97–106 (2018). https://doi.org/10.3233/jae-172286

12. Aldo, C., Luca, G.: High performance magnetic shielding solution for elf sources. 24th International Conference On Electricity Distribution Glasgow, 12–15 June 2017. Paper 1029, pp. 1–5
13. Celozzi, S.: Active compensation and partial shields for the power-frequency magnetic field reduction. In: Proceedings of IEEE International Symposium on Electromagnetic Compatibility, pp. 222–226. Minneapolis, USA (2002)
14. Celozzi, S., Garzia, F.: Active shielding for power-frequency magnetic field reduction using genetic algorithms optimization. IEE Proc. Sci. Meas. Technol. **151**(1), 2–7 (2004). https://doi.org/10.1049/ip-smt:20040002
15. Aldo, C., Luca, G., Vincenzo, C.: Active and passive shield for aerial power lines. 25th International Conference on Electricity Distribution Madrid, 3–6 June 2019. Paper n°1096, pp.1 –5
16. Rozov, V., Grinchenko, V.S., Yerisov, A.V., Dobrodeyev, P.N.: Effective circuit shielding of a magnetic field of three-phase cable lines with a limited thermal influence on power cables. Electr. Eng. Electromechanics **6**, 50–54 (2019)
17. Rozov, V.Y., Reutskyi, S.Y., Pelevin, D.Y., Kundius, K.D.: Approximate method for calculating the magnetic field of 330–750 kV high-voltage power line in maintenance area under voltage. Electr. Eng. Electromechanics **5**, 71–77 (2022). https://doi.org/10.20998/2074-272X.2022.5.12
18. Rusanov, A., Chugay, M., Rusanov, R.: Advanced computer technologies in the new flow part development for reactive type HPC steam turbine of T-100 series. Advances in Mechanical and Power Engineering. Selected Papers from The International Conference on Advanced Mechanical and Power Engineering (CAMPE 2021), pp. 55–63 (18–21 Oct. 2021)
19. Tarasova, V., Kostikov, A., Kuznetsov, M.: Thermodynamic analysis and optimization of the cycle of a CHP plant power unit operating with ultra-supercritical steam parameters. Advances in Mechanical and Power Engineering. Selected Papers from The International Conference on Advanced Mechanical and Power Engineering (CAMPE 2021), pp. 44–54 (18–21 Oct. 2021)
20. Ugrimov, S., Smetankina, N., Kravchenko, O.: A study of the dynamic response of materials and multilayer structures to shock loads. Advances in Mechanical and Power Engineering. Selected Papers from The International Conference on Advanced Mechanical and Power Engineering (CAMPE 2021), pp. 304–313 (18–21 Oct. 2021)
21. Solovey, V., Rusanov, A.: Integrated energy technologies when using natural gas in utility heat power engineering. Advances in Mechanical and Power Engineering. Selected Papers from The International Conference on Advanced Mechanical and Power Engineering (CAMPE 2021), pp. 78–87 (18–21 Oct. 2021)
22. Sushchenko, O., et al.: Algorithms for design of robust stabilization systems. In: Computational science and its applications – ICCSA 2022. Lecture Notes in Computer Science, vol. 13375, pp. 198–213. Springer, Cham (2022). https://doi.org/10.1007/978-3-031-10522-7_15
23. Hashim, F.A., Hussain, K., Houssein, E.H., Mabrouk, M.S., Al-Atabany, W.: Archimedes optimization algorithm: a new metaheuristic algorithm for solving optimization problems. Appl. Intell. **51**, 1531–1551 (2021). https://doi.org/10.1007/s10489-020-01893-z
24. Working group C4.204, Int. Council Lucio Ippolito, Pierluigi Siano. Using multi-objective optimal power flow for reducing magnetic fields from power lines. Electr. Power Syst. Res. **68**, 93–101 (2004)

Magnetic Field Active Silencing in One-Storey Old Buildings Closed to Power Line

Ihor Bovdui[1], Borys Kuznetsov[1](✉), Olena Voloshko[1], Tetiana Nikitina[2], and Borys Kobylianskyi[2]

[1] Anatolii Pidhornyi Institute of Power Machines and Systems of the National Academy of Sciences of Ukraine, Komunalnykiv Str. 2/10, Kharkiv 61046, Ukraine
kuznetsov.boris.i@gmail.com

[2] Bakhmut Education Research and Professional Pedagogical Institute V.N. Karazin Kharkiv National University, 9a, Nosakov Str., Bakhmut, Donetsk Region 84511, Ukraine
tatjana55555@gmail.com

Abstract. In this paper reduction of magnetic field generated by a single-circuit power line with a triangular arrangement of wires in a one-story residential old building by means of active silencing is proposed. The vector of the desired design parameters of the active silencing robust system, the components of which are the geometric parameters of the silencing windings and the desired parameters of the regulators for open and closed current control of the silencing windings calculated as vector minimax optimization problem. The vector objective function of this optimization problem calculated based on Biot-Savart's law. The solution of this optimization problem calculated based on nonlinear multiswarm stochastic optimization algorithms. Results of experimental studies of robust system of active silencing for a magnetic field generated by a single-circuit power line with a triangular arrangement of wires in a one-story residential old building are given. Based on results of experimental studies the possibility of reducing of initial magnetic flux density level to the living safe level by means of active silencing shown.

Keywords: Overhead Power Lines · Magnetic Field · System of Active Silencing · Field Experimental Study

1 Introduction

Near one-story houses of old buildings, single-circuit power lines with a triangular arrangement of wires often pass. These power lines generate a magnetic field of industrial frequency inside residential buildings, the level of which exceeds the safe level of 0.5 μT for life, adopted in Europe for residential buildings. The stay of the population for a long time in such a magnetic field of industrial frequency can lead to cancer [1]. One of the safest for the health of the population living in such houses and the most economically feasible way for the further operation of such houses is the use of active silencing methods for the magnetic field generated by overhead power lines [2–4].

H. Altenbach et al. (Eds.): CAMPE 2023, pp. 73–81, LNME 2025.
https://doi.org/10.1007/978-3-031-82979-6_8

2 Literature Review

The problem of reducing the level of the magnetic field in residential buildings located near overhead power lines is relevant for many countries [5–7]. Active shielding systems have long been successfully used in many developed countries [8–10]. The system of active silencing includes a power source, silencing coils, a magnetic field sensor and a control system [11–13]. The number of silencing windings determined by the initial parameters of the magnetic field, the size of the residential building in which it is necessary to keep level the magnetic field and transmission by the silencing efficiency. The most widely used in world practice are single-circuit active silencing systems containing one silencing coil. However, in the event of detection, two, three, six, and even 24 silencing windings are used to obtain the initial silencing of the magnetic field in a given silencing space. However, these works [14–17], when synthesizing the system, uncertainties in the model of the initial magnetic field were not taken into account. But, designed system of active silencing must be robust. In addition, the works [8–17] did not consider the issues of reducing the level of the magnetic field generated by a single-circuit power line with a triangular arrangement of wires in one-story residential buildings.

The aim of the work is to design, theoretical and experimental studies of system of active silencing for a magnetic field generated by a single-circuit power line with a triangular arrangement of wires in a one-story residential old building.

This research contributes the development of method for design of robust system for increased the efficiency of active silencing of magnetic field generated by a single-circuit power line with a triangular arrangement of wires in a one-story residential old building. This improvement consists possibility to increase the value of the silencing factor of the initial magnetic field to the population safe level.

The paper consists of fifth sections. The first section is introduction. The second section is literature review and concentrates on analysis of related works and problem statement. Third section describes research methodology. Fourth section describes the results. Fifth section is conclusion.

3 Research Methodology

First, we calculate the vector of the instantaneous value of the induction of the initial magnetic field generated by the power transmission line at a point Q_j in the silencing space. Let us introduce the uncertainty vector δ of the parameters of the initial magnetic field model. In particular, the components of the uncertainty vector δ of the model parameters are the inaccurately specified geometric parameters of the power transmission line wires that change during operation, as well as the wires currents. Let us set the instantaneous value $I_k(\delta, t)$ of the current in k wire of the power transmission line at time t in the form of the following sinusoidal dependence

$$I_K(\delta, t) = A_K(\delta)\sin(\omega(t) + \phi_K) \tag{1}$$

Here, A_K - the amplitude of the current in that wire of the power transmission line, ω - the frequency of the current of the power transmission line, ϕ_K - the phase of the current to that wire of the power transmission line. Then the value of the elementary vector

$\boldsymbol{dB}_{pk}(Q_j, \delta, t)$ of the instantaneous value of the induction generated by the elementary section dl_k of conductor k of the power transmission line with current $I_k(\delta, t)$ at the point Q_j calculated based on the Biot-Savart's law [18–19]

$$d\dot{\vec{B}}_{pk}(Q_j, \delta, t) = \mu_0 \cdot \dot{I}_k(\delta, t) \frac{\vec{r} \times d\vec{l}_k}{2\pi \cdot r^3} \tag{2}$$

where r is the point Q_j coordinate vector, r^3 is the cube of the distance to the point Q_j. Then the vector $\boldsymbol{B}_p(Q_j, \delta, t)$ of the instantaneous value of the induction of the initial magnetic field generated by the power transmission line is calculated as the sum of the elementary vectors $\boldsymbol{dB}_{pk}(Q_j, \delta, t)$ of the instantaneous value of the induction generated at the point Q_j by all elementary sections dl_k of all conductors k of the power transmission line with current $I_k(\delta, t)$.

Based on the calculated value of the induction vector $\boldsymbol{B}_p(Q_j, \delta, t)$, we calculate the spatio-temporal characteristic of the initial magnetic field generated by the power transmission line at the point Q_j. The spatio-temporal characteristic of the magnetic field is a line that describes the end of the vector of the instantaneous value of the induction when the time changes over time equal to the period of the current change in the wires of the power transmission line. The space-time characteristic of the magnetic field is the hodograph of the vector of the instantaneous value of the magnetic field induction when the instantaneous value of time changes during the period of change in the current of the power transmission line. The space-time characteristic of the magnetic field calculated for a given point Q_j in the silencing space and for a given value of uncertainty vector δ of the parameters of the initial magnetic field model.

Let us set the vector X of the desired design parameters [20–23] of the active silencing system, the components of which are the geometric parameters of the silencing windings and the desired parameters of the regulators for open and closed current control of the silencing windings of the active silencing system with two degrees of freedom.

We also introduce into the vector δ of uncertainty parameters [24] of the model (1)–(2) of the initial magnetic field generated by the power transmission line, the uncertainty parameters of the designed system, the components in the form of deviations of the real parameters of the designed system from their values adopted in the design of the system.

Let us set the instantaneous value $I_{wk}(X, \delta, t)$ of the current in k compensation winding at time t in the form of the following expression

$$I_{WK}(X, \delta, t) = A_{WK} \sin(\omega(t) + \phi_{WK}) + K_{WK} B_R(Q_J, X, \delta, t) \tag{3}$$

Here both the amplitude A_{WK} and phase ϕ_{wK} of the instantaneous value of the current in k compensation winding, which implements open-loop control. K_{WK} is the gain of the current controller in k silencing winding, which implements closed-loop control with feedback of the vector $\boldsymbol{B}_R(Q_j, X, \delta, t)$ of the instantaneous value of the induction of the resulting magnetic field at a point Q_j in the shielding space at time t.

The amplitude A_{WK} and phase ϕ_{wK} of the instantaneous value of the current as well as the gain K_{WK} of the current regulator in that compensation winding (3) are the components of vector X of the desired design parameters of the active silencing system.

Then the vector $\boldsymbol{B}_w(Q_j, X, \delta, t)$ of the instantaneous value of the induction of the silencing magnetic field generated by the silencing windings calculated likewise (2) as the sum of the elementary vectors of the instantaneous value of the induction generated at the point Q_j by all elementary sections of all conductors of the silencing windings with current $I_{wk}(X, \delta, t)$.

The vector $\boldsymbol{B}_R(Q_j, X, \delta, t)$ of the instantaneous value of the induction of the resulting magnetic field calculated as the sum of the vectors and the induction of the magnetic field generated by the power transmission line $\boldsymbol{B}_P(Q_j, X, \delta, t)$ and compensation windings $\boldsymbol{B}_w(Q_j, X, \delta, t)$. Using the vector $\boldsymbol{B}_R(Q_j, X, \delta, t)$ of instantaneous values of the magnetic field induction, we calculate the effective value $\boldsymbol{B}(Q_j, X, \delta)$ of the induction of the resulting magnetic field at the point Q_j in the silencing space.

Then the problem of designing a robust system of active silencing of the magnetic field reduced to solving the minimax optimization problem [24]

$$B(X^*, \delta^*) = \underset{X}{MIN}\ \underset{\delta}{MAX}\ \underset{Q_i}{MAX}\ B(X, \delta, Q_I) \tag{4}$$

The solution of this problem is the vector X^*, which minimizes the maximum value of the induction $\boldsymbol{B}(Q_j, X, \delta)$ over all points Q_j of the silencing space, and the vector δ^*, which maximizes the same maximum value $\boldsymbol{B}(Q_j, X, \delta)$ of the induction over all points Q_j of the silencing space. This approach is the standard to robust control systems designing.

This minimax optimization problem (4) reduced to solving a minimax vector optimization problem [24–25] with real X and δ vector variables. The solution of this minimax vector optimization problem calculated based on nonlinear multiswarm stochastic optimization algorithms [26–28].

4 Results

As an example, Fig. 1a shows the location of a one-story house in the immediate vicinity of a single-circuit power line with a triangular arrangement of wires. A full-scale model of the active silencing system has been developed for field experimental studies in field conditions. On Fig. 1b shows the spatial arrangement of the location of the model of this system relative to a single-circuit power line with a triangular arrangement of wires.

The active silencing system is double-circuit and contains two silencing windings. The spatial arrangement of these two silencing windings is shown in Fig. 2a. To implement closed-loop control by two silencing windings, two magnetic field sensors are used. Both of these sensors are located inside the silencing space and measure the resulting magnetic field generated in the silencing space by the transmission line and by two silencing windings. On Fig. 2.b shows the spatial arrangement of both magnetic field sensors inside the silencing space.

The control system contains two control channels for two silencing windings. Each control channel contains a power amplifier for supplying the silencing winding with a given current, an internal current regulator and an external regulator of the resulting magnetic field induction. On Fig. 3a shows the control part of the system of active silencing.

Let us now consider the results of field experimental studies. To conduct field experimental studies, a special measuring system developed, with the help of which the spatio-temporal characteristic of the magnetic field is quickly measured. The system contains

a) b)

Fig. 1. The location of a one-story buildings a) closed to power line and model b) of system of active silencing.

a) b)

Fig. 2. Two windings a) and two magnetic field sensors b) of system of active silencing.

two magnetic field sensors, the axes of which are orthogonal to each other. The voltages from the outputs of these magnetic field sensors are fed to the vertical and horizontal scan inputs of the oscilloscope. With the help of such a system, the space-time characteristic of the magnetic field is measured at the installation point of the magnetic field sensors of the measuring system. Before measurements, the spatial and angular positions of the sensors of the measuring system are adjusted, and the controllers of the channels of the measuring system are adjusted for the identity of signal delays in the channels of the vertical and horizontal scans of the oscilloscope.

On Fig. 3b shows the experimentally measured space-time characteristic of the initial magnetic field generated by a single-circuit power transmission line with a triangular arrangement of wires. The spatio-temporal characteristic of this initial magnetic field has the form of a circle.

We adjust the external regulators of the magnetic field induction of one-channel channels during their operation. The parameters of the regulators of individual channels adjusted in such a way that the spatio-temporal characteristic of the resulting magnetic field generated by the power transmission line and only single first channel is orthogonal to the spatiotemporal characteristic of the resulting magnetic field generated by the power transmission line and only single second channel.

Figure 4a shows the experimentally measured spatio-temporal characteristic of the resulting magnetic field generated by the power transmission line and only single first channel, and Fig. 4b shows the experimentally measured spatio-temporal characteristic of the resulting magnetic field generated by the power transmission line and only single second channel. These experimentally measured space-time characteristics are practically orthogonal to each other.

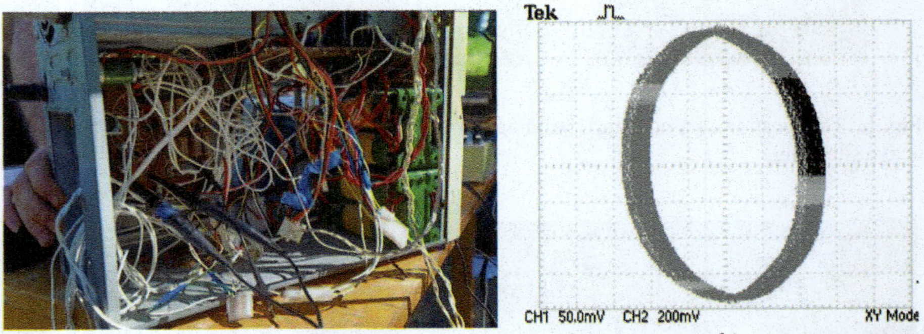

a) b)

Fig. 3. The control system a) and space-time characteristic b) of the initial magnetic field.

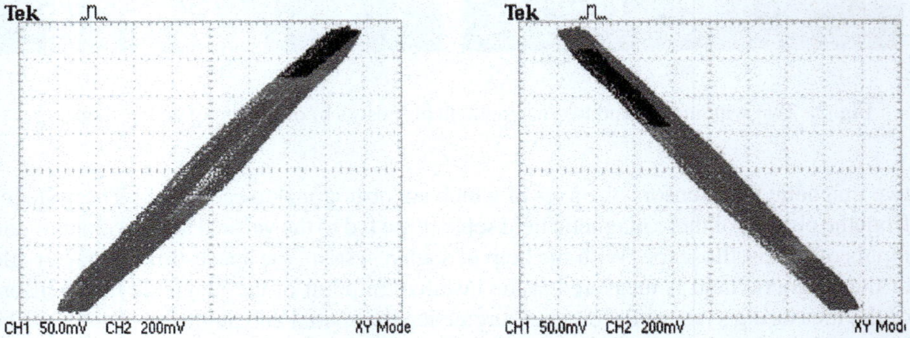

a) b)

Fig. 4. Spatio-temporal characteristic of the resulting magnetic field generated by only single first a) and only single second b) channels.

With this adjusting of the regulators of individual channels, during their independent operation with simultaneous operation of both channels, the experimentally measured spatio-temporal characteristic of the resulting magnetic field generated by the power transmission line and both channels, during their joint operation, practically turns into a point, the size of which is due to the noise of measuring the magnetic induction fields.

Such adjustment of the regulators of individual channels makes it possible to effectively compensate for the initial magnetic field generated by the power transmission line, with the joint operation of both control channels. The experimental value of the silencing factor is more than 5 units.

Based on theoretical and experimental studies, it has been shown, that the designed system of active silencing has a higher silencing factor compared to existing systems [12, 15]. In addition, the designed robust system has less sensitivity to changes in the parameters of the initial magnetic field and, therefore, this system is robust.

5 Conclusions

1. In this paper, we proposed reduction of magnetic field generated by a single-circuit power line with a triangular arrangement of wires in a one-story residential old building by means of active silencing.
2. The problem of design of two degree of freedom robust two circuit system of active silencing of initial magnetic field comes down to a solution of the vector minimax optimization problem. The solution of this problem calculated based on nonlinear multi-swarm stochastic optimization algorithms. Objective vector function of this minimax optimization problem is nonlinear vector function calculated based on Biot-Savart's law.
3. Based on field experimental measurements space-time characteristics the level of the magnetic field generated by overhead transmission lines with a triangular arrangement of wires in one-storey residential building is reduced by more than 5 times.

Acknowledgment. The authors express their gratitude to the engineers Sokol A.V. and Shevchenko A.P. of the department of magnetism of technical object of Anatolii Pidhornyi Institute of Power Machines and Systems of the National Academy of Sciences of Ukraine for the creative approach and courage shown during the creation under fire, under martial law, of an experimental installation and successful testing of model of the system of active silencing.

References

1. Global Cancer Statistics: GLOBOCAN Estimates of Incidence and Mortality Worldwide for 36 Cancers in 185 Countries. Enhanced Reader (2020). Accessed 28 Feb 2023
2. Bravo-Rodríguez J., Del-Pino-López J., Cruz-Romero P.A Survey on Optimization Techniques Applied to Magnetic Field Mitigation in Power Systems. Energies, 2019, vol.12, no.7, p. 1332. https://doi.org/10.3390/en12071332
3. Salceanu, A., Paulet, M., Alistar, B.D., Asiminicesei, O. Upon the contribution of image currents on the magnetic fields generated by overhead power lines. 2019 International Conference on Electromechanical and Energy Systems (SIELMEN) (2019). https://doi.org/10.1109/sielmen.2019.8905880
4. Ghanim Thiab Hasan, Kamil JaduAli, Ali Hlal Mutlaq. The Influence of the Mixed Electric Line Poles on the Distribution of Magnetic Field. Indonesian Journal of Electrical Engineering and Informatics (IJEEI) **10**(2), June 2022, pp. 292~301 ISSN: 2089-3272. https://doi.org/10.52549/ijeei.v10i2.3572

5. Victoria Mary, S., Pugazhendhi Sugumaran, C.: Investigation on magneto-thermal-structural coupled field effect of nano coated 230 kV busbar. 2020, IOP Publishing Ltd., Physica Scripta **95**(4). https://doi.org/10.1088/1402-4896/ab6524
6. Del Pino Lopez, J.C., Cruz, R.P.: Influence of different types of magnetic shields on the thermal behavior and ampacity of underground power cables. IEEE Trans. Power Delivery **26**(4), 2659–2667 (2011). https://doi.org/10.1109/tpwrd.2011.2158593
7. Ghanim Thiab Hasan, kamil JaduAli, Ali Hlal Mutlaq. The Influence of the Mixed Electric Line Poles on the Distribution of Magnetic Field. Indonesian Journal of Electrical Engineering and Informatics (IJEEI), 10(2), 292–301 (June 2022). ISSN: 2089-3272. https://doi.org/10.52549/ijeei.v10i2.3572
8. Victoria Mary, S., Pugazhendhi Sugumaran, C.: Investigation on magneto-thermal-structural coupled field effect of nano coated 230 kV busbar. 2020, IOP Publishing Ltd., Physica Scripta, **95**(4). https://doi.org/10.1088/1402-4896/ab6524
9. Barsali, S., Giglioli, R., Poli, D.: Active shielding of overhead line magnetic field: Design and applications. Electr. Power Syst. Res. **110**, 55–63 (2014). https://doi.org/10.1016/j.epsr.2014.01.005
10. Bavastro, D., Canova, A., Freschi, F., Giaccone, L., Manca, M.: Magnetic field mitigation at power frequency: design principles and case studies. IEEE Trans. Ind. Appl. **51**, 2009–2016
11. Beltran, H., Fuster, V., García, M.: Magnetic field reduction screening system for a magnetic field source used in industrial applications. 9 Congreso Hispano Luso de Ingeniería Eléctrica (9 CHLIE), Marbella (Málaga, Spain), 2005, pp. 84–99
12. Canova, A., del-Pino-Lopez, J.C., Giaccone, L., Manca, M.: Active Shielding System for ELF Magnetic Fields. IEEE Transactions on Magnetics **51**(3), 1–4 (2015). https://doi.org/10.1109/tmag.2014.2354515
13. Canova, A., Giaccone, L.: Real-time optimization of active loops for the magnetic field minimization. Int. J. Appl. Electromagnet Mech **56**, 97–106 (2018). https://doi.org/10.3233/jae-172286
14. Canova Aldo, Giaccone Luca. High performance magnetic shielding solution for elf sources. 24th International Conference On Electricity Distribution Glasgow, 12–15 June 2017. Paper 1029. pp. 1–5
15. Celozzi, S. Active compensation and partial shields for the power- frequency magnetic field reduction. Proc. IEEE Int. Symp. On Electromagnetic Compatibility, Minneapolis, USA, pp. 222–226 (2002)
16. Celozzi S., Garzia F. Active shielding for power-frequency magnetic field reduction using genetic algorithms optimization. In IEE Proceedings – Science, Measurement and Technology **151**(1), 2–7 (2004). https://doi.org/10.1049/ip-smt:20040002
17. Canova Aldo, Giaccone Luca, Cirimele Vincenzo: Active and passive shield for aerial power lines. 25th International Conference on Electricity Distribution Madrid, 3–6 June 2019. Paper n°1096, pp. 1–5
18. Rozov, VYu., Grinchenko, V.S., Yerisov, A.V., Dobrodeyev, P.N.: Effective circuit shielding of a magnetic field of three-phase cable lines with a limited thermal influence on power cables. Electrical engineering & electromechanics **6**, 50–54 (2019)
19. Rozov, V. Yu., Reutskyi, S. Yu., Pelevin, D. Ye., Kundius, K. D.: Approximate method for calculating the magnetic field of 330–750 kV high-voltage power line in maintenance area under voltage. Electr. Eng. & Electromechanics **5**, 71–77 (2022). https://doi.org/10.20998/2074-272X.2022.5.12
20. Rusanov, A., Chugay, M., Rusanov, R.: Advanced Computer Technologies in the New Flow Part Development for Reactive Type HPC Steam Turbine of T-100 Series. Advances in Mechanical and Power Engineering. Selected Papers from The International Conference on Advanced Mechanical and Power Engineering (CAMPE 2021), October 18–21, 2021, pp. 55–63

21. Tarasova, V., Kostikov, A., Kuznetsov, M.: Thermodynamic Analysis and Optimization of the Cycle of a CHP Plant Power Unit Operating with Ultra-supercritical Steam Parameters. Advances in Mechanical and Power Engineering. Selected Papers from The International Conference on Advanced Mechanical and Power Engineering (CAMPE 2021), October 18–21, 2021, pp. 44–54
22. Ugrimov, S., Smetankina, N., Kravchenko, O.: A Study of the Dynamic Response of Materials and Multilayer Structures to Shock Loads. Advances in Mechanical and Power Engineering. Selected Papers from The International Conference on Advanced Mechanical and Power Engineering (CAMPE 2021), October 18–21, 2021, pp. 304–313
23. Solovey, V., Rusanov, A.: Integrated energy technologies when using natural gas in utility heat power engineering. Advances in Mechanical and Power Engineering. Selected Papers from The International Conference on Advanced Mechanical and Power Engineering (CAMPE 2021), October 18–21, 2021, pp. 78–87
24. Sushchenko, O., et al.: Algorithms for Design of Robust Stabilization Systems. In Computational Science and Its Applications—ICCSA 2022. Lecture Notes in Computer Science, vol. 13375, pp. 198–213. Springer, Cham (2022). https://doi.org/10.1007/978-3-031-10522-7_15
25. Hashim, F.A., Hussain, K., Houssein, E.H., Mabrouk, M.S., Al-Atabany, W.: Archimedes optimization algorithm: a new metaheuristic algorithm for solving optimization problems. Appl. Intell. **51**, 1531–1551 (2021). https://doi.org/10.1007/s10489-020-01893-z
26. Working group C4.204, Int. Council Lucio Ippolito, Pierluigi Siano. Using multi-objective optimal power flow for reducing magnetic fields from power lines. Electr. Power Syst. Res. **68**, 93–101 (2004)
27. Bravo-Rodríguez, J., Del-Pino-López, J., Cruz-Romero, P.: A Survey on Optimization Techniques Applied to Magnetic Field Mitigation in Power Systems. *Energies* **12**(7), 1332 (2019). https://doi.org/10.3390/en12071332
28. Xiaohui Hu., Eberhart R.C., Yuhui Shi: Particle swarm with extended memory for multiobjective optimization. *Proceedings of the 2003 IEEE Swarm Intelligence Symposium. SIS'03* (Cat. No. 03EX706). https://doi.org/10.1109/sis.2003.1202267

Modeling of Temperature Distributions in Axisymmetric Elements of Complex Metallurgical Systems on the Basis of Nonlocal Problems

Viktor Lyashenko[1], Michail D. Todorov[2], Elena Kobilskaya[1](✉), Olga Demyanchenko[3], and Dimitrios Statharas[4]

[1] Kremenchuk Mykhailo Ostrohradskyi National University, Kremenchuk, Ukraine
kobilskaya1983@gmail.com
[2] Technical University, Sofia, Bulgaria
[3] Azov Maritime Institute, National Universit "Odessa Maritime Academy", Mariupol, Ukraine
[4] University of Leicester, Leicester, England

Abstract. During the study and control of temperature distributions in moving and stationary axisymmetric elements of complex systems, mathematical models of heating processes play an important role. Unlike natural measurements of temperature in a finite number of points, these models help to analyze the temperature field in the entire heating area. The goal of this paper is to build a mathematical model of temperature distributions in moving and stationary axisymmetric elements of complex systems in the form of a nonlocal problem for the heat conduction equation with an integral condition, to develop a method for solving the nonlocal inverse problem. For the practical use and application of the mathematical model in the design of technological equipment, it is important to determine the control parameters of the temperature distribution. By introducing an integral condition to the mathematical model of the thermal process that occurs in a moving axisymmetric area, the control parameters of the temperature field are determined. In the paper, as additional information for formulating the inverse problem, a nonlocal integral condition is used, which is constructed based on the thermal energy balance condition in the heating area. This made it possible to increase the efficiency of temperature control in complex systems. A method for solving the nonlocal inverse problem of thermal conductivity was proposed.

Keywords: mathematical model · control parameters · nonlocal integral condition · temperature distributions

1 Introduction

Nonlocal integral conditions arise in the study of various physical phenomena in the case when the boundary of the process flow region is not available for direct measurement [1]. As an example, we can cite problems associated with the study of particle diffusion in a turbulent plasma, associated with the processes of heat propagation, the process of

H. Altenbach et al. (Eds.): CAMPE 2023, pp. 82–92, LNME 2025.
https://doi.org/10.1007/978-3-031-82979-6_9

moisture transfer in capillary-porous media. During the study and control of temperature distributions in moving and stationary axisymmetric elements of complex metallurgical systems, mathematical models of heating processes play an important role [2, 3]. Unlike natural measurements of temperature in a finite number of points, these models help to analyze the temperature field in the entire heating area. For the practical use and application of the mathematical model in the design of technological equipment, it is important to determine the control parameters of the temperature distribution [4]. These parameters make it possible to determine the temperature distribution both inside the heating area and at its borders. As is known, most mathematical models of heat transfer processes are based on boundary value problems for the heat conduction equation [2]. With the help of boundary and other additional conditions, a complete description and identification of a specific temperature distribution takes place. It should be noted that mathematical models based on boundary value problems fairly accurately reflect the temperature distributions inside the heating area and do not always accurately reflect the temperature at its boundaries [2]. This is due to the peculiarity of constructing a mathematical model in the form of a boundary value problem. Thus, when the temperature field is known at all points of the region, then the boundary value problem perfectly describes the temperature field. And if the temperature distribution is known only at the initial moment of time and/or at a part of the boundary, then additional conditions related to the balance of the energy allocated in the heating zone are needed to determine the temperature distribution in the entire heating area [6]. Knowing the amount of energy released in the entire heating area allows you to fully identify the temperature distribution. From a mathematical point of view, determining the temperature distribution in the heating area, under such conditions, leads to the solution of not a boundary value problem, but a nonlocal problem for the heat conduction equation. By combining the obtained solutions of the boundary value and nonlocal problem, it is possible to solve the inverse problem, in particular, the inverse problem for the heat conduction equation with an additional integral condition, which allows determining the unknown parameters of the thermal process. Such problems arise during the production of wire and tape [7], where in many cases preliminary heating occurs before plastic deformation [8] and during the production of products using 3D printing technology. Due to the mentioned reasons, there is a need to modify standard methods of studying nonlocal problems and develop new methods.

The goal of this paper is to build a mathematical model of temperature distributions in moving and stationary axisymmetric elements of complex systems in the form of a nonlocal problem for the heat conduction equation with an integral condition, to develop a method for solving the nonlocal inverse problem.

As additional information, a nonlocal integral condition is used to formulate the inverse problem, which is constructed based on the thermal energy balance condition in the heating area, which made it possible to determine the temperature field control parameters. Thus, the topic of research is relevant from both a theoretical and a practical point of view.

2 Literature Review

Recently, the class of problems with nonlocal integral conditions, which are a natural generalization of discrete nonlocal conditions, has attracted the increasing attention of specialists in the field of partial differential equations and specialists in mathematical physics. Most of the works are devoted to the study of the existence of and finding solutions to nonlocal problems. In [9], the two-dimensional nonlinear elliptic equation with the boundary integral condition depending on two parameters is solved by finite difference method. In [10] considered new results on the spectrum on complex plane for discrete Sturm–Liouville problem with one integral type nonlocal boundary condition depending on three parameters. The integral condition is approximated by the trapezoidal rule. In [11] the two-dimensional elliptic equation with integral boundary conditions is solved by finite difference method. In [11] the convergence of a finite difference scheme for two-dimensional nonlinear elliptic equation in the rectangular domain with the integral boundary condition is considered. However, these works do not describe the physical interpretation of the problem and the integral condition itself. As for the formulation of the inverse problem of heat conduction and the search for control parameters of the thermal process, it should be noted that the approach based on solving inverse problems by switching to extremal formulations of heat conduction problems and further using iterative minimization algorithms is quite well known. For example, in [13] the inverse bio-heat transfer nonlinear problem to determine the space-dependent perfusion coefficient from final time or time-average temperature measurements, which are themselves space-dependent quantities is considered. The most interesting thing in setting the inverse problem is the search for an additional condition, which largely determines the method for solving the inverse problem, it is here that the integral condition can be applied. In [14] the inverse source problem of heat conduction equation with time-dependent diffusivity on a spherical symmetric domain is considered.. To solve this problem, an iterative regularization method is proposed.

The analysis showed that the questions of the existence and uniqueness of the solution of nonlocal problems with integral conditions are sufficiently well studied [15], but their possibilities, in particular, the application of integral conditions in inverse problems, are not sufficiently studied. These works do not show the effectiveness of using nonlocal problems as mathematical models.

3 Research Methodology

Unlike nonlocal problems, boundary value problems have been sufficiently studied. When formulating a nonlocal problem, an integral condition that has a clear physical meaning is used as additional information.

In this work, as a nonlocal condition, the condition of energy balance [16] allocated in the entire heating area or in its part is proposed

$$Q_1 = Q_2 + Q_3 \tag{1}$$

where Q_1, Q_3, Q_2 are the amount of heat released per unit time and in an element of a complex system in the heating area; the amount of heat used to increase the temperature of the region and the amount of heat lost through its borders, respectively. During the formation of a nonlocal problem, this condition is the most complete. By adding to it the appropriate number of local conditions and a heat conduction equation, we obtain a nonlocal problem.

Determining the main control parameters of the temperature field of a moving wire or fixed bars before plastic deformation is one of the problems that arises during the creation of modern systems for automatic control of technological processes of heat treatment or plastic deformation of metals [17]. Similar problems arise during 3-D printing of products made of low-melting metals or thermoplastics. Control parameters can be obtained both experimentally and with the help of mathematical models that take into account the conditions of interaction of the wire surface with the environment. In a number of cases, determining the control parameters for heat treatment of metals is associated with large financial costs, much greater than obtaining parameter values using mathematical models. If a wire or rod is heated by an electric current, the amount of heat released in the heating area is determined by the Joule's law

$$Q_1 = \frac{I^2 \rho_0}{S}(1 + \beta T) \tag{2}$$

where S is the cross-sectional area of the heating zone, ρ_0 is the resistivity of the region at an 0°C, β is the temperature coefficient of resistance.

The following amount of heat is lost from a unit of side surface l per unit of time according to Newton and Stefan-Boltzmann laws

$$Q_2 = l\left[\alpha(T - T_c) + \varepsilon\sigma\left(T^4 - T_c^4\right)\right] \tag{3}$$

where α is the convective heat transfer coefficient, ε is the emissivity, the degree of blackness of the filament, σ is the Stefan–Boltzmann constant, T_c is the ambient temperature. In the case when all the heat released in an element of a complex system is lost, its temperature does not change and we have a case of thermodynamic equilibrium. Thermodynamic equilibrium can occur at different temperature values T, depending on the characteristics of the thermal process (I, α, ε). With a known value of the current I and the heat transfer coefficient α, the thermodynamic equilibrium temperature T of the axisymmetric element is determined from the equation

$$Q_1 = Q_2 \tag{4}$$

or

$$\frac{I^2 \rho_0 (1 + \beta T)}{\pi r_0^2} = 2\pi r_0\left[\alpha(T - T_c) + \varepsilon\sigma\left(T^4 - T_c^4\right)\right] \tag{5}$$

$$Q_3 = c\rho_n \Delta T \tag{6}$$

In this case, the relation (1) is valid.

As an additional condition for setting the inverse heat conduction problem, an integral condition for $v(t) > 0$ is chosen, which determines the energy balance of the heating zone, which has the form

$$\int_{\varepsilon+}^{t_0}\int_{0+}^{r_0}\int_{0}^{l}\frac{I(t)^2\rho_0 l+\beta I(t)^2\rho_0 lT(r,z,t)}{v(t)r_0^4\pi^2}dzdrdt=c\rho_n\int_{\varepsilon+}^{t}\iint_{G}(T(r,z,t)-T_0)dgdt+\frac{\alpha l}{r_0}\int_{\varepsilon+}^{t_0}\int_{0}^{r_0}\int_{0}^{l}\frac{T(r,z,t)-T_c}{v(t)}dzdrdt \quad (7)$$

where $G-$ considered surface, $\iint_G dg-$ surface area G.

The mathematical model of the temperature field of an axisymmetric moving element of a system with constant thermophysical characteristics is considered in the form of a boundary value problem (the function $I(t)$ is unknown) for the non-stationary heat conduction equation in the region.

$\Omega_{1t}=\{(z,r,t)|0<z<L, 0<r<r_0, 0<t\le t_0\}$[2, 3]

$$\lambda\frac{1}{r}\frac{\partial}{\partial r}\left(r\frac{\partial T}{\partial r}\right)+\lambda\frac{\partial^2 T}{\partial^2 z}-v(t)c\rho_n\frac{\partial T}{\partial z}-c\rho_n\frac{\partial T}{\partial t}=-W(T) \quad (8)$$

$$T(r,z,0)=T_0 \quad (9)$$

$$\left.\frac{\partial T}{\partial r}\right|_{r=0}=0,\ \lambda\left.\frac{\partial T}{\partial r}\right|_{r=r_0}=-\alpha(T-T_c)-\sigma\varepsilon\left(T^4-T_c^4\right) \quad (10)$$

$$T(r,0,t)=T_0,\ T(r,l,t)=T \quad (11)$$

where $W(T)=\frac{I^2(t)\rho_0(1+\beta T)}{\pi^2 r_0^4}$.

Control of the temperature field can be done using any parameter presented in the equation or in the boundary condition. In particular, it can be the current when the area is heated by internal heat sources. A control parameter can be either a constant or a variable value, for example, a control parameter can be a function of time. When applying condition (7), we assume that the temperature distribution $T(r,z,t)$ is known, so by substituting it, we can determine any temperature field control parameter. The deviation of equality (7) is used to refine the value of the parameter I_0 at each time step. So, the question is to find the function $T(r,z,t)$, which is reflected in the mathematical model. In the case of a thermally thin element, we turn to the consideration of the average integral temperature in the region $\Omega_{2t}=\{(z,t)|0<z<L, 0<t\le t_0\}$.

To solve the inverse problem of thermal conductivity, we use the search method, which is based on finding a temperature function that satisfies Eq. (8), boundary conditions (10)–(11), and initial condition (9). Condition (7) is satisfied at all points of the region $\Omega_{2t}=\{(z,t)|0<z<L, 0<t\le t_0\}$. The control $I(t)$ is carried out according to the requirement of minimizing deviations in equality (7) after substituting the obtained model temperature function that satisfies (8)–(11).

Consider the inverse problem of thermal conductivity as an optimization problem. The solution of the inverse problem of thermal conductivity as an optimization problem is connected with the introduction of some parameters of the system $a_j (j = 1, 2...J)$, which are controlled according to the requirement of minimizing the deviations of the model (calculated) temperature function from (7).

In order to express the desired function $I(t)$ using the control parameters of the system, we expand it in a Taylor series $I(t) = I_0 + a_1 t + a_2 t^2 + a_3 t^3 + ...$ The process of finding the control parameter $I(t)$ begins with an a priori specification of the initial approximation of the function $I(t)$ as a constant value of I_0. It can be found by taking in (8) $\lambda = 0, \ \sigma = 0, \ \frac{\partial u}{\partial t} = 0$.

This leads to the solution of the Cauchy problem for the ordinary differential equation

$$\frac{du}{dz} - (\frac{2}{r_0 v c \rho_n} + \frac{\beta \rho_0 I(t)^2}{\pi^2 r_0^4 v c \rho_n})u = \frac{\rho_0 I(t)^2}{\pi^2 r_0^4 v c \rho_n} - \frac{2T_c}{r_0 v c \rho_n}, \ 0 < z < l \tag{12}$$

$$u(0) = T_0 \tag{13}$$

Its solution can be written in the form

$$u(z) = \frac{\chi_1}{\theta_1} + (T_0 - \frac{\chi_1}{\theta_1})e^{\theta_1 z} \tag{14}$$

where $\chi_1 = \frac{\rho_0 I(t)^2 - 2T_c \pi^2 r_0^3}{\pi^2 r_0^4 v c \rho_n}$, $\theta_1 = \frac{2\pi^2 r_0^3 + \beta \rho_0 I(t)^2}{\pi^2 r_0^4 v c \rho_n}$.

We determine the value of I_0 from it. A more accurate approximation of I_0 can be obtained by assuming in (8) $\lambda = 0, \ \sigma \neq 0, \ \frac{\partial u}{\partial t} = 0$. Under such conditions, the solution of problem (8)–(11) is sought by the predictor-corrector method.

If the coefficients a_1, a_2 in $I(t)$ are found, then we come to the direct problem of thermal conductivity and solve it by the finite-difference method. The model temperature function obtained as a result of the solution of the direct problem will be the desired solution of the inverse problem of thermal conductivity, if $I(t)$ is selected in such a way that condition (16) is satisfied. Consider the discrete analogue of (7).

$$\int_0^l \frac{I_j^2 \rho_0 l + \beta I_j^2 \rho_0 u^j(z) l}{r_0^4 \pi^2 v_j} dz = c\rho_n \int_0^l (u^j(z) - T_0) dz + \frac{\alpha l}{r_0} \int_0^l \frac{u^j(z) - T_c}{v_j} dz \tag{15}$$

As a result of the substitution in (15) of the model temperature function, which corresponds to the function I_k in the k-th approximation, we get the deviation $\eta_{j(k)}$ in the k-th approximation. The totality of the discontinuities obtained for each time layer j forms a vector of discontinuities $\overline{H}$, all of whose components must be reduced to zero during the solution process.

Each parameter a_j is associated with a certain deviation η_j. A small test step Δa_1 is made for the first parameter and the direct problem (8)–(11) by is solved using the finite-difference method. To solve system (8)–(11), the Crank-Nicholson scheme was used, which is absolutely stable, as a result of which the increment of the corresponding deviation $\Delta \eta_1$ and the approximate value of the derivative are found $\frac{\partial \eta_1}{\partial a_1} = \frac{\Delta \eta_1}{\Delta a_1}$.

Further, several working steps along the parameter a_1 with fixed values of other parameters are taken until the condition $|\eta_1| \leq \delta$ for the deviation η_1 is satisfied, where δ is the required calculation accuracy.

The size of the working step can be determined under the condition $A_j = 1$ according to the formula

$$\Delta a_j = -A_j \eta_j \frac{\partial a_j}{\partial \eta_j} \tag{16}$$

The derivative $\frac{\partial \eta_j}{\partial a_j}$ is refined after each work step β, i.e.

$$\left(\frac{\partial a_j}{\partial \eta_j}\right)^{(\beta)} = \frac{a_j^{(\beta)} - a_j^{(\beta-1)}}{\eta_j^{(\beta)} - \eta_j^{(\beta-1)}} \tag{17}$$

The calculation stops when all $\eta_j (j = 1, 2, ...s, s < J)$ satisfy the condition $\left|\eta_j\right| \leq \delta$.

4 Results

A comparative analysis of temperature measurements using temperature sensors during 3D printing and numerical calculations obtained using a mathematical model during 3D printing using fdm (fusion deposition modeling) technology was carried out. As well as analysis of the temperature distributions of copper wire during its annealing.

The softened polymer thread is extruded onto the heated surface of the platform with gradual layering of layers of material on top of each other, creating a finished part. The filament is heated over a limited area in the extruder zone by an external heat source. The temperature distributions of a polymer filament product were constructed for various conditions of the 3d printing process. A technological experiment for the production of a product from low-melting filament was carried out on an Ultimaker 3 Extended 3D printer.

A 2.85 mm PLA thermoplastic filament was used to make the product on the printer. The sample had the form of a thin wall. calculations were performed at the following values of filament and medium parameters.

During the numerical experiment, the speed of movement of the filament through the extruder was $v = 60\frac{mm}{\text{min}}$, the temperature of the medium surrounding the filament $T_c = 30$°C, the temperature of the nozzle $T_l = 200$°C, the temperature of the platform on which the filament is laid out $T_0 = 60$°C, the radius of the filament thread $r_0 = 0, 2$ 10^{-3} m.

The printer, during the technological experiment, was installed in a thermal chamber. The temperature inside the chamber was controlled using a thermocouple installed in it. The filament temperature was monitored through a viewing window with a radiation pyrometer (Fig. 1).

According to Fig. 2, the speed of movement of the filament through the extruder, with other parameters constant, significantly affects the temperature in the heating zone. The same can be said about temperature distributions depending on heat transfer coefficients in the extruder.

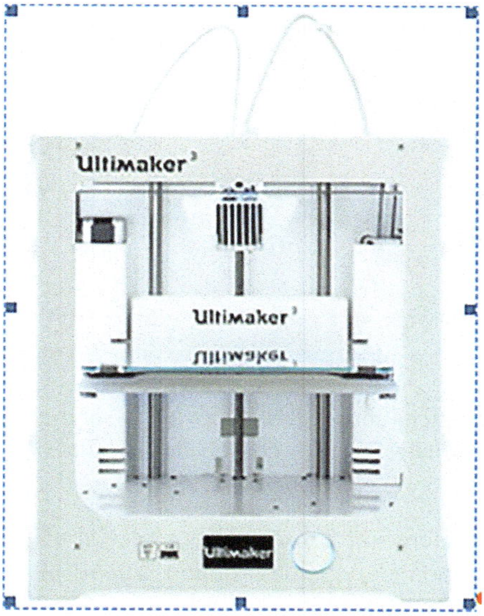

Fig. 1. 3D printer ULTIMAKER 3 extended

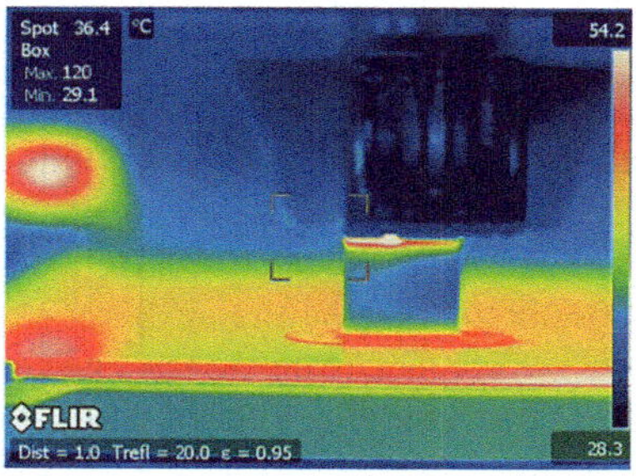

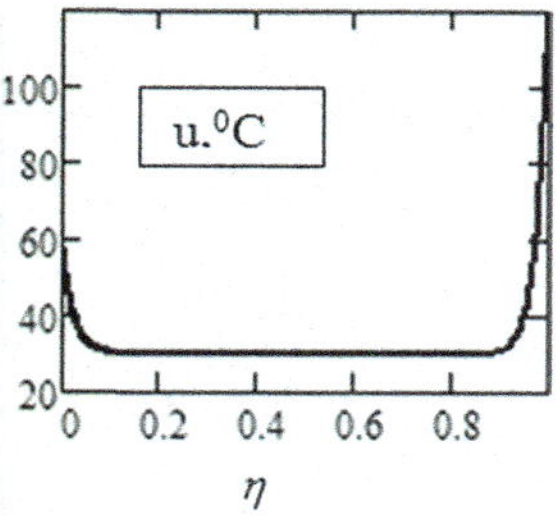

Fig. 2. Temperature distribution curves of the heating zone of PLA filaments moving through the heating zone $v = 60\frac{mm}{\min}$

The temperature distributions of copper wire during its annealing.

The conducted studies show that the temperature distribution obtained from the solution of the nonlocal-local problem coincides with the temperature distribution obtained by measuring the temperature during the technological process of annealing copper wire (point 3 in Fig. 3).

In Fig. 3 shows the temperature distributions obtained from the solution of the boundary value problem (8)–(11)—dashed and solid lines, curve 1, the nonlocal problem (7)–(11)—dashed lines, curve 2 and the values obtained experimentally at three control points—point 3. In Fig. 3 shows the results of three experiments at the beginning of the heating zone z = 0.1 m, inside the heating zone z = 0.5 m, and at the end of the heating zone z = 0.95 m. The results of the experiment agree with the solutions

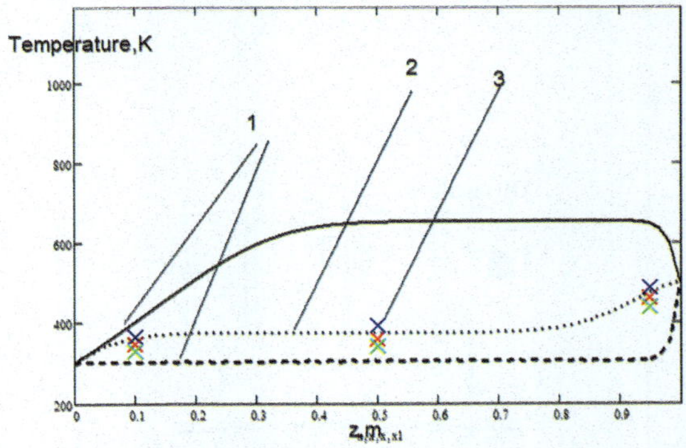

Fig. 3. Temperature distributions obtained from the solution of the boundary value problem (8)–(11)—curve 1, nonlocal problem (7)–(11)—curve 2 and the values obtained experimentally –3.

of the nonlocal-local problem (the error does not exceed 8%) and these results are far from the solutions of the boundary value problem. The results of three experiments are summarized in Table 1.

$$\delta T = \frac{|T_s - T_n|}{T_s} \cdot 100\%$$

T_s– the values of temperature are obtained from the solution of the boundary value problem (8)–(11), T_n– the values of temperature are obtained from the solution nonlocal problem. And the solutions obtained from the solution of the boundary value problem using a constant value of the current are far from the real ones and obtained from the nonlocal problem, which makes it possible to control the temperature field. These distributions indicate that the boundary value problem cannot always be considered as a mathematical model of the temperature field of a moving axisymmetric element (tape, wire, filament), since its solution does not always correspond to the physical content of the problem. As a mathematical model of the thermal process in axisymmetric elements of complex systems, it is advisable to use a nonlocal problem with an integral condition of heat balance.

5 Conclusions

In this work, a mathematical model of the thermal process during plastic deformation that occurs in a moving axisymmetric element in the form of a nonlocal problem for the heat conduction equation with an integral condition is constructed. It is shown that mathematical models based on boundary value problems fairly accurately reflect the temperature distributions inside the heating area and do not always accurately reflect the temperature at its boundaries. A method for solving the nonlocal inverse problem of thermal conductivity was proposed, which made it possible to determine the unknown parameter of the thermal process in an axisymmetric element of a complex system.

Table 1. Comparison of the results of numerical and natural experiments

	Experimental data 1	Experimental data 2	Experimental data 3	The values are obtained from the solution of the boundary value problem	The values are obtained from the solution nonlocal problem	δT (%)
Temperature value at the beginning of the heating zone z = 0.1 m	332 K	368 K	351 K	400 K	362 K	8
The temperature inside the heating zone z = 0.5 m	345 K	495 K	362 K	651 K	375 K	8
Temperature value at the end of the heating zone z = 0.95 m	440 K	490 K	465 K	647 K	474 K	7

The integral condition is constructed based on the thermal energy balance condition in the heating area. A comparative analysis of the temperature distributions obtained from the solutions of the boundary value, nonlocal-local problem and the temperature distributions obtained experimentally was carried out. The comparative analysis showed that it is appropriate to use a nonlocal problem as a mathematical model of the thermal process. For the practical use and application of the mathematical model during the design of technological equipment, it is important to determine the temperature distribution control parameters. The temperature field control parameters are determined from the solution of a nonlocal problem. The results obtained in the work make it possible to increase the efficiency of temperature control in complex systems.

References

1. Assanova, A.T.: On the theory of nonlocallocal problems with integral conditions for systems of equations of hyperbolic type. Ukr. Math. J. **70**, 1514–1525 (2019). https://doi.org/10.1007/s11253-019-01587-x
2. Lyashenko, V., Kobilskaya, E.: Methods for solving of inverse heat conduction problems. AIP Conf. Proc. **1773**(1), 040005-1–040005-7. AIP Publishing, United States (2016)

3. Lyashenko, V., Kobilskaya, E., Zagirnyak, M., Demyanchenko, O.: Mathematical model of thermal process in an infinite cylinder heated by a moving heat source. AIP Conf. Proc. **2522**, 040003–040009. AIP Publishing, United States (2022)
4. Zhang, B.: Temperature distribution model and parameter optimization of multilayer thermal protective clothing. AIP Conf. Proc. **2066**, 020035-1–020035-6. AIP Publishing, United States (2019). https://doi.org/10.1063/1.5089077
5. Mykola, Z., Mykhailo, S.: Mathematical model of thermal processes during the fermentation of biomass in a biogas reactor. 2020 IEEE KhPI Week on Advanced Technology, pp. 227–231 (2020). (KhPIWeek), Kharkiv, Ukraine https://doi.org/10.1109/KhPIWeek51551.2020.9250160
6. Shin, D.-U., Jeong, C.-H.: Energy savings of simultaneous heating and cooling system according to indoor set temperature changes in the comfort range. Energies **14**, 7691–7699 (2021). https://doi.org/10.3390/en14227691
7. Baier, D., Weckenmann, T., Wolf, F., Wimmer, A., Zaeh, M.F.: Underlying methodology for a thermal process monitoring system for wire and arc additive manufacturing. J. Manuf. Mater. Process **7**, 10–15(2023). https://doi.org/10.3390/jmmp7010010
8. Hwang, J.-K.: Comparison of temperature distribution between TWIP and plain carbon steels during wire drawing. Materials **15**, 8696–8699 (2022). https://doi.org/10.3390/ma15238696
9. Pupalaigė, K., Sapagovas, M., Čiupaila, R.: Nonlinear elliptic equation with nonlocal integral boundary condition depending on two parameters. Math. Model. Anal. **27**(4), 610–628 (2022)
10. Bingelė, K. Bankauskienė, A. Štikonas, A.: Spectrum curves for a discrete Sturm-Liouville problem with one integral boundary condition. Nonlin. Anal. Model. Control **24**, 755–774 (2019). https://doi.org/10.15388/NA.2019.5.5
11. Čiupaila, R., Pupalaigė, K., Sapagovas, M.: On the numerical solution for nonlinear elliptic equations with variable weight coefficients in an integral boundary conditions. Nonlin. Anal. Model. Control **26**(4), 738–758 (2021). https://doi.org/10.15388/namc.2021.26.23929
12. Sapagovas, M., Stikoniene, O., Jakubeliene, K., Ciupaila, R.: Finite difference method for boundary value problem for nonlinear elliptic equation with nonlocal conditions. Bound. Value Probl. **94**, 1–16 (2019). https://doi.org/10.1186/s13661-019-1202-4
13. Cao, K., Lesnic, D.: Reconstruction of the space-dependent perfusion coefficient from final time or time-average temperature measurements. J. Comput. Appl. Math. **337**, 150–165 (2018). https://doi.org/10.1016/j.cam.2018.01.010
14. Geng, X., Cheng, H., Mian, L.: Inverse source problem of heat conduction equation with time-dependent diffusivity on a spherical symmetric domain. Inverse Probl. Sci. Eng. **29**, 1–16 (2021). https://doi.org/10.1080/17415977.2021.1899172
15. Sadybekov, M., Dildabek, G., Ivanova, M.: Direct and inverse problems for nonlocal heat equation with boundary conditions of periodic type. Bound. Value Probl. **53**, (2022). https://doi.org/10.1186/s13661-022-01632-y
16. Stikonas, A., Sen, E.: Asymptotic analysis of Sturm-Liouville problem with nonlocal integral type boundary condition. Nonlin. Anal. Model. Control **26**(5), 969–991 (2021). https://doi.org/10.15388/namc.2021.26.24299
17. Han, S., Yang, H., Wang, Y. L.: Temperature control and numerical analysis for mass concrete pile cap of Haihuang bridge. In: IOP Conference Series: Earth and Environmental Science, vol. 153, pp. 052046–052051. IOP Publishing, Bristol, United Kingdom (2018)

Experimental Study of the Porosity of Ni + Zn + Al_2O_3 Coatings Deposited by Low-Pressure Cold Spraying

Oleksandr Shorinov[1(✉)] and Kostyantyn Balushok[2]

[1] Kharkiv Aviation Institute, National Aerospace University, 17, Chkalova Street, Kharkiv 61070, Ukraine
o.shorinov@khai.edu
[2] Motor Sich JSC, 15 Motorobudivnykiv Avenue, Zaporizhzhia 69068, Ukraine

Abstract. The goal of the article is to understand the effect of temperature and air pressure at the inlet to the supersonic nozzle, as well as the distance from the nozzle to the substrate (stand-off distance), on the microstructure and porosity of coatings obtained from Ni + Zn + Al_2O_3 powder mixture during low-pressure cold spraying. In the study, a nickel-based powder mixture was used as a powder material, in which the proportion of Al_2O_3 is approximately 30% mass. VT9 titanium alloy plates were used as the substrate material. Based on the multifactor planning of the experiment, the influence of complex parameters of the low-pressure cold spraying on the porosity of coatings was studied. Porosity estimation was performed on prepared microsamples using electron microscopy. The scientific novelty is the following: the dependences of the influence of complex parameters of the spraying process on the porosity of Ni + Zn + Al_2O_3 coatings are constructed from the analysis of the obtained statistical data. The results show that the air temperature has the greatest influence on the studied parameter, and by changing the spraying modes it is possible to control the porosity values in a rather wide range, which ranged from 2.7% to 6.6% in this study. The practical value is that obtained results can be used for development of technological recommendations for cold spraying of nickel-based coatings with minimum porosity for surface protection and repairing.

Keywords: Particle Velocity · Gas Flow · Computational Simulation · Coating · Optimization

1 Introduction

The technology of cold spraying (CS) is a promising method of forming coatings due to their high-quality characteristics. The method is based on powder particles acceleration in the solid phase with a supersonic gas flow, and their subsequent collision with the substrate surface and the coating formation. The main physical mechanism of CS is high-speed deformation of sprayed particles on impact, which leads to intense adiabatic shear instability of the material along the contact boundaries, and the formation of

H. Altenbach et al. (Eds.): CAMPE 2023, pp. 93–102, LNME 2025.
https://doi.org/10.1007/978-3-031-82979-6_10

adhesive-cohesion bonds. The process is characterized by low operating temperatures and high kinetic energy of particles in the gas stream [1]. For the purpose of adhering metal particles to the substrate, they must reach a certain velocity, which is called the critical velocity [2]. The critical velocity depends on both characteristics of deposited material and particle size [3]. The particle velocity affects the quality of the coating, porosity level, and the powder utilization factor. One of the important properties of coatings is the microgeometry of the coating surface, structure of internal areas and transition area from the coating to the substrate. These properties determine main physical characteristics of coatings, such as adhesion, cohesion, coating density, thermal conductivity, microhardness, etc. The selection of optimal process parameters is important from practical point of view – formation of coatings with pre-defined properties, ensuring the maximum coating adhesion, porosity level, thickness, deposition efficiency, etc.

The idea of the present work is to use nickel-based coatings to restore damaged surfaces of titanium alloy parts by low-pressure cold spraying. This paper aims to establish the dependences of the influence of the flow parameters at the nozzle inlet (temperature and pressure), as well as the distance from the nozzle section to the substrate on the porosity of $Ni + Zn + Al_2O_3$ powder mixture during low-pressure cold spraying (LPCS).

2 Literature Review

A low-pressure cold spraying (operating pressure up to 1.0 MPa) is the most widely used for repair and restoration of light alloy parts in the aerodrome and repair plants, as the equipment is portable and spraying is done manually. This is primarily the repair of aircraft parts made of aluminum [4] and magnesium alloys [5]. In addition to the above-mentioned alloys, the production of restorative coatings by low-pressure cold spraying on titanium alloys is also of interest from the point of view of technology implementation [6]. Among all the existing methods of thermal spraying of coatings, CS has a number of advantages, namely the absence of high temperatures, oxidation, phase transformations of the coating material and the substrate, and it can be successfully used for forming coatings on titanium alloys.

The CS process is characterized by strong deformation of particles in the coating and their dense packaging, which evenly covers the surface. For nickel-based coatings obtained by LPCS, the microstructure study clearly shows the shape of the deformed particles, their orientation in the coating layers and on the contact surface [7]. The coatings have a "scale-like" structure with a dense package without visible pores and "discontinuities". No traces of open porosity are observed [8].

In addition to the nature of particle deformation and their packaging in the sprayed layer, microstructure studies allow assessment of the amount of porosity and the nature of the pores in the coating material. On the cross sections, it was determined that the porosity is relatively evenly distributed in volume, the average size of individual pores is much smaller than the average powder particle size. Depending on the spraying modes, the porosity value amounted to 1.0…5.0% [9].

The porosity of coatings under LPCS can be reduced by increasing the particle velocities before contact with the substrate. This can be achieved by improving spraying parameters (pressure and temperature of the flow supplied to the nozzle, stand-off distance, powder feed rate) [10], ensuring the correct shape [11] and length [12] of the internal channel of the nozzle, selecting the correct fraction and shape of powder particles [13], etc. Particular attention shall be paid to the correct content of the ceramic component in the powder mixture [14]. In this case it is necessary to understand that the increase in the velocity of powder particles has a certain upper limit, exceeding which negative phenomena are observed, such as the effect of abrasive surface treatment and lack of adhesion of particles to the latter, reduced coating quality (reduced adhesion, increased porosity, etc.).

The analysis of literature sources established that the previous positive results on the spraying of titanium coatings were obtained using expensive CS equipment of medium (operating pressure 2.5 MPa) and high pressure (operating pressure up to 5.0 MPa) [15]; a low-pressure cold spraying with standard nozzle [16] also can be used for this purpose. It is concluded that coating properties significantly depend on spaying parameters which need to be carefully selected in order to obtain the required coating quality. The complexity of cold spray process determines that recommendations for spraying conditions have to be developed for each particular powder and spraying facility mainly by an experiment. Despite the availability of previously conducted one-factor numerical and experimental investigations, the effect of complex spraying parameters on crucial coating characteristic needs to be done.

3 Research Methodology

4 Materials

A commercially available powder mixture based on nickel Ni+Zn+Al_2O_3 was used as a powder material (see Fig. 1). The composition and fraction size of Ni + Zn + Al_2O_3 powder mixture are shown in Table 1.

VT9 titanium alloy was chosen as the substrate material. Before spraying, the samples were sandblasted with corundum fraction 200...350 μm.

5 Equipment

The coatings were sprayed using the DYMET-405 low-pressure cold spraying equipment.

Acceleration of the gas stream with particles occurred using a standard supersonic Laval nozzle of the DYMET-405 machine with a throat diameter of 2.7 mm and an outlet diameter of 6.0 mm. The powder entry point is located behind the critical section of the nozzle. Air was used as the operating gas. The temperature and air pressure at the nozzle inlet and the spray distance were set according to the experiment plan. Constant spray parameters are shown in Table 2.

Metallographic studies were performed using a Selmi REM-106 scanning electron microscope.

Fig. 1. Micrograph of Ni+Zn+Al_2O_3 powder mixture.

Table 1. Composition and fraction size of the Ni + Zn + Al_2O_3 powder mixture.

Powder characteristics	Powder composition		
	Ni	*Zn*	*Al2O3*
Content in powder mixture, mass %	60.0	10.0	30.0
Average particle size, um	10.0	20.0	25.0

Table 2. Constant deposition parameters.

Parameter	Value
Powder feed rate *G*, g/s	5.0
Traverse speed *v*, mm/s	25.0
Number of passes	3

6 Experiment Planning

A series of experiments was conducted in order to study the influence of the parameters of the low-pressure CS process on the porosity of the coatings and to find rational parameters to achieve its maximum values. The theory of experimental planning was used to obtain the necessary data with a minimum number of experiments, and a full factorial experiment (FFE) was performed to implement all possible combinations of non-repeating factor levels. From the analysis of literature sources and the results of one-factor experiments, the following factors influencing the coating porosity were identified for further study: temperature (T_0) and air pressure (P_0) at the nozzle inlet, as well as stand-off distance (spraying distance, S) (see Table 3). A type 2^3 experiment was conducted, where the number of factors k = 3, the number of levels P = 2, the number of experiments N = 8, the number of repeated experiments n = 2.

The stages of planning and implementation of FFE are as follows: selection of optimization parameters and their variation levels; factor coding; drawing up an experiment planning matrix; randomization of experiments; implementation of the experiment plan; checking the uniformity of variances of parallel experiments, reproducibility of results; calculation of regression equation coefficients, their errors and significance; checking the model adequacy [17].

Table 3. Values of factor variation levels and intervals.

Factor	Variation levels			Variation intervals
	–1	*0*	*1*	
T0, °C	300	450	600	150
P0, MPa	0.7	0.85	1.0	0.15
S, mm	10.0	15.0	20.0	5.0

After determining of cold spraying parameters and their maximum and minimum values (see Table 3), the experiment matrix was developed. In this research, the response (porosity) is a function of the initial gas temperature at the nozzle inlet T_0 (x_1), the gas pressure at the nozzle inlet P_0 (x_2), the spraying distance S (x_3), and can be written as following.

$$\text{Porosity}\ (\%) = f(T_0, P_0, S). \tag{1}$$

A second-order polynomial was used to describe the dependence of spraying parameters on coating porosity and its prediction. The regression equation for an experiment with three variables has the form as

$$\begin{aligned} y(x_1x_2x_3) = &\beta_0 + \beta_1x_1 + \beta_2x_2 + \beta_3x_3 + \beta_{12}x_1x_2 + \beta_{13}x_1x_3 + \beta_{23}x_2x_3 \\ &+ +\beta_{11}x_1^2 + \beta_{22}x_2^2 + \beta_{33}x_3^2, \end{aligned} \quad y \tag{2}$$

where β_0 is constant coefficient; β_1, β_2, β_3 are coefficients of variables x_1, x_2, and x_3, respectively; β_{11}, β_{22}, β_{33} are coefficients of the squares of the variables x_1, x_2, and x_3, respectively; β_{12}, β_{12}, β_{13} are coefficients of interaction of variables x_1, x_2, and x_3, respectively.

As can be seen from Eq. 2, the model includes the influence of the main factors and their interaction.

7 Results and Discussion

According to the developed experimental program, samples with coatings from Ni + Zn + Al_2O_3 powder mixture were obtained. The total thickness of 1.0...2.0 mm of the investigated coatings was reached. Cross sections of the samples were prepared by a standard grinding and polishing procedure according to ASTM Standard E3 – 11. The porosity of obtained coatings was determined on microsections to obtain a microstructure at magnifications of × 400. ImageJ software was used to analyze and process the obtained images.

Figure 2 shows an example of a microsample of the Ni + Zn + Al_2O_3 coating on a VT9 titanium alloy substrate.

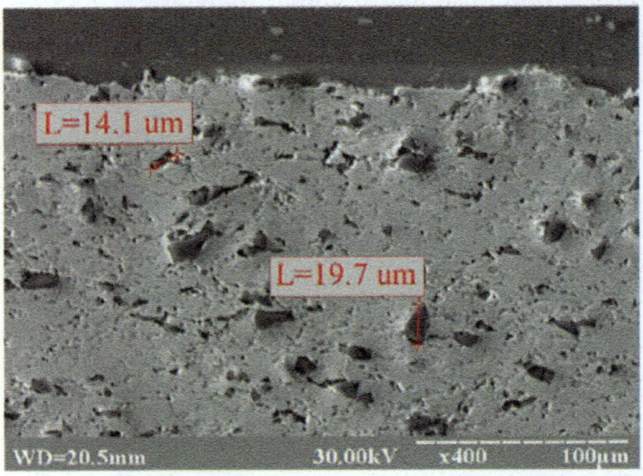

Fig. 2. SEM image of Ni + Zn + Al2O3 coating.

Results of determining the porosity of coatings are shown in Table 4.

Table 4. Results of coating porosity studies.

Exp. No	Spray parameters			Average po rosity value, %
	T0, °C	*P0, MPa*	*S, mm*	
1	600	1.0	10	5.8
2	300	1.0	10	2.7
3	600	0.7	10	3.1
4	300	0.7	10	3.8
5	600	1.0	20	6.6
6	300	1.0	20	3.6
7	600	0.7	20	4.5
8	300	0.7	20	5.6

The influence of the temperature and air pressure at the nozzle inlet, as well as the spraying distance on the porosity values of the coatings from Ni + Zn + Al_2O_3 powder mixture is shown in Fig. 3.

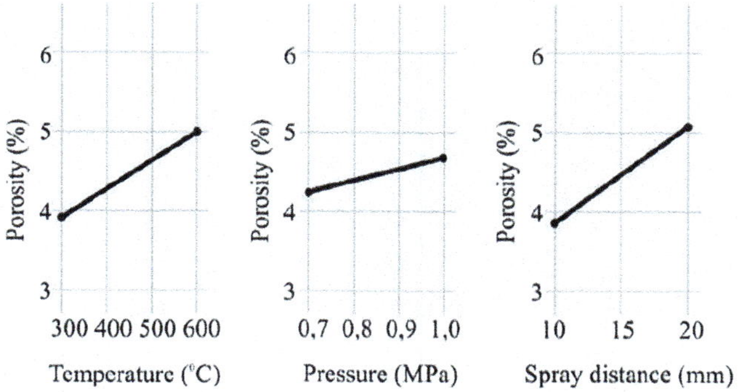

Fig. 3. Main effects plot for porosity.

The dependences of the combinations of the studied parameters of the spraying process on the optimization parameter – porosity of Ni + Zn + Al_2O_3 powder – are shown in Fig. 4–6.

As can be seen from the results obtained, among the studied parameters, the air temperature has the greatest influence on the value of the porosity of the coatings. For Ni + Zn + Al_2O_3 powder mixture, an increase in temperature from 300 to 600 °C leads to an increase in porosity from 2.7 to 5.8% at a constant pressure of 1.0 MPa and a spraying distance of 10 mm. The effect of pressure change on the porosity is as follows: with increasing pressure from 0.7 to 1.0 MPa, the porosity changes from 3.1 to 5.8% at constant temperatures of 600 °C and a spraying distance of 10 mm. On the contrary, distance reduction from 20 to 10 mm leads to a drop in porosity from 6.6 to 5.8% at a constant temperature of 600 °C and a pressure of 1.0 MPa. Changing the spray modes, namely the temperature and pressure of the air, as well as the spray distance, leads to a change in the coating porosity. The studied parameters affect the speed and temperature of the sprayed particles, which is of great importance for the CS process.

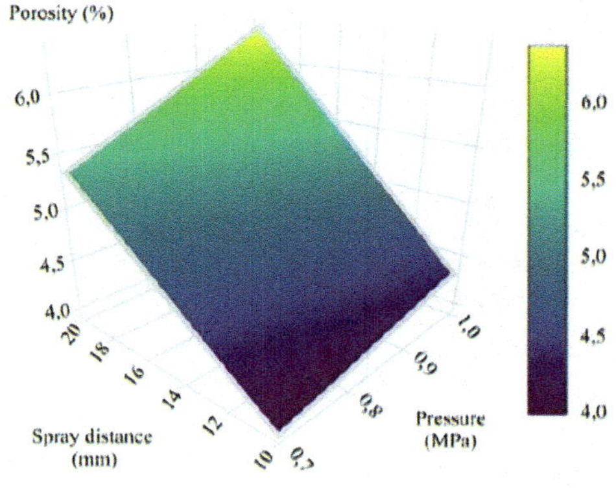

Fig. 4. Graph of dependence of porosity on P0 and T0.

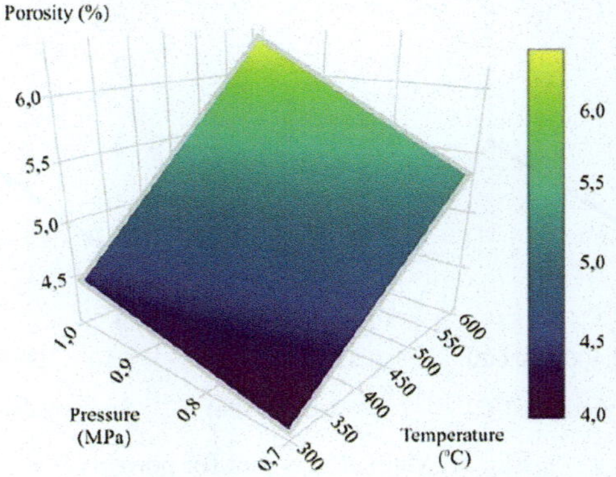

Fig. 5. Graph of dependence of porosity on P0 and S.

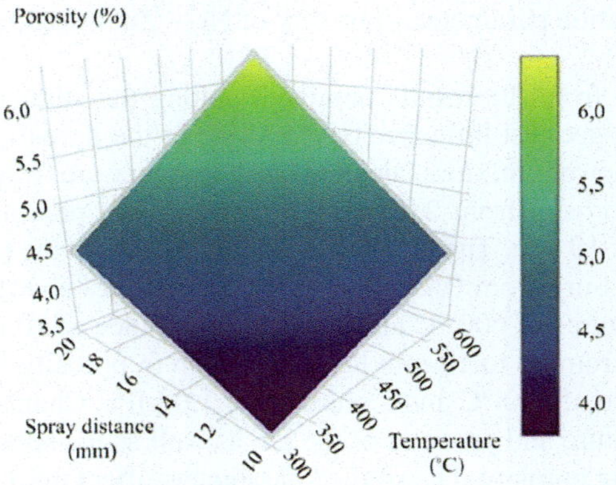

Fig. 6. Graph of dependence of porosity on S and T0.

When the speed of the gas flow increases, the speed of the powder particles presents in this flow increases. High particle velocities and temperatures lead to severe plastic deformations when particles collide with the substrate and the appearance of the phenomenon of unstable shear instability, which is associated with the adhesion mechanism during CS.

Verification of the obtained dependences of the porosity of the coating based on nickel Ni + Zn + Al_2O_3 consisted in conducting an additional series of experiments (3 runs) with arbitrarily selected spraying modes in the studied range of values. According to the research results, it was established that deviations do not exceed ± 5%. This testifies to the sufficiently high accuracy of the obtained dependencies and the possibility of using them to predict the porosity of the Ni + Zn + Al_2O_3 powder mixture coating in the studied ranges of values of spraying modes.

8 Conclusions

Based on the multi-factor experiment planning, the paper studied the influence of complex parameters of the process of low-pressure cold spraying, namely temperature and air pressure at the nozzle inlet, as well as spraying distance on the porosity of coatings obtained from $Ni + Zn + Al_2O_3$ powder mixture.

It is established that the air temperature has the greatest impact. Based on the results of experimental research and analysis of the obtained data, the dependences of the selection of optimal spraying parameters to achieve the minimum values of porosity (2.7%) were investigated.

The results obtained are the basis for the development of technological recommendations and processes for the formation of protective and restorative coatings on titanium-alloy parts during their repair as a result of corrosion, wear, mechanical damage, etc. Further optimization of the nozzle geometry, composition of the powder mixture and particle size is necessary to improve the quality of the obtained coatings at the maximum values of the powder utilization factor.

Acknowledgment. O. S. acknowledges the support from the project funded by National Research Foundation of Ukraine, under grant agreement No. 2020.02/0119.

References

1. Champagne, V. K.: Introduction. In: Champagne, V. K., Ozdemir, O. C., Nardi, A. (eds.) Practical Cold Spray, pp. 1–25. Springer, Cham (2021). https://doi.org/10.1007/978-3-030-70056-0_1
2. Assadi, H., Gärtner, F., Klassen, T., Kreye, H.: Comment on "Adiabatic shear instability is not necessary for adhesion in cold spray." Scripta Mater. **162**, 512–514 (2019). https://doi.org/10.1016/j.actamat.2018.07.065
3. Hu, W., Tan, K., Markovych, S., Cao, T.: A simple method for estimate critical velocity of cold spraying via SP. In: 3rd International Conference on Artificial Intelligence and Ad- vanced Manufacture (AIAM), pp. 215–221. IEEE, New York City (2021). https://doi.org/10.1109/AIAM54119.2021.00052
4. Han, X., et al.: A repair method for dam- age in aluminum alloy structures with the cold spray process. Materials **14**(22), 6957 (2021). https://doi.org/10.3390/ma14226957
5. Champagne, V., Jr., Kaplowitz, D., Champagne, V.K., III., Howe, C., West, M.K.: Dissimi- lar metal joining and structural repair of ZE41A-T5 cast magnesium by the cold spray pro- cess. Material Manufacturing Processes **33**, 130–139 (2018). https://doi.org/10.1080/10426914.2016.1257137
6. Volkov, A., Shorinov, O., Polyviany, S.: Deposition of titanium-containing coatings by low-pressure cold spraying. In: Nechyporuk, M., Pavlikov, V., Kritskiy, D. (eds.) Integrat- ed Computer Technologies in Mechanical Engineering - ICTM 2021, Lecture Notes in Networks and Systems, vol. 367, pp. 565–574. Springer, Cham (2022). https://doi.org/10.1007/978-3-030-94259-5_48
7. Koivuluoto, H., Lagerbom, J., Vuoristo, P.: Microstructural studies of cold sprayed copper, nickel, and nickel-30% copper coatings. J. Therm. Spray Technol. **16**, 488–497 (2007). https://doi.org/10.1007/s11666-007-9060-5

8. Koivuluoto, H., Milanti, A., Bolelli, G., Lusvarghi, L., Vuoristo, P.: High-pressure cold-sprayed Ni and Ni-Cu coatings: Improved structures and corrosion properties. J. Therm. Spray Technol. **23**, 98–103 (2014). https://doi.org/10.1007/s11666-013-0016-7
9. Koivuluoto, H., Vuoristo, P.: Structural analysis of cold-sprayed nickel-based metallic and metallic-ceramic coatings. J. Therm. Spray Technol. **19**, 975–989 (2010). https://doi.org/10.1007/s11666-010-9481-4
10. Mohankumar, A., et al.: Optimization of cold spray process in- puts to minimize porosity and maximize hardness of metal matrix composite coatings on AZ31B magnesium alloy. J. Nanomater. **2022**, 7900150 (2022). https://doi.org/10.1155/2022/7900150
11. Dolmatov, A.I., Polyviany, S.A.: Interaction of solid particles from a gas stream with the surface of a flat nozzle. Metallofiz. Noveishie Tekhnol. **43**(3), 319–328 (2021). https://doi.org/10.15407/mfint.43.03.0319
12. Dolmatov, A.I., Bilchuk, O.V.: Modelling of gas flow with solid particles in a short nozzle. Metallofiz. Noveishie Tekhnol. **40**(9), 1257–1271 (2018). https://doi.org/10.15407/mfint.40.09.1257
13. Shorinov, O.V., Polyviany, S.A.: Simulation of gas flow with nanocomposite carbon- containing powders in supersonic nozzle. Metallofiz. Noveishie Tekhnol. **44**(5), 601–611 (2022). https://doi.org/10.15407/mfint.44.05.0601
14. Cavaliere, P., et al.: Microstructural and fatigue behavior of cold sprayed Ni-based superalloys coatings. Sur- face and Coatings Technology **324**, 390–402 (2017). https://doi.org/10.1016/j.surfcoat.2017.06.006
15. Li, W., Cao, C., Yin, S.: Solid-state cold spraying of Ti and its alloys: A literature review. Prog. Mater. Sci. **110**, 100633 (2020). https://doi.org/10.1016/j.pmatsci.2019.100633
16. Dolmatov, A.I., Bilchuk, O.V.: Simulation of process of solid particles sputtering with dimet nozzle. Metallofiz. Noveishie Tekhnol.Tekhnol. **41**(7), 927–940 (2019). https://doi.org/10.15407/mfint.41.07.0927
17. Rodrigues, M. I., Emma, A. F.: Experimental design and process optimization. CRC Press, Boca Raton, FL, USA (2014). https://doi.org/10.1201/b17848

System of Automatic Notification of Personnel About Damage to Power Supply Facilities

Sergiy Panchenko, Valentyn Moiseienko, Olha Ananieva(✉), Mykhailo Babaiev, and Vasyl Sotnyk

Ukrainian State University of Railway Transport, Kharkov 61001, Ukraine
{info,mojseenko,ananeva,babaev,sotnyk.va}@kart.edu.ua

Abstract. Modern power supply systems have a direct impact on the development of industry, transport, economic and social aspects. Being one of the most complex and powerful consumers of electricity, electric power complexes of railway transport are constantly faced with emergency and non-working situations.

Changes in weather conditions, natural disasters, lightning strikes, overloads, and terrorist attacks often lead to their shutdown. Therefore, the study of the issues of early automatic notification of technical personnel about damage to railway power supply facilities is relevant. The report shows that despite the new information technologies used in the operation of modern automated dispatch control systems for railway power supply devices, the process of interaction with technical personnel has not changed significantly.

The presence of a human operator integrated into the mechanism for transferring information from the control system to technical personnel leads to additional, sometimes significant, unproductive losses of time. They can occur due to errors, distraction, or operator fatigue. Therefore, it becomes necessary to eliminate intermediate circuits in the process of transmitting information and withdrawing a person from this technological operation.

This will not only reduce ineffective losses of time but also improve fault tolerance. The paper proposes a new approach to building a system for organizing the maintenance and repair of power supply devices based on the principle of interactive cooperation between the control system and technical personnel and the introduction of information computer-integrated technologies.

Keywords: Power supply system · Operator · Energy dispatcher · Database · Information · Human factor · Monitoring

1 Introduction

Recently, significant changes have been observed in the implementation of computer-integrated information technologies in Ukrainian railway transport. Automated control systems (ACS) and control devices for railway power supply became the basis for the implementation of this process. It should be noted that if the first systems of the type "Granite" and "Lisna" actually copied the logic of the old relay system, then the new development of "Loza" is based on modern information technologies.

H. Altenbach et al. (Eds.): CAMPE 2023, pp. 103–111, LNME 2025.
https://doi.org/10.1007/978-3-031-82979-6_11

The relevance of the work has increased significantly over the past year due to damage to the energy infrastructure caused by enemy shelling. This has led to a sharp increase in the volume of repair and restoration work on railway power supply devices. The main criteria used to evaluate the work of the staff are the speed of restoration and the quality of work. These indicators ensure the continuity of the transportation process, which is a key factor in the conditions of a state of war in the Ukrainian economy.

In case of damage to specific objects of the energy infrastructure, the system becomes unbalanced, leading to losses and reduced power supply efficiency. Therefore, improving the efficiency of restoring power supply devices will certainly lead to a reduction in losses and an increase in energy efficiency. The authors did not consider the environmental aspects of the issue, as the list of tasks performed by the staff and their content remain unchanged.

The introduction of modern microprocessor systems opens up new opportunities for expanding the functions of dispatch control systems for power supply devices, but unfortunately, the vast majority of developments actually repeat the functionality of old systems with minor additions of functional capabilities.

According to the authors, significant progress in this direction can be achieved due to the systematic introduction of the latest computer-integrated information technologies using the experience gained in related fields.

2 Principles of Constructing an Automatic System for Alerting Personnel About Damage to Power Supply Facilities

The development of industry, transport, economic and social aspects are extremely dependent on modern power supply systems. Being one of the most complex, they are constantly faced with emergency and non-working situations. The power grid is affected by changes in weather conditions, natural disasters, lightning strikes, overloads, and cascading failures. The vulnerability of power grids often leads to outages.

A significant number of publications in most countries of the world are devoted to the issues of studying the operating conditions of electrical networks, the impact of overloads, climate change, natural disasters, the human factor, security and political stability of a number of energy producing countries [1–5].

The analysis of power outages and the human factor in the operation of equipment in electrical networks are considered in [6]. Measures are proposed to prevent accidents due to personnel errors, depending on their objectivity and subjectivity. The issues of introducing innovations in automated dispatch control systems with a human operator, built into the technological process of energy systems, are described in the article [7].

The system for monitoring accidents in medium voltage electrical networks, which allows you to determine the local places of their damage in real time, is able to adapt in accordance with changes in the electrical network [8]. A review and multidimensional analysis of the reliability criteria for power systems is presented in research papers [9, 10]. An integrated approach to assessing the functional reliability of a system using a dynamic Bayesian network and a hidden Markov model is considered in the article [11]. In research [12], in order to achieve minimization of maintenance costs for distribution feeders, it is proposed to use a new model of the reliability index of the weighted importance of components and the corresponding method for determining the priorities of the elements of the distribution system. The issues of assessing the cascading impact on energy networks of the most important elements of energy and transport infrastructure at a fundamental level, the failure rate in distribution networks and the reliability of power supply systems, including fault tolerance and adaptability, are considered in articles [13–15]. Works [16, 17] are devoted to methods of early warning about the quality of power supply services based on the theory of making three-way decisions and the neural network LSTM, cloud prediction of failures and monitoring the state of devices using machine learning.

Despite the latest elemental base and modern information technologies used in the work of modern automated dispatching control systems of railway power supply devices, the technology of interaction with technical personnel has not changed significantly.

Let us examine this matter using the example of detecting damage to technical means of power supply. If the power supply devices of the site are connected to the dispatch control system, the damage will be displayed on the monitor of the automated workplace of the power dispatcher and the person on duty at the traction substation. The energy dispatcher uses dispatch communication to contact the person on duty at the traction substation and coordinate his actions with him.

However, it is necessary to emphasize some essential features that are of fundamental importance:

- any damage identified by the system is clearly linked to a specific element of the identified device;
- in addition to the device and the nature of the damage, its territorial and administrative affiliation is determined;
- technical personnel are clearly assigned to specific devices within predetermined limits of the railway territory.

Let us examine how the information interaction of management systems with technical personnel is organized.

Information about damage to technical equipment and further identification of the event by the control system appears in the database of the automated workplace (AW) of the personnel (see Fig. 1).

The human operator, while monitoring the information on the AW screen, records the occurrence of a malfunction in the power supply devices. According to the information about the location of the damage, a decision is made to inform a specific person of the technical personnel. Previously, for this, a business telephone connection was used, and later - a mobile connection. After receiving the information, the technical personnel began to carry out work on restoring devices after damage, spending some time on this T_w. So, the total recovery time of devices after damage T_r has two main components

$$T_r = T_i + T_w \qquad (1)$$

where T_i is the time for informing the technical personnel about the damage.

It should be noted that at this stage we will not analyze the components of the restoration work time T_r, meaning the time spent on arriving at the work site, disassembling and reassembling the device structure, etc.

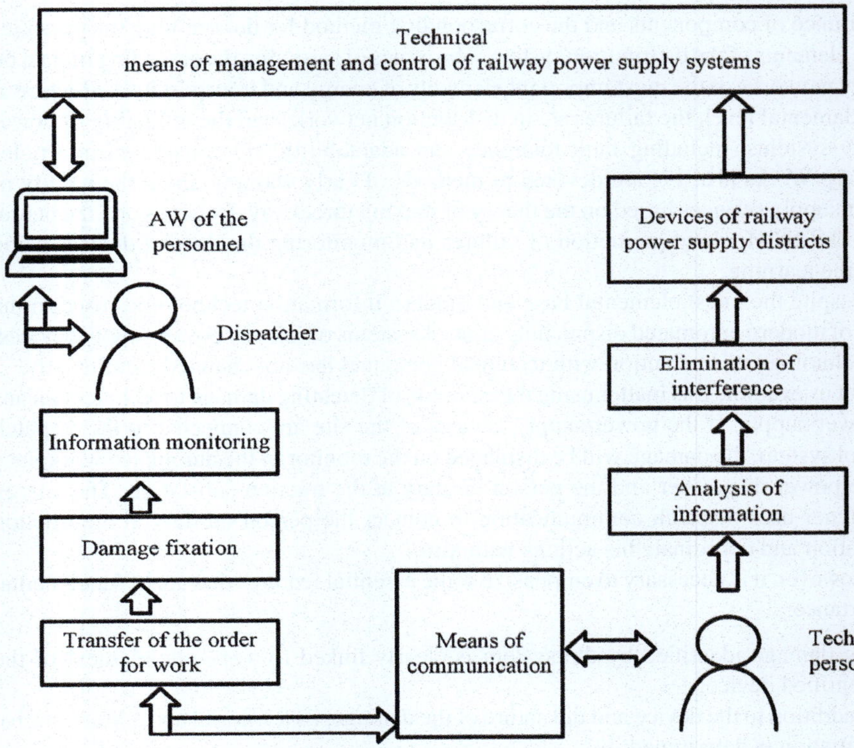

Fig. 1. Block diagram of the automated workplace of the personnel of the dispatch control system

It is important to have a person-operator integrated into the mechanism of information transfer from the ACS to the technical personnel. This leads to additional, sometimes significant, unproductive losses of time. They can occur due to operator errors, distraction, or fatigue.

The scientific works of specialists investigating the work of a human operator testify:

- the vast majority of unusual situations in railway transport occur as a result of wrong human actions;
- nowadays it is impossible to significantly improve the quality of work of a human operator and reduce the number of errors.

In this sense, the most productive is the elimination of intermediate chains in the process of information transfer and the withdrawal of a person from this technological operation. This will not only reduce unproductive time spent but also improve fault tolerance.

These types of approaches are quite actively used in the field of providing information services, for instance, by Internet service provider companies. Also in this way, it could be quite interesting to experience the service for ordering various services, especially taxis.

In the context of this ideology, the interaction of modern ACS with power supply devices and technical personnel changes somewhat. For its implementation, each employee must have an official personal mobile gadget, with the help of which information interaction with the control system is carried out (see Fig. 2). The easiest way is to send SMS messages to the relevant persons with the control of their reading.

All information about damage and personnel involved in this process is stored in a database that is accessible only to predetermined categories of industry workers (see Fig. 2). The perspective of such an approach becomes obvious with the active use of outsourcing mechanisms during the maintenance and repair of power supply devices.

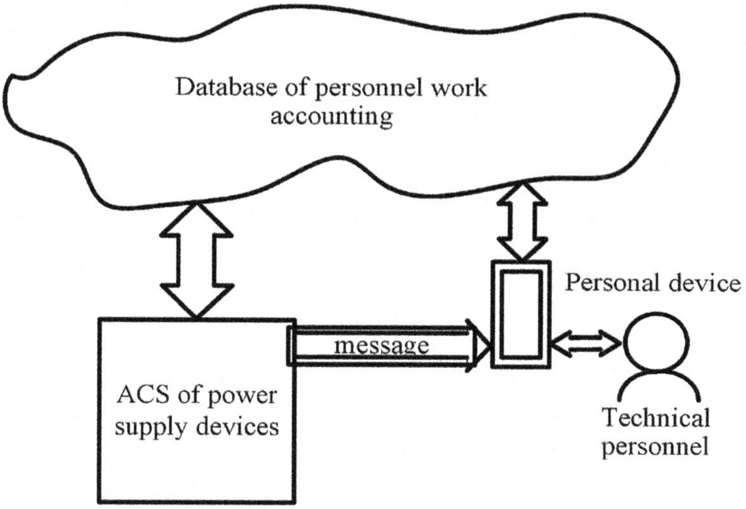

Fig. 2. Organization of interaction between the ACS and the technical personnel

In order to ensure personnel motivation, it is advisable to classify damages according to their level of impact on the process of train movement, as well as the necessary time spent on recovery after the damage. Each damage can be assessed by a number of P points with a corresponding critical time to recover T_{cr}.

Then the recovery time coefficient K_p

$$K_{pi} = \frac{T_{aci}}{T_{cri}} \tag{2}$$

where K_{pi} is the recovery time coefficient for the i-th damage;

T_{aci} is the actual recovery time after the i-th damage;

T_{cri} is a predetermined recovery time after the i-th damage (defined as the upper limit of the weighted average value of the recovery time).

Taking into account the K_B, the actual value of the score of P_{aci} can be determined

$$P_{aci} = P_i \cdot K_{pi} \tag{3}$$

where P_i is some normative value of the points awarded for the i-th damage.

In fact, P_{aci} are penalty points, according to the numerical value of T_{aci}, they can decrease when $T_{aci} < T_{cri}$, or increase when $T_{aci} > T_{cri}$.

For practical usage, it is advisable to introduce some scaling coefficient of K_{sc}, which will ensure the reduction of P_i losses to a point indicator, then

$$P_i^* = K_{sc} \cdot \mathrm{P}_{ri} \cdot T_{cri} \tag{4}$$

After the damage is identified by the system, an order for work is automatically generated (see Fig. 3).

- the time of appearance of damage is determined;
- the priority of execution is established and the values of P_i and T_{cri} indicators are determined;
- the circle of persons involved in the liquidation of the consequences of damage is defined, namely:
- direct executor;
- the person who is their manager;
- an official at the enterprise who is responsible for the maintenance of power supply devices.

So, after determining all the components of the order structure, it is sent to the database to the direct executor of the work, if necessary, to the direct manager. In case of non-acceptance of the order by the executor within the set period, the list of persons to whom the notification is sent is adjusted by involving higher-ranking officials. After accepting the order for execution, the start time of work is fixed and, as in the previous case, the information is transferred to the database using cloud technologies (see Fig. 3).

After completing the work, the performer makes an appropriate mark in the database about the successful completion of the task. The system automatically records the time spent on the work, after which the order is removed and the P_{aci} indicator is calculated, which is entered into the storage.

Due to this approach, over time, statistical material is collected about device failures, as well as the work of personnel with assessments of their work quality. This information will be important, especially in the conditions of transition to a new system of organizing maintenance and repair of power supply devices based on outsourcing procedures.

3 Conclusions

The report discusses a new approach to building a system for organizing the maintenance and repair of power supply devices based on the principle of interactive cooperation between the control system and technical personnel. The use of such an approach will make it possible to increase the efficiency of response to damage, reduce the time spent on repair and regulatory works, and reduce the downtime of trains due to damage to the railway's power supply. Conditions are being created for the introduction of innovations in the system of maintenance and repair of power supply devices.

Further development of scientific research is associated with the problems of reforming the power supply sector, which requires a very large amount of research-based,

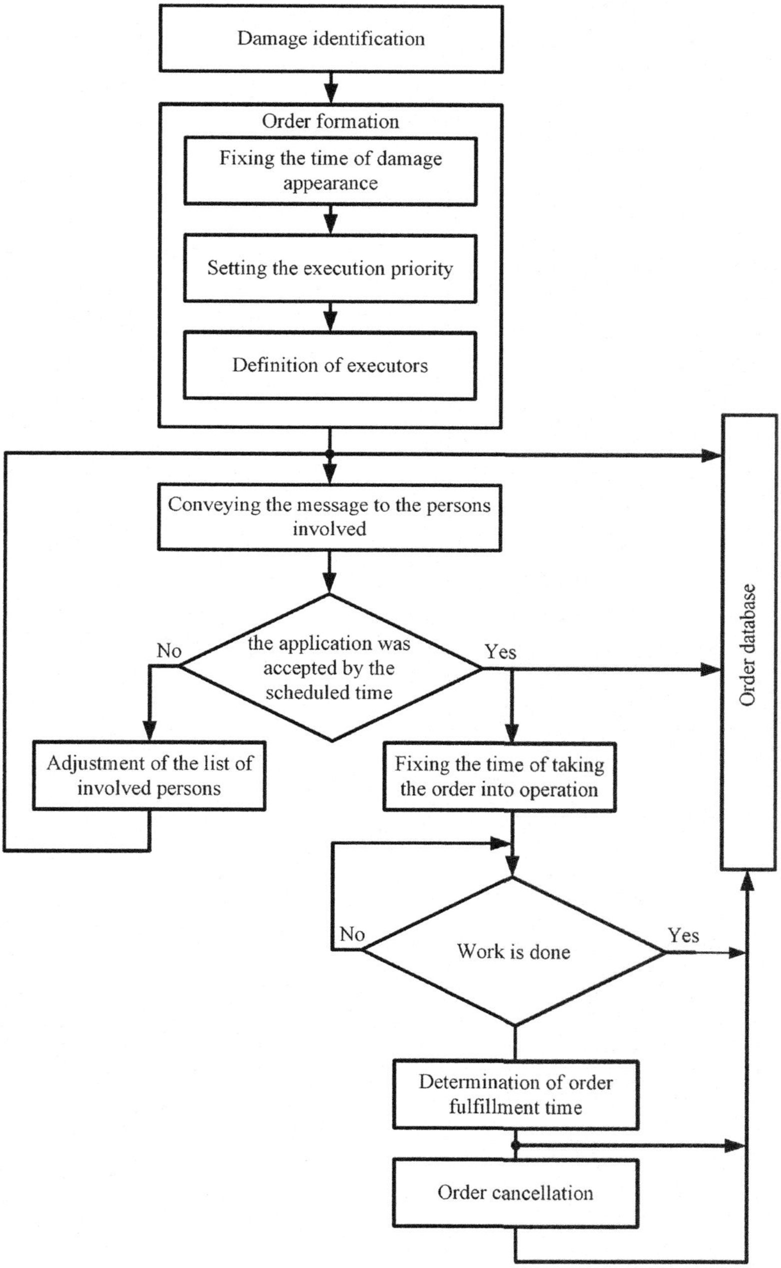

Fig. 3. Block diagram of an implementation of orders for restoration in case of damage

economic and engineering work. In particular, this means the introduction of leasing technologies for fixed assets for their maintenance and repair. It is necessary to develop a methodology for assessing the quality of work by the performer, enhance methods and technical means of monitoring energy infrastructure, and develop and enact relevant regulatory documents.

References

1. Fonseca, F. R., Craig, M., Jaramillo, P., Bergés, M., Severnini, E., Loew, A., Zhai, H., Cheng, Y., Nijssen, B., Voisin, N., and Yearsley, J.: Effects of climate change on capacity expansion decisions of an electricity generation fleet in the Southeast U.S. Environ. Sci. Technol. **55**(4), 2522–2531 (2021). https://doi.org/10.1021/acs.est.0c06547
2. Tomaszewski, M.: Ruszczak, Bogdan, Michalski, Paweł, Zator, Sławomir: The study of weather conditions favourable to the accretion of icing that pose a threat to transmission power lines. Int. J. Crit. Infrastruct. Prot. **25**, 139–151 (2019). https://doi.org/10.1016/j.ijcip.2019.04.001
3. Waseem, M., Manshadi, S.D.: Electricity grid resilience amid various natural disasters: Challenges and solutions. Electr. J. **33**(10), 106864 (2020). https://doi.org/10.1016/j.tej.2020.106864
4. Tina, G.M., Nicolosi, C.F.: Assessment of the impacts of climate change on power systems: the Italian case study. Appl. Sci. **11**, 11821 (2021). https://doi.org/10.3390/app112411821
5. Bloomfield, H.C., et al.: Quantifying the sensitivity of european power systems to energy scenarios and climate change projections. Renewable Energy **164**, 1062–1075 (2021). https://doi.org/10.1016/j.renene.2020.09.125
6. Ivanova, M., Dimitrova, R., Filipov, A.: Analysis of power outages and human errors in the operation of equipment in power grids. 12th Electrical Engineering Faculty COFERENCE (BulEF). pp. 1–5. Varna, Bulgaria (2020). https://doi.org/10.1109/BulEF51036.2020.9326058
7. Prostejovsky, A.M., Brosinsky, C., Heussen, K., Westermann, D., Kreusel, J., Marinelli, M.: The future role of human operators in highly automated electric power systems. Electric Power Systems Research **175**, 105883 (2019). https://doi.org/10.1016/j.epsr.2019.105883
8. Zhanlei Gu, Jun Wang and Linfeng Han: The application of an integrated monitoring system for power distribution and consumption. 8th International Symposium on Sensors, Mechatronics and Automation System. Journal of Physics: Conference Series 2246, 012050 (2022). https://doi.org/10.1088/1742-6596/2246/1/012050
9. Han, T., Gao, Z., Wenxin, D., Shizhao, H.: Multi-dimensional evaluation method for new power system. Energy Rep. **8**(6), 618–635 (2022). https://doi.org/10.1016/j.egyr.2022.03.150
10. Heylen, E., Ovaere, M., Proost, S., Deconinck, G., Hertem, D.: A multi-dimensional analysis of reliability criteria: From deterministic $N - 1$ to a probabilistic approach. Electr. Power Syst. Res. **167**, 290–300 (2019). https://doi.org/10.1016/j.epsr.2018.11.001
11. Li, Y., Zio, E., Pan, E.: An MEWMA-based segmental multivariate hidden Markov model for degradation assessment and prediction. Proceedings of the Institution of Mechanical Engineers, Part O: Journal of Risk and Reliability **235**(5), 831–844 (2021). https://doi.org/10.1177/1748006X21990527
12. Afzali, P.: Keynia, Farshid, Rashidinejad, Masoud: A new model for reliability-centered maintenance prioritisation of distribution feeders. Energy **171**, 701–709 (2019). https://doi.org/10.1016/j.energy.2019.01.040
13. Dai, Y., Preece, R., Panteli, M.: Risk assessment of cascading failures in power systems with increasing wind penetration. Electric Power Systems Research **211**, 108392 (2022). https://doi.org/10.1016/j.epsr.2022.108392

14. Clavijo-Blanco, J.A., Rosendo-Macías, J.A.: Failure rates in distribution networks: Estimation methodology and application. Electric Power Systems Research **185**, 106398 (2020). https://doi.org/10.1016/j.epsr.2020.106398
15. Felder, F.A.: Petitet, Marie: Extending the reliability framework for electric power systems to include resiliency and adaptability. Electr. J. **35**(8), 107186 (2022). https://doi.org/10.1016/j.tej.2022.107186
16. Lin, Z.: Shi, Yunchao, Chen, Bo, Liu, Shengyuan, Ge, Yuejun, Ma, Jien, Yang, Li, Lin, Zhenzhi: Early warning method for power supply service quality based on three-way decision theory and LSTM neural network. Energy Rep. **8**(5), 537–543 (2022). https://doi.org/10.1016/j.egyr.2022.02.243
17. Uppal, Mudita, Gupta, Deepali, Juneja, Sapna, Dhiman, Gaurav, Kautish, Sandeep: Cloud-Based fault prediction using IoT in office automation for improvisation of health of employees. Journal of Healthcare Engineering, vol. 2021, Article ID 8106467, 13 (2021). https://doi.org/10.1155/2021/8106467

Effective Management of Water Resources Cooling System Discharges at Nuclear Power Plants

Pavlo Kuznietsov[1,2](✉), Olesya Yaroschuk[1,2], Olha Biedunkova[1], Alla Pryshchepa[1], and Oleksandr Antonyuk[1,2]

[1] National University of Water and Environmental Engineering, 11, Str. Soborna, Rivne 33028, Ukraine
kuznetpavel@gmail.com
[2] SS Rivne, NPP 1, Str. Promyslova, 34400 Varash, Ukraine

Abstract. Water use at nuclear power plants is essential for the safe, reliable, efficient, and sustainable operation of power facilities. The purpose of the research is to analyze the compliance with water use, temperature influence, and pollutant discharge into the surface waters of the Styr River at the Rivne Nuclear Power Plant. The research was based on long-term average data of environmental and operational factors of the power plant, using classification and generalization approaches, computational mathematical methods, and assessment of compliance with current water quality standards. The positive effect of lime softening on the reduction of cooling water replenishment consumption by up to 30% in different seasons has been demonstrated. The absence of temperature effects of return cooling water discharges on the environment was proved, the value of which was 1.07 ± 0.64 °C. Cooling water quality was monitored and it was found that there is no negative impact on the structural materials of the cooling system and low-scale formation, which are explained by the fact that the actual values of water quality indicators to the results of monitoring over long periods are 1.0 times lower than the standard. The elements of an effective water management system for the cooling system of the Rivne Nuclear Power Plant are proposed to solve environmental and technological problems. The practical value of the research lies in the possibility of applying the methods to other nuclear power plants.

Keywords: Temperature and Chemical Effects · Optimization · Water Use · Corrective Treatment · Pollutant Discharges

1 Introduction

Nuclear power is recognized as a low-cost, clean energy resource with a low carbon trace [1]. Recent studies have confirmed that nuclear power contributes to environmental excellence and that increased investment in nuclear power is critical to improving energy efficiency and achieving economic sustainability [2, 3]. The use of nuclear energy is more efficient in reducing carbon dioxide (CO_2) emissions, and the introduction of

H. Altenbach et al. (Eds.): CAMPE 2023, pp. 112–122, LNME 2025.
https://doi.org/10.1007/978-3-031-82979-6_12

innovations in the technological cycles of nuclear power plants (NPPs) increases energy efficiency and reduces environmental impact [4]. The negative impact of nuclear energy on the quality of the environment is possible as a result of an inefficient operation, and technical incompetence associated with limitations of infrastructure and process control in the operation of NPPs [5].

The purpose of the research is to monitoring the quality of discharge water, to identify the safety of contact of cooling water with materials of power plant process equipment, and to identify elements of an effective water management system for the Rivne NPP cooling system by blocks of solving environmental and technological problems. The subject of the study is water consumption at NPPs, methods are the subject of research is water management. Which was studied in the example of the Rivne Nuclear Power Plant (Rivne NPP). The research was conducted on the example of the Rivne NPP, the operation of which is connected with the use of surface waters of the Styr River and requires the solution of the tasks of the Water Strategy of Ukraine for the period up to 2050 to assess the non-radiative environmental impact of NPPs with wastewater discharge.

2 Literature Review

NPPs consume large amounts of water, ranging from 20% to 83% more than thermal power plants [6]. Water scarcity may limit the possibility of siting NPPs in some areas, which provides an incentive to introduce innovative water use practices and technologies and efficient methods and techniques for managing water resources at NPPs related to design, construction, and operation [7].

Water in NPPs is used for cooling with heat removal. NPPs water consumption depends on condenser cooling and is determined by the efficiency of electricity conversion [8]. In terms of water consumption, a nuclear power plant is a large water consumer with an average specific water consumption of $2.67 \cdot 10^4$ dm^3/MW [9]. Compared to thermal power plants, NPPs produce steam at lower temperatures and pressures, are less thermally efficient, and consume more cooling water per unit of electricity [10].

Effective management of water resources at nuclear power plants is essential to ensure the safe and sustainable operation of energy facilities [11]. The implementation of measures to protect water resources from pollution and their rational use at NPPs must meet several of environmental standards [12], which is also essential for the sustainable development of the entire energy sector [13].

3 Research Methodology

Elements of effective water management systems are explored in the example of the cooling system of the Rivne NPP. The Rivne NPP is the first nuclear power plant in Ukraine with a power water-water reactor (VVER), and located in western Polissia, near the town of Varash in the Rivne region. The water of the Styr River is used as a source of water supply for the Rivne NPP. The water of the Styr River belongs to the carbonate-calcium type, the water treatment of the cooling system the Rivne NPP is carried out by lime softening in the hydrocarbonate mode is used. Lime softening takes place in the ВТИ-1000 clarifies (material Ст.20) and sulphuric acid is dosed after the clarifier, so there is no corrosive effect of sulphuric acid on the water treatment equipment. The make-up water, which has undergone clarification and sulphuric acid treatment, is not stored (tanks, reservoirs), but is immediately fed to the waste water treatment plant.

The research is based on a model that takes into account the environmental and operational factors of NPPs exploitation (Fig. 1). Heated cooling water in the cooling system flows from the condenser 3 (Fig. 1b) through cooling tower 2 (Fig. 1b), where it is cooled by evaporation, cooled water returns to condenser 3 (Fig. 1b) and is heated again, the cycles of cooling water concentration are determined by the evaporation coefficient (φ) and are regulated by blowdown with discharge of return water into the environment. Return water from a cooling system is discharged into the body of water from which water is taken and water consumption due to evaporation and blowdown is replenished with additional water 1 (Fig. 1a). In cooling towers, water heat is released through evaporation, the components of the cooled water are concentrated, and part of the water is removed by blowing off the return cooling water, the chemical composition of which forms the discharge. The discharge of waste cooling water can lead to a deterioration in the chemical state of the receiving water body [8].

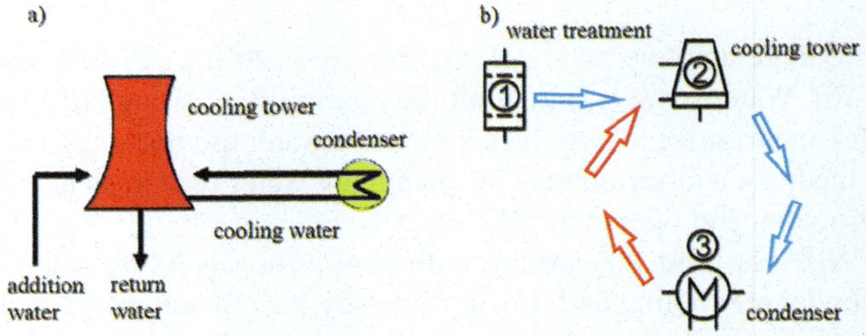

Fig. 1. Schematic representation of the model of water resources use by NPPs.

The research included the processing and systematic analysis of long-term observational data on the parameters of water consumption and cooling water discharge from the cooling systems of the Rivne NPP. The data were analyzed by means of classification and generalization, computational mathematics methods, and assessment of compliance with current water quality standards.

When implementing the technology of water treatment for the cooling system blowdown water, a model of recirculation of a part of the cooling system blowdown water was used to ensure the permissible concentrations (PC) to ensure discharge standards and, accordingly, the calculated permissible cycle of concentration φ (1) was determined.

$$\varphi = C_{system}/C_{makeup} \tag{1}$$

where: φ is the cycle of concentration; C_{system} is the concentration of components of the system water; C_{makeup} is the concentration of components of the make-up water.

The thermal effect of water consumption was defined as the difference in temperature before and after wastewater discharge NPPs. The permissible temperature increase was assumed to be no more than 3 °C in accordance with the Permit for Special Water Use of the Rivne NPP No. 53/PV/49d-20.

The monitoring of the quality of the surface water of the Sty River, which receives the effluent from the cooling systems of the Rivne NPP, was carried out by comparing the actual concentrations of substances and water quality indicators in the effluent of the power plant with the regulatory indicators. The parameters controlled in the study during monitoring and their normalized values are determined in accordance with the standard in Ukraine SOU NAEC 067:2013, they were measured by standard control methods.

4 Results and Discussion

4.1 Optimization of Water Use

Power characteristics and installed capacity utilization factors (ICUF) of units of the Rivne NPP are shown in Fig. 2. The cooling water consumption of power units No. 1, 2 the Rivne NPP (type VVER-440) is 91 thousand m^3/h for each power unit, power units No. 3, 4 the Rivne NPP (type VVER-1000) – 188.9 thousand m^3/h each power unit. The amount of water required for the power generation process is also highly dependent on the cooling technologies used in the power plant [11, 12]. Optimization of a NPP water use is realized through the use of appropriate design solutions, minimization of leaks and losses, water recycling and reuse [13]. The choice between optimising the use of methods depends on several factors, including specific site requirements, water availability, climatic conditions, and economic considerations, and a careful evaluation of these factors is necessary to determine the most appropriate cooling system for a particular application when building or retrofitting cooling systems for NPPs [14].

After treatment, the blowdown water is reused to recharge the cooling system, which minimizes water consumption [15]. Using the example of the cooling system of the Rivne NPP, the possibility of reducing water consumption by re-using purge water through liming and correcting water acidification with sulfuric acid (Table 1, Fig. 3) was assessed.

Maintaining allowable evaporation coefficient φ_{AEC} (Table 1) during the operation will allow to reduce the Rivne NPP water consumption by 1000 m^3/h in the warm season and up to 1500 m^3/h in the cold season, which depends on the quality of the cooling water, the load of the unit, recycling and reuse of the Rivne NPP cooling water by liming will allow reducing the cooling water make-up costs by 10–30%. Compared to the operation mode without recycling of the Rivne NPP cooling water and at the φ_{AEC} (Table 1), the increase of cooling water consumption will be from 500 to 750 m^3/h, and accordingly, the consumption of additional water for make-up will increase by the same amount, depending on the season (warm or cold), the make-up water consumption will be in the range of 2500 to 5600 m^3/h, compared to the consumption of 4000 to 8000 m^3/h without recycling of treated cooling water.

In general, water purification during recycling or reuse can be implemented using the following purification methods: lime softening, reverse osmosis, ion exchange and distillation desalination plants [16]. Implementation of measures for optimization of water use by means of water treatment with lime softening and sulfuric acidification allows to minimize water consumption and water discharge at the Rivne NPP (Table 1, Fig. 3). The current state of water treatment at Rivne NPP (Fig. 1b) includes lime

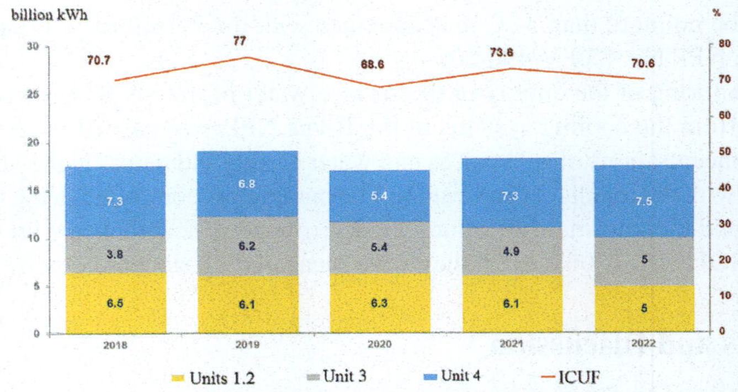

Fig. 2. Power characteristics and ICUF of units of the Rivne NPP.

Table 1. Calculation of the allowable evaporation coefficient (φ_{AEC}) with recirculation of cooling blowdown water (RBW) and make-up water (MW) according to values of permissible concentrations (PC).

Components	PC*, mg/dm^3	RBW, mg/dm^3	MW, mg/dm^3	φ_{AEC}
Mineralization	1000	198	913	4.6
Nitrates	40	7.77	35.7	5.1
Sulfates	250	47.11	353	5.3
Chlorides	200	25.19	115.8	7.9

Note: * - in accordance with the Permit for the Permit for Special Water Use of the Rivne NPP No. 53/PV/49d-20

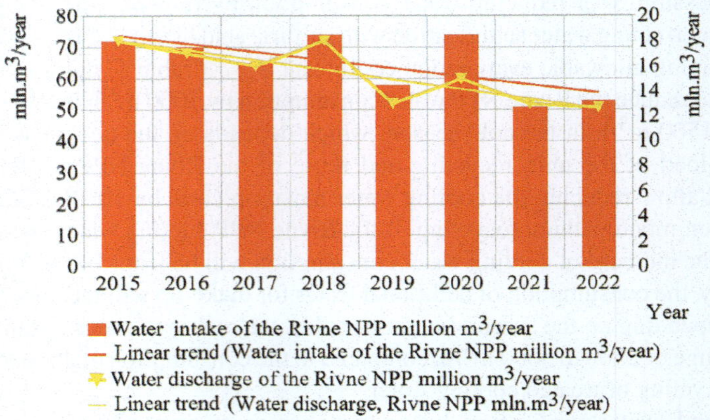

Fig. 3. Water consumed and discharged into the Styr River by the Rivne NPP.

softening and sulphuric acid treatment, and part of the return water will be recycled to the existing water treatment system.

Clearly, water recycling and reuse can reduce water consumption without affecting the cooling capacity of the system by operating at higher concentration cycles and evaporation rates. Based on the current state of the art in water treatment technology, if optimization methods are applied, it is possible to increase the concentration cycles and evaporation coefficients maintained in the cooling system, hereby reducing water consumption.

4.2 Environmental Impact

The difference in water temperature of the Styr River before water intake and after the Rivne NPP cooling water discharge does not exceed the established temperature increase effect under the conditions of special water use of 3 °C and amounts to an average value of 1.07 °C, standard deviation ± 0.64 °C (Fig. 4).

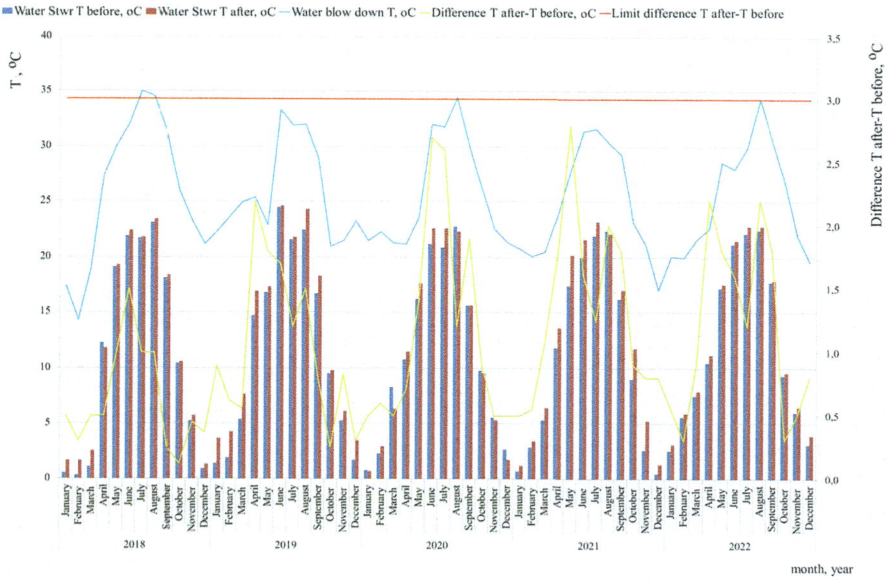

Fig. 4. Temperature influence of return water discharges from the cooling system of the Rivne NPP on water of the Styr River.

Research show that climate changing can increase uncertainty about water resources, including increases in their temperature [17]. The impact of increased water temperature of the cooling water reservoirs on a power plant includes reduced production, plant shutdown, and changes in regulatory deviation requirements [18]. It is noted that responding to climate change involves the development of adaptation approaches to reduce potential risks, including the implementation of adaptation plans for operating a nuclear power plant during drought or heat [19].

Since the temperature fluctuations in the water of the Styr River, where the effluent of the Rivne NPP is discharged, reflect the seasonal fluctuations in the ambient temperature, we assume that this may be due to the absence of an increase in temperature effects.

4.3 Monitoring of Water Quality

NPPs must have robust monitoring programs in place to regularly measure water quality parameters in water samples. Results of the Rivne NPP effluent monitoring indicate that quality indicators are maintained within the range of acceptable values, and ensure efficient resource management because the multiplicity of components content increase of monitoring results does not exceed 1.0 (Table 2). Regulation of the water-chemical regime of the cooling system at Ukrainian NPPs is carried out in compliance with SOU NAEC 067:2013 "Management of chemical technologies. The water-chemical regime of the technical water supply system of responsible consumers of NPPs with VVER. General requirements" (Table 2).

Table 2. List of chemical control parameters of cooling water and their normalized values.

Components and units of measurement	Normalized values*	Multiplicity of components content increase*	Impact on the cooling system of a nuclear power plant
pH, units	from 6.5 to 8.5	0.84**	Formating acid or alkaline conditions and affect corrosion
Total hardness, $mmol/dm^3$	no more 7,0	0.79	Indicates scale formation
Carbonate hardness, $mmol /dm^3$	no more 2,5	0.65	Indicates the ability to neutralize acid
Chloride ions, mg/dm^3	no more 150	0.58	Maintains proper chemical composition of water
TSS***, mg/dm^3	no more 50	0.21	Creates sludge deposits on equipment
TSC***, mg/dm^3	no more 800	0.75	Affects metal corrosion and salt deposits

Note: * - the normalized values in accordance with SOU NAEC 067:2013, monitoring results of the cooling system of the Rivne NPP for 2019–2022, ** - for the upper limit of the normalized value, *** - TSS is the total suspended solids, TSC is the total salt content

Taking into account the fact that the chemical composition of water affects all materials in contact with cooling water, and at the same time it is affected by the materials themselves [20], it can be assumed that the water quality in the cooling system of the Rivne NPP does not contain potential risks to the safety of the equipment. To put it another way, the results of water chemical composition monitoring meet the conditions for maintaining the integrity of target materials, elements and systems of NPPs.

4.4 Analysis of Emergency Preparedness

NPPs should have well-defined emergency response plans for water-related incidents or natural disasters that may affect water resources [21]. These plans should include provisions for alternative water sources, containment measures, and rapid response actions to prevent potential releases of radioactive materials. NPPs should cooperate with regulatory agencies responsible for water management to ensure compliance with applicable regulations and standards [22]. Regular interaction with these agencies can help resolve issues, obtain necessary permits, and implement best practices for water management [23].

Having conducted a theoretical and applied analysis of the problems related to the reliability of the water supply systems at the Rivne NPP, the authors consider it appropriate to present a scheme of possible environmental and operational problems caused by the water-chemical regime of the cooling system NPPs (Fig. 5).

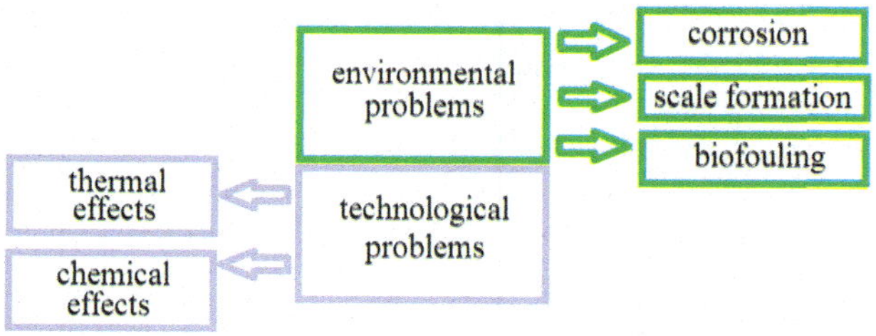

Fig. 5. Environmental and operational problems of using water resources at NPPs.

To minimize the impact on aquatic ecosystems, nuclear power plants must comply with standards and protocols to ensure the temperature and chemical composition of cooling water discharged to water bodies. The physical, chemical and biological characteristics of the cooling water are considered when optimizing environmental impact [24, 25]. In particular, physical characteristics such as seasonal water temperature, intake design, condenser cooling efficiency, effluent plume, chemical and biological characteristics such as the presence of contaminants, fish in the plume, and the purpose of the water body [26]. Physical, chemical, and biological characteristics are critical to identifying opportunities to optimize water use at NPPs.

The need to protect the environment by reducing the negative impact on the environment obliges [27, 28] power plants to take measures to regulate the discharge of pollutants to permissible levels. Therefore, power plants take the necessary measures to improve the quality of technological processes, control and reduce pollutant discharge [29, 30].

Thus, the stages of the emergency preparedness analysis should be strengthened by research related to operational problems of corrosion, scale formation and biofouling. Despite the identified favorable environmental parameters within the study area of the Rivne NPP, the temperature and chemical load during water discharge should be monitored regularly.

5 Conclusions

Effective water management at NPPs is critical to ensuring the safe and sustainable operation of energy facilities. According to the results of studies of water use in the cooling system of the Rivne NPP, a positive impact of liming was found to reduce cooling water recharge by 30%, both in warm and cold seasons.

The temperature effect of the return cooling water discharged into the Styr River was 1.07 ± 0.64 °C, which indicates compliance with the regulatory values and no temperature impact on the environment. The monitoring of the quality of the discharged water showed that it exceeded the standards for nuclear power plants by a factor of 0.21 to 0.84, which guarantees the safety of the contact of the cooling water with the materials of the technological equipment. Improvement of the water management system at the Rivne NPP is possible by expanding the emergency prevention scheme to include the stages of corrosion, scale formation and biofouling assessment. The proposed cooling system blowdown water recirculation scheme makes use of the existing water treatment system and requires economic costs for implementation only for the installation of a blowdown pipeline into the water treatment system and will reduce water consumption by up to 40%.

Improvement of technological processes, modification of systems and elements of a nuclear power plant, cooperation with regulatory authorities and scientific institutions allow increasing the efficiency of water management in NPPs.

References

1. Lau, L.-S., Choong, C.-K., Ng, C.-F., Liew, F.-M., Ching, S.-L.: Is nuclear energy clean? Revisit of Environmental Kuznets Curve hypothesis in OECD countries. Econ. Model. **77**, 12–20 (2019). https://doi.org/10.1016/j.econmod.2018.09.015
2. Azam, A., Rafiq, M., Shafique, M., Zhang, H., Yuan, J.: Analyzing the effect of natural gas, nuclear energy and renewable energy on GDP and carbon emissions: A multi-variate panel data analysis. Energy **219**, 119592 (2021). https://doi.org/10.1016/j.energy.2020.119592
3. Sadiq, M., Shinwari, R., Usman, M., Ozturk, I., Maghyereh, A.I.: Linking nuclear energy, human development and carbon emission in BRICS region: Do external debt and financial globalization protect the environment? Nucl. Eng. Technol. **54**(9), 3299–3309 (2022). https://doi.org/10.1016/j.net.2022.03.024
4. Çakar, N.D., Erdoğan, S., Gedikli, A., Öncü, M.A.: Nuclear energy consumption, nuclear fusion reactors and environmental quality: The case of G7. Nucl. Eng. Technol. **54**(4), 1301–1311 (2022). https://doi.org/10.1016/j.net.2021.10.015
5. Mahmood, N., Wang, D.Z., Zhang, B.: The role of nuclear energy in the correction of environmental pollution: Evidence from Pakistan. Nucl. Eng. Technol. **52**(6), 1327–1333 (2020). https://doi.org/10.1016/j.net.2019.11.027
6. Ren, K., Tang, X., Jin, Y., Wang, J., Feng, C., Höök, M.: Bi-objective optimization of water management in shale gas exploration with uncertainty: A case study from Sichuan. China. Resour. Conserv. Recycl. **143**, 226–235 (2019). https://doi.org/10.1016/j.resconrec.2019.01.003
7. Jin, Y., Behrens, P., Tukker, A., Scherer, L.: Water use of electricity technologies: A global meta-analysis. Renew. Sust. Energ. Rev. **115**, 109391 (2019). https://doi.org/10.1016/j.rser.2019.109391

8. Kuznietsov, P., Tykhomyrov, A., Biedunkova, O., Zaitsev, S.: Improvement of methods for controlling power oil of cooling tower recycling water supply units at Rivne nuclear power plant. Sci. Horiz. **25**(12), 69–79 (2022). https://doi.org/10.48077/scihor.25(12).2022.69-79
9. Argüelles, R., Toledo, M., Martín, M.A.: Study of the Tagus River and Entrepeñas reservoir ecosystem around the Trillo nuclear power plant using chemometric analysis: Influence on water, sediments, algae and fish. Chemosphere **279**, 130532 (2021). https://doi.org/10.1016/j.chemosphere.2021.130532
10. Ali, B., Hedayati-Dezfooli, M., Gamil, A.: Sustainability assessment of alternative energy power generation pathways through the development of impact indicators for water, land, GHG emissions, and cost. Renew. Sust. Energ. Rev. **171**, 113030 (2023). https://doi.org/10.1016/j.rser.2022.113030
11. Kang, J. S., Lee, S. J.: Concept of an intelligent operator support system for initial emergency responses in nuclear power plants. Nucl. Eng. Technol. **54**(7), 2453–2466 (2022). https://doi.org/10.1016/j.net.2022.02.010
12. Zargar, A., Kodkani, A., Vickers, B., Flynn, M.R., Secanell, M.: A hybrid cooling tower model for plume abatement and performance analysis. Appl. Therm. Eng. **219**(25), 119593 (2022). https://doi.org/10.1016/j.applthermaleng.2022.119593
13. Batista, N.E., Carvalho, P., Fernández-Ramírez, L.M., Braga, A.P.S.: Optimizing methodologies of hybrid renewable energy systems powered reverse osmosis plants. Renew. Sust. Energ. Rev. **182**, 113377 (2023). https://doi.org/10.1016/j.rser.2023.113377
14. Zhou, G., Tan, D.: Review of nuclear power plant control research: Neural network-based methods. Ann. Nucl. Energy **181**, 109513 (2023). https://doi.org/10.1016/j.anucene.2022.109513
15. Redfoot, E.K., McKellar, M.G., Borrelli, R.A.: Allocating heat and electricity in an integrated energy system coupled with a water purification system. Nucl. Eng. Des. **397**, 111902 (2022). https://doi.org/10.1016/j.nucengdes.2022.111902
16. Ray, S.S., Gusain, R., Kumar, N.: Carbon Nanomaterial-Based Adsorbents for Water Purification. Elsevier, 37–66 (2020). URL: https://doi.org/10.1016/b978-0-12-821959-1.00003-9Su ganya, P., Swaminathan, G., Anoop, B.: Assessing the factors affecting the water chemistry parameters in the auxiliary water system of a nuclear power plant. SN Appl. Sci. **2**(11), 1889 (2020). https://doi.org/10.1007/s42452-020-03693-z
17. Hyvärinen, J., Vihavainen, J., Ylönen, M., Valkonen, J.: An overall safety concept for nuclear power plants. Ann. Nucl. Energy **178**, 109353 (2022). https://doi.org/10.1016/j.anucene.2022.109353(dateofaccess:28.07.2023)
18. Abdin, I.F., Fang, Y.P., Zio, E.: A modeling and optimization framework for power systems design with operational flexibility and resilience against extreme heat waves and drought events. Renew. Sust. Energ. Rev. **112**, 706–719 (2019). https://doi.org/10.1016/j.rser.2019.06.006
19. Fan, G., Hou, B., Dong, X., Ding, X.: Technical points of water-draw and discharge impact analysis in guidelines for water resource assessment of coastal nuclear power plants. Sustainability. **13**(11), 6308 (2021). https://doi.org/10.3390/su13116308
20. Gursel, E., et al.: Using artificial intelligence to detect human errors in nuclear power plants: A case in operation and maintenance. Nucl. Eng. Technol. **55**(2), 603–622 (2023). https://doi.org/10.1016/j.net.2022.10.032
21. Alrammah, I.A.: Analysis of nuclear accident scenarios and emergency planning zones for a proposed Advanced Power Reactor 1400 (APR1400). Nucl. Eng. Des. **407**, 112275 (2023). https://doi.org/10.1016/j.nucengdes.2023.112275
22. Jyotish, N.K., Singh, L.K., Kumar, C.: A state-of-the-art review on performance measurement petri net models for safety critical systems of NPP. Ann. Nucl. Energy **165**, 108635 (2022). https://doi.org/10.1016/j.anucene.2021.108635

23. Kuznietsov, P., Biedunkova, O.: Experimental tests of biocidal treatment for cooling water of safety systems at Rivne NPP units. Nucl. Radiat. Saf. **1**(97), 30–40 (2023). (in Ukrainian) https://doi.org/10.32918/nrs.2023.1(97).04
24. Kurek, K., Młyński, D., Bugajski, P., Nowobilska-Majewska, E.: The impact of treated sewage on water quality in Mordarka stream. J. Ecol. Eng. **20**(1), 39–45 (2019). https://doi.org/10.12911/22998993/93874
25. Kuznietsov, P.N., Biedunkova, O.O., Yaroshchuk, O.V.: Experimental study of transformation of carbonate system components cooling water of Rivne Nuclear Power Plant during water treatment by liming. Probl. Atomic Sci. Technol. №2(144), 69–73 (2023) https://doi.org/10.46813/2023-144-069
26. Al-Rawajfeh, A.E., Araj, K.: Treatment of wastewater to meet the requirements for cooling water systems in Jordan's nuclear plants. Chem. Process. Eng. **34**(2), 253–267 (2013). https://doi.org/10.2478/cpe-2013-0021
27. Mavredaki, E., Stathoulopoulou, A., Neofotistou, E., Demadis, K.D.: Environmentally benign chemical additives in the treatment and chemical cleaning of process water systems: Implications for green chemical technology. Desalination **210**(1–3), 257–265 (2007). https://doi.org/10.1016/j.desal.2006.05.050
28. Parasuraman, S., Ganapathiraman, S., Bhargavan, A.: Multivariate regression studies for investigating and setting the action levels for the system water quality parameters of a nuclear plant. Water Environ. J. **36**(3), 553–563 (2022). https://doi.org/10.1111/wej.12786
29. Siraj, G., Khan, H.H., Khan, A.: Dynamics of surface water and groundwater quality using water quality indices and GIS in river Tamsa (Tons), Jalalpur, India. HydroResearch, **6**(89–107), (2023). https://doi.org/10.1016/j.hydres.2023.02.002
30. Kuznietsov, P.M., Biedunkova, O.O.: Technological and environmental problems in the stabilization treatment of the main condenser cooling circuit by sulfuric acid. J. Eng. Sci. (Ukraine), **10**(2), (2023). https://doi.org/10.21272/jes.2023.10(2).h1

Modelling Safety System of Waveguides Made of Regenerated Hard Alloy

Oleksandr Shapoval[1(✉)], Iurii Savchenko[2], Yurii Parshyn[3], Oksana Khrebtova[1], and Viktoriia Kulynych[1]

[1] Kremenchuk Mykhailo Ostrohradskyi National University, 20, Pershotravneva Street, Kremenchuk 39600, Ukraine
alexshap.as@gmail.com
[2] University of Customs and Finance, Vladimir Vernadsky Street, 2/4, Dnipro 49000, Ukraine
[3] Dnipro State University of Internal Affairs, 26, Gagarina, Dnipro 49005, Ukraine
parshin22@ukr.net

Abstract. If obsolete munitions are no longer needed, then their components, in particular high-quality metals, are quite suitable for use. Hard alloys of the tungsten carbide-nickel-cobalt system have been developed with different content of components compared to the traditionally used tungsten-cobalt alloys, which, in the presence of the same binder percentage, have increased hardness and flexural strength and, due to this, increased wear resistance. It has been established that in the regenerated mixtures the cobalt content corresponds to its content in the original scrap, the content of bound carbon corresponds to the stoichiometric content for each grade of alloy, the content of free carbon does not exceed the norm, and the physical and mechanical properties of sintered alloys meet the requirements for standard alloys of the corresponding grades. To increase the flexural strength, it is necessary to choose a binder in the nickel to cobalt ratio of 0.5–0.6. The obtained WCNi alloy powders are not in high commercial demand, but they can be used to manufacture new designs of projectile cores. Compositions with new properties have been developed for mass use of the WC + Ni type powders for cutters of rock-cutting tools, mandrels, blanks, and dies for drawing wire and pipes from ferrous and non-ferrous metals.

Keywords: tungsten · cobalt · tool · cermet · recycling

1 Introduction

If obsolete munitions are no longer needed, then their components, in particular high-quality metals, are quite suitable for use (Fig. 1). An industrial high-performance technology has been developed for the production of waveguides and tools from tungsten-cobalt and tungsten-nickel alloys by direct regeneration of secondary raw materials without the use of thermochemical and metallurgical methods [1].

H. Altenbach et al. (Eds.): CAMPE 2023, pp. 123–132, LNME 2025.
https://doi.org/10.1007/978-3-031-82979-6_13

2 Literature Review

Material science studies of tungsten carbide alloys with cobalt, nickel, and iron have been carried out since 1926 when this class of materials appeared. Using cobalt as a binder phase, turned out to be most successful. Therefore, the largest number of works on hard alloys focus on the cobalt bond [2]. Hard alloys of the carbide system WC + Ni have been of heightened interest lately. This is due to the need to replace the cobalt, ever-rising in cost, with cheaper and less scarce nickel [3]. It should be noted that, despite the relatively poorer mechanical properties of WC + Ni hard alloys [4], they have found application in technical fields where corrosion resistance is required [5] and in precision instrumentation due to their non-magnetic nature [6] (Fig. 2).

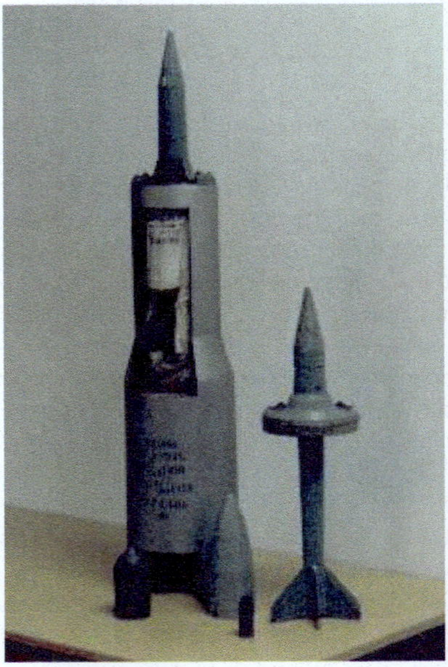

Fig. 1. Obsolete Munitions

The obtained WC + Ni alloy powders are not in high commercial demand, but they can be used to manufacture new designs of projectile cores. Compositions with new properties have been developed for mass use of the WC + Ni type powders for cutters of rock-cutting tools, mandrels, blanks, and dies for drawing wire and pipes from ferrous and non-ferrous metals [7].

The industrial high-efficiency technology of production of cutting, breaking and forming tools from WC + Co and WC + Ni alloys by direct regeneration of secondary raw materials without the use of thermochemical and metallurgical methods with the use of shock waves has been developed. The technology fundamentally differs from the existing ones by high technical and economic indicators, productivity, low energy consumption and environmental friendliness.

Fig. 2. Alloy scrap WCNi5, WCNi8 и WCNi10

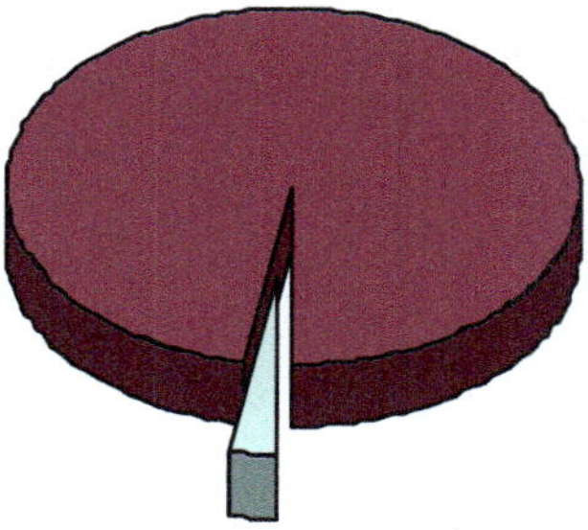

☐ 1- shock-wave treatment;

■ 2- without shock-wave treatment.

Fig. 3. Alloy grinding time WC + Ni:

Figure 3 shows the data illustrating the results of industrial tests of wet milling of WC + Ni powder after shock-wave treatment. The same figure shows the duration of grinding the powder to the same dispersity, which has not undergone shock-wave treatment. The diagram shows that the duration of the grinding process of powder processed by explosion is more than 30 times less compared to the existing technology of grinding, which helps to reduce energy consumption of the grinding process by 25–30%.

The technology combines technological flows of utilization of carbide scrap, production of reaction ampoules, preparation of explosive charges, regeneration, classification, fine grinding and sintering, the implementation of which provides the finished product. Recovery of the tungsten-containing mixture is carried out under conditions of high-pressure gradients and loading rates in reaction cylindrical ampoules subjected to explosive load generated by detonation of the explosive charge.

It should be noted that in contrast to classical mechanical-thermal methods of influence, which have limited possibilities both in terms of mechanical and physical-chemical influence, shock-wave treatment introduces changes in the structure at all scale levels.

In this regard, the idea of stimulating the processes of defect formation and chemical activation of tungsten-containing alloys by shock waves proved to be vital and fruitful.

3 Research Methodology

To save cobalt and to improve the physical and mechanical characteristics of finely dispersed hard alloys, a new tungsten-containing hard alloy is to be obtained, cobalt and nickel to be used simultaneously in different proportions as a binder [8].

Conversion scrap from WCNi5, WCNi8, and WCNi10 alloys is used as a feedstock (Fig. 2). The chemical composition of scrap is given in Table 1.

Grinding of grains was carried out in a ball mill MMP 70/110 380 l and in a vibrating mill. The grain size was classified into fractions −1 mm, −2 + 1 mm, −3 + 2 mm. Grinding modes of WNi10 grains in a vibrating mill are presented in Table 2.

Table 1. Chemical composition of scrap.

Element	Content	Element	Content
C	4.5	Ni	8.9
O_2	0.24	Mo	0.09
V	0.012	Sn	0.03
Fe	0.04	W	rest

The dried mixture was sifted through a nylon sieve № 58-61 GOST 4403-77. Screenings were rounded carbide particles with a fraction of up to 0.9 mm [9]. 35×5×5 mm rods were pressed from the resulting mixtures to determine the physical and mechanical properties according to GOST 3882-74 [10]. The grain size was controlled by the optical method using the MIM-8 microscope [11]. The granularity of the mixture is shown in Table 3. Then the resulting mixture was additionally charged with cobalt metal according to TI 48-4203-01-11-86 based on the production of an alloy with a total nickel and cobalt content of 14.5%, similar to the WCo15 alloy according to GOST 3882-74 [12].

The pressed products were dried in an electrical cabinet at 120 ÷ 150°C for 12 h. The dried products were sintered in a vacuum furnace in the mode indicated in Table 4 [13].

4 Results and Discussion

The physical and mechanical properties of the sintered products were determined according to the parameters specified in GOST 3882-74 [14]. The test results of the rods are shown in Table 5.

Metallographic studies were carried out by GOST 9391-80.

The alloy is two-phase. Porosity is 0.15–0.2%. Pores size is up to 10 mkm. No η_1 phase (complex carbide W_3Co_3C). The nickel + cobalt bond is evenly distributed. The

Table 2. WNi10 grain grinding modes in a vibrating mill.

Equipment	Mixture type	Balls weight, [kg]	Charge weight, [kg]		Ethanol amount, [l]	Grinding duration, hours	Mixture yield by mass, [%]
			WNi	Co			
Ball mill V = 380 l	WCNi8	1200	320	3	170	20	97
Horizontal vibrating mill, V = 1 l	WCNi10	3	0,4		0,5	48	60
	WCNi15	3	0,36	0,017	0,5	30	60
Vertical vibrating mill, V = 1 l	WCNi8	5	0,5		0,5	1	75

Table 3. Mixture granularity.

Equipment	Mixture type	Cobalt, %	Nickel, %	Grinding duration, hours	Grain size, mkm			Notes
					Grains weight	1,5–2, before	more 2, before	
Ball mill V = 380 l	WCCoNi8	5,54	2,25	20	0,5–1,0	3%	0.5%	WCo6 base
Horizontal vibrating mill, V = 1 l	WCNiCo15	5,6	8,9	30	0,3–0,8	5%	0.5%	
	WCNi10	-	8,9	48	0,5–1,2	10%	3%	
Vertical vibrating mill, V = 1 l	WCNi10	-	8,9	1	0,5–1,0	10%	2%	

Table 4. Sintering modes in a vacuum furnace.

Equipment	Residual pressure, mm Hg	Temperature rise rate, °C/min	Sintering temperature, °C	Holding time at the sintering temperature, mins
Vacuum furnace	1×10–2	3	1450	50

thickness of the interlayer is 0.5–1.5 mkm, at times found up to 3 mkm. The grain size is

Table 5. Physical and mechanical properties of regenerated alloys

Alloy type	Flexural strength, H/mm^2 (kG/mm^2)	Density, g/cm^3	Hardness, HRA	Type of vibrogrinding
WCNi10	1519 (155)	14,02	85,0	horizontal
WCNiCo15	1627 (166)	13,45	87,5	
WCNi10	1559 (159)	14,04	84,5	vertical

uniform, the bulk of the grains is 1–3 mkm, individual grains are up to 15 mkm. GOST 4872-75 does not allow grains larger than 20 mkm for WCo type alloys [15].

Metallographic analysis shows that although the alloys were produced according to different technological modes and differ in binder content, they meet the requirements for hard alloys. The positive effect of the addition of cobalt to the mixture is confirmed by the increased physical and mechanical properties of the WNiCo alloy, shown in Table 5 [16].

The mixture obtained as a result of 40-h grinding in a ball mill in ethanol medium with a cobalt content of 5.6% ($d_{cp} = 0,75$ mkm) was additionally charged with nickel powder (NPC) to a total binder content of 8% [17]. This was done to allow comparison with the standard WCo8 type. The results of the chemical analysis of the resulting mixture are given in Table 6 [18].

Table 6. Chemical analysis of regenerated mixture

Ctotal	Cfree	Co	Ni	Fe
5,49	0,08	5,54	2,25	0,25

The sintering of rods and cutting molds was carried out in the following modes: holding temperature at the maximum sintering temperature - 1385°C, duration - 45 min - series 1, holding temperature at the maximum sintering temperature - 1395°C, duration - 40 min - series 2.

The physical and mechanical properties of the resulting alloy were determined according to GOST 3882-91, 20017 (determination of hardness by Rockwell), 20018 (determination of density), 20019 (determination of flexural strength), 9391-80 (determination of microstructure). The results are presented in Table 7.

The macrostructure of the alloy is satisfactory [19, 24].

It should be noted that the conditions of grinding mixtures have a significant effect on the properties of regenerated hard alloys. So, the hardness of the alloy increases by about 1 HRA unit after 20 h of dry grinding, however, low cobalt alloys have a porosity of up to 0.4% (by volume) [20]. An increase in the duration of dry grinding does not contribute to a decrease in porosity.

Applying wet grinding for 20 h ensures the production of alloys with a porosity of 0.1 - 0.2%, which is allowed by GOST, while there is no increase in hardness compared to dry grinding [21].

The cobalt content in regenerated alloys corresponds to its content in the original scrap, the content of bound carbon corresponds to the stoichiometric content for each type of alloy, the content of free carbon does not exceed 0.1%, and the physical and mechanical properties of sintered alloys meet the requirements of GOST 3882-74 for standard alloys of the corresponding types [22].

To select the optimal ratio of cobalt to nickel, tungsten-containing alloys with different binder contents were produced [23].

Table 7. Physical and mechanical properties of regenerated alloy

Alloy type	Flexural strength, [kgs/mm^2]	Hardness, [HRA]	Density, [g/sm^3]	Cutting coefficient	Notes
WCCo6Ni2	145	96,5	14,7	2,6	series 1
WCCo6Ni2	188 (max 200)	89	14,4–14,7	1,9	series 2
WCCo8	170	88	14,5–14,8	1	GOST 3882–91

To find the optimal modes, a statistical method of experiment planning was applied. The following variables were chosen as input variables: x_1 – percentage of cobalt 7.5 $\pm$ 7.5, x_2 – percentage of nickel 6 $\pm$ 4; x_3 – alloy density 14.05 g/cm^3 $\pm$ 0.81; x_4 - hardness HRA 91.25 $\pm$ 5.25.

The matrix of the planned experiment was compiled according to the full factorial experiment 2^4. The flexural strength kg/mm^2 (У) was chosen as an optimization criterion. Experiments were carried out in a vacuum furnace. Reproducibility dispersion resulted in $D = 7.41$, mean square error of experience $S_y = 2{,}72$; regression coefficients: $B_0 = 66{,}68$, $B_1 = 6{,}525$, $B_2 = -12{,}85$; $B_3 = -7{,}125$, $B_4 = -6{,}7$, $B_5 = -2{,}8$; confidence interval at $\alpha = 0{,}05$, $N = 8$, $\Delta bi = \pm 2.22$. Linear model (code) $У = 66.68 + 6.525\ x_1 - 12.85\ x_2 - 7.125\ x_3 - 6.7\ x_4 - 2{,}8\ x_5$.

The adequacy hypothesis is not rejected since $F_{calc} = 9.1 < F_{table} = 19.0$.

Natural model: $P = 318.52 + 3.263Co - 0.257Ni - 0.2375\rho - 0.983HRA$, where P—flexural strength limit, kg/mm^2; Co—percentage of cobalt, %; Ni—percentage of nickel, %, ρ—alloy density, g/sm^3; HRA—hardness HRA.

New technology is resistant towards negative factors. The center can operate at no loss, selling over 11 tons of products a year. In case of price drop by 10–20% or material cost increase by 20–50%, product profitability will reach 60–70%, which is still rather high.

Economical indices of the project are as follows:

net present value (NPV) for the settlement period of project realization (5 years) is over 1265 thousands USD, which corresponds to the effectiveness condition, where NPV should be > 0;
investments profitability index (JR) – 6,0 (JR > 1);
recoupment of capital investments is 8 months from the issuance of the loan.

Industrial development of the above-described technology, within short timeframe, will allow to solve several important issues:

- hard alloys scrap utilization, its further reduction and consequent significant cut down of raw material usage;
- making use of new technology which omits low productive ways of deficit raw material utilization and manufacturing products of international quality standard;
- subsequent import refusal in favor of hard alloy cermet tools;
- reduction and subsequent termination of strategic materials and alloys scrap export, in favor of exporting ready product
- powders and tools, which will increase foreign currency earnings.

5 Conclusions

It has been established that the cobalt content in regenerated mixtures corresponds to its content in the original scrap, the content of bound carbon corresponds to the stoichiometric content for each type of alloy, the content of free carbon does not exceed 0.1%, and the physical and mechanical properties of sintered alloys meet the requirements of GOST 3882-74, applied to standard alloys of the respective grades. To increase the flexural strength, it is necessary to choose a binder in the nickel to cobalt ratio of 0.5–0.6. Hard alloys of the WC–Ni–Co carbide system with different content of components compared to traditionally used WC + Co alloys using the same binder percentage have been developed to ensure increased hardness and flexural strength and, due to this, increased wear resistance. Environment friendly technology of super hard material recycling, cermet components of out aged ammunition and different types military equipment was suggested.

References

1. Kulynych, V., Shapoval, A., Dragobetskii, V.: Hard alloys recycling as a promising direction of technological equipment for machine-building production. Mater. Sci. Forum (MSF) **1052**, 423–428 (2022)
2. Savchenko, I., Shapoval, A., Kuziev, I.: Modeling of high module power sources systems safety processes. Mater. Sci. Forum (MSF) **1052**, 399–404 (2022)
3. Jagadeesh, P., Puttegowda, M., Rangappa, S.M., Alexey, K., Gorbatyuk, S., Khan, A., et al.: A comprehensive review on 3D printing advancements in polymer composites: technologies, materials, and applications. Int. J. Adv. Manuf. Technol. **121**(1–2), 127–169 (2022)
4. Markov, O.E., Oleshko, M.V., Mishina, V.I.: Development of energy-saving technological process of shafts forging Weighing more than 100 tons without ingot upsetting. Metall. Min. Ind. **3**(7), 87–90 (2011)

5. Allieva, L., Hrudkina, N., Aliiev, I., Zhbankov, I., Markov, O.: Effect of the tool geometry on the force mode of the combined radial-direct extrusion with compression. East Eur. J. Enterp. Technol. **2**(1–104), 15–22 (2020)
6. Trembach, B., Grin, A., Turchanin, M., Makarenko, N., Markov, O., Trembach, I.: Application of Taguchi method and ANOVA analysis for optimization of process parameters and exothermic addition (CuO-Al) introduction in the core filler during self-shielded flux-cored arc welding. Int. J. Adv. Manuf. Technol. **114**(3–4), 1099–1118 (2021)
7. Karnaukh, S.G., Markov, O.E., Aliieva, L.I., Kukhar, V.V.: Designing and researching of the equipment for cutting by breaking of rolled stock. Int. J. Adv. Manuf. Technol. **109**(9–12), 2457–2464 (2020)
8. Zagirnyak, M., Chornyi, O., Nykyforov, V., Sakun, O., Panchenko, K.: Experimental research of electromechanical and biological systems compatibility. Prz Elektrotech **92**(1), 128–131 (2016)
9. New methods of obtaining materials and structures for light armor protection. In: ICMT 2015—International Conference on Military Technologies 2015 (2015)
10. Analysis of operation of power components compensation systems at harmonic distortions of mains supply voltage. In: Joint International Conference—ACEMP 2015: Aegean Conference on Electrical Machines and Power Electronics, OPTIM 2015: Optimization of Electrical and Electronic Equipment and ELECTROMOTION 2015: International Symposium on Advanced Electromechanical Motion Systems (2016)
11. A model of the assessment of an induction motor condition and operation life, based on the measurement of the external magnetic field. In: 2018 IEEE 3rd International Conference on Intelligent Energy and Power Systems, IEPS 2018—Proceedings (2018)
12. Pavlenko, D.V., Tkach, D.V., Danilova-Tretyak, S.M., Evseeva, L.E.: Thermophysical properties and temperature of the start of titanium recrystallization in different structural states. J. Eng. Phys. Thermophys. **90**(3), 685–696 (2017)
13. Bykov, I.O., Ovchinnikov, A.V., Pavlenko, D.V., Lechovitzer, Z.V.: Composition, structure, and properties of sintered silicon-containing titanium alloys. Powder Metall. Met. Ceram. **58**(9–10), 613–621 (2020)
14. Olshanetskii, V.E., Stepanova, L.P., Greshta, V.L., Pavlenko, D.V., Tkach, D.V.: Deformation behavior of titanium VT1-0 with submicrocrystalline structure formed by the method of helical extrusion. Met. Sci. Heat Treat. **55**(11–12), 603–607 (2014)
15. Shmelov, Y., Bazyk, A., Kitsel, N.: Research of parametric influence of light-emitting diodes on the multicomponent module near-field illuminated zone formation. Lect. Notes Mech. Eng. 801–809 (2022)
16. Yakovliev, R., Shmelov, Y., Petchenko, M., Honchar, S., Kovalskyi, V.: Research of the synthesis of radiant intensity indicatrix of multicomponent beam diode module. In: E3S Web of Conferences (2021)
17. Pavlenko, D.V.: Assessment of gas saturation of titanium alloys synthesized from powders using twist extrusion. Powder Metall. Met. Ceram. **56**(5–6), 273–282 (2017)
18. Markov, O.E., Aliiev, I.S., Aliieva, L.I., Hrudkina, N.S.: Computerized and physical modeling of upsetting operation by combined dies. J. Chem. Technol. Met. **55**(3), 640–648 (2020)
19. Vlasenko, K., Hrudkina, N., Reutova, I., Chumak, O.: Development of calculation schemes for the combined extrusion to predict the shape formation of axisymmetric parts with a flange. East Eur. J. Enterp. Technol. **3**(1–93), 51–59 (2018)
20. Hrudkina, N.S.: Process modeling of sequential radial-direct extrusion using curved triangular kinematic module. FME Trans. **49**(1), 56–63 (2020)
21. Potapkin, V.F., Levkin, A.N., Satonin, A.V., Romanov, S.M., Vorob'yev, Y.A., Gribkov, E.P.: Stressed state and kinematics in the rolling of powder materials on a metal substrate. Powder Metall. Met. Ceram. **39**(1–2), 11–17 (2000)

22. Kostikov, A.A., Perig, A.V., Larichkin, O.V., Stadnik, A.N., Gribkov, E.P.: Research into payload swaying reduction through cable length manipulation during boom crane motion. FME Trans. **47**(3), 464–476 (2019)
23. Tarasov, A.F., Altukhov, A.V., Gribkov, E.P., Abdulov, A.R.: Development and FEM modeling of a new severe plastic deformation process according to the reverse shear scheme. Model. Simul. Eng. (2019)
24. Hrushko, O.V., Hutsalyuk, O.V., Andreev, I.V., Mel'nychenko, V.V., Studenets', S.F.: Mechanical characteristics of alloys of the W–Ni–Fe system. Mater. Sci. **54**(1), 88–95 (2018)

Increasing the Life of Lance Tips with Extended Laval Nozzles

Iurii Savchenko[1], Oleksandr Shapoval[2](✉), Sergei Panteikov[1], Inna Zhulkovska[3], and Oleg Zhulkovskyi[3]

[1] University of Customs and Finance, Volodymyr Vernadskyi Str., 2/4, Dnipro 49000, Ukraine
ser_pant_in@ukr.net

[2] Kremenchuk Mykhailo Ostrohradskyi National University, 20, Pershotravneva Street, Kremenchuk 39600, Ukraine
alexshap.as@gmail.com

[3] Dniprovsky State Technical University, Dniprobudivska Str., 2, Kamianske 51918, Ukraine
olalzh@ukr.net

Abstract. The purpose of this work is to develop high-strength mechanical designs for feeding oxygen into the cavity of converters. On the basis of the conducted set of industrial experiments, for the first time the hypothesis of significant increase of lance tip durability at providing cooling of nozzle zones of their copper bowls was confirmed and high degree of efficiency of this method was established. A new design (patent-protected) of lance tip with five extended Laval nozzles was developed, which due to a more rational distribution of a coolant in it, allowed increasing its resistance by 2.205 and 4.195 times in comparison with two similar designs. It was found that the external welding seams around the Laval nozzles at the end of the copper bowl are the limiting link that prevents significant durability of welded lance tip structures. This is in favor of forged and/or solid-steel lance tip designs with higher durability, which would be improved to a greater extent by providing cooling in the back-to-back (with water flow) zones of the copper bowls. The results of experimental evaluation of effectiveness degree of lance tip design, reflected in the work, testify to the possibility of more effective solution of the problem of low durability of lance tip of this type, which is always present at the converter shops in operation.

Keywords: BOF process · cooper bowl · cooler · blowing device

1 Introduction

The engineering industry is heavily dependent on the industrial production of many metals and alloys, particularly the production of various grades of steel, which are smelted by various methods: oxygen-converter, electric arc, induction, etc.

BOF production is the main steelmaking process [1], which involves conversion of pig iron into steel by reducing the carbon content in a high-carbon melt that is blown with high-purity technical oxygen [2]. Converter technology accounts for more than 60% of

H. Altenbach et al. (Eds.): CAMPE 2023, pp. 133–142, LNME 2025.
https://doi.org/10.1007/978-3-031-82979-6_14

world steel production due to its high productivity and greater environmental efficiency [2].

The main mechanical structures for oxygen supply to the converter are top oxygen lances—copper multi nozzle lance tips at the end, however the construction of the latter is mostly far from being perfect and this causes different difficulties for top oxygen lance operation, negatively influencing operation rhythm and productivity of oxygen converters.

It is known that the location of blow nozzles on the way of water movement in multi-nozzle lance tip constructions lead to the formation of stagnant coolant zones behind each nozzle (in the way of water movement). It is obvious, that elimination of stagnant water zones in lance tip constructions will increase their durability and prolong the top lances' lifetime. In this case, the number of Laval nozzles, which should be located in a blowing device from the point of view of providing the rational blowing and slagging regimes of the converter melting, will not be limited by the stamina of lance tips.

The work objective is to develop a new design of highly resistant lance tips with extended Laval nozzles on the basis of developed theoretical positions and further evaluation of efficiency degree of suggested methods for elimination of the problem of low resistance of blowing devices of such type at the existing BOF shops.

2 Literature Review

Improving the technical and economic performance of BOF Shop and increasing the operating characteristics of the process equipment (including top blowing lances) involves the development and use of top multi-nozzle (six and higher) top blowing devices.

High melt temperature, presence of flame flames as a result of melt carbon oxidation when blowing with pure oxygen, splashing of liquid metal and slag, occurrence of CO to CO_2 afterburning flames creates high-temperature atmosphere in the converter cavity [3], which has an aggressive (thermal and oxidizing) effect on materials of top lance, and especially on its lance tip [4]. Operation of the top lances in such unfavorable conditions may result in the following: changes of microstructure of initial materials (copper, steel) of its components (bowls, nozzles, pipes), eventually causing metal fatigue; softening of welding (copper and steel) seams, which will lead to the destruction, usually, of external (and often internal) copper welding seams [3].

In order to avoid these phenomena, it is necessary to arrange the necessary water velocity in the water-cooling path of the top lances (and especially in its lance tip) in order to ensure proper cooling of the entire blowing device. The area of the nozzle space increases with the number of nozzles. This explains the dramatic drop in yield of lance tips with the increase in the number of nozzles. Therefore, in everyday practice of BOF shops they try to apply top lances with a minimum number of lance tubes (usually not more than five), that have comparatively simple construction and relatively high resistance of lance tips, however, it leads to unsatisfactory process of melting in oxygen converter—bad slag formation, metal carry-over from converter and notification of technological equipment, reduction of finished steel output, increase of harmful admixtures in it (sulphury, phosphorus) and others.

At present time in development of top lances the main attention is always given to increase of resource-saving efficiency of converter process due to use of its own reserves, therefore different designs of lance tips are constantly developed, adapted for afterburning of CO to CO_2 and allowing to increase heat balance of melting, to improve conditions of melt mixing, to reduce blowing time, to increase speed of dephosphorization, to reduce rejection rate [5]. Waste gas temperature has been found to increase from 1800 to 2000 K under CO afterburning conditions at $P_{CO_2}/P_{CO} = 1$ [6]. In order to effectively utilise the chemical and physical heat of the waste converter gas, blowdown processes with and without afterburning were analyzed [2].

In order to increase the mass transfer rate between slag and melt, a self-rotating lance using a gas jet as an energy source has been developed; the mass transfer rate has increased by more than 30% [7].

Study of influence of cross-sectional area of the main blowing orifice of the top lances and blowing rate on characteristics of the outgoing jet allowed to design and optimize the central nozzle of the top oxygen lances for blowing of $O_2 + CO_2$ gas mixture in converter production [8]. An influence of consumption of this mixture (80% O_2 + 20% CO_2) on distribution of gas-spark jets carrying powdered lime is investigated [1].

In order to design and develop the oxygen lance nozzle design, effect of its outlet wear on characteristics of the supersonic jet flowing out of the lance nozzle has been studied. It was found [9] that increasing blowout nozzle wear leads to faster jet velocity reduction, which also leads to earlier convergence point for each gas jet. At the same time, the causes of blow-out nozzle blow-out have not been investigated and no measures have been taken to eliminate them.

For projecting and manufacturing of oxygen lance with high Mach number (oxygen pressure 1,01 MPa, Mach number at the nozzle outlet 2,15), characteristics of lance jet at different pressures of oxygen supply and its influence on the melting bath have been analyzed and investigated [10].

Characteristics of blowing-down dynamic oxygen lance of the converter are investigated - impact of its design (including angle of twist and speed of rotation) on melt diphosphonate technology is determined [11].

In order to design upper plunging lance, we studied the following parameters: gas injection in melt [12], bubbles formation and growth in the bath [13], as well as by means of X-ray visualization [14], rotation of splashing waves in a metallic bath [15]. The influence of viscous and interfacial forces on the bath hydrodynamics was established [16].

Recently, in connection with considerable spreading of protective skull liner technology at refractory converter lining, different activities at blow-out and foaming of liquid slag by upper blowing tuyere have been widely spread. Thus, impact of upper lance blow effect on slag foaming in a converter with combined upper and lower blowing was studied in paper [17].

As we can see from the analysis of literature on this subject, cooling systems of multi-nozzle lance heads are not studied, and there are no publications on this issue, though it is one of the important problems, and it always takes place at all BOF shops, especially if out-of-date lance heads of welded structure are used. Thus, the urgency of improving the durability of tuyere heads, particularly with extended Laval nozzles, is unquestionable and needs to be effectively addressed.

3 Research Methodology

The influence of the cooling arrangement of the copper bowl nozzle zones on the service life of the lance tips was investigated. The object of the study was the developed design of a five-nozzle lance tip with extended Laval nozzles; their profile and arrangement were determined by high-temperature modelling of the blowing process. Experiments were carried out on 250-ton converters, structured observation was used, causes of lance tip failures were recorded by photography.

In order to verify the previously put forward hypothesis about the influence of the degree of cooling of the inner surface of top lance nozzle zones on its durability, a new design (Patent 40296 UA) of five-nozzle lance tip with extended laval nozzles (Fig. 1, a) for the top lances of 250-ton converters of PJSC "Dneprovsky Metallurgical Combine" (now PrJSC "KAMET-STAL"", Kamianske, Ukraine), in which intensive cooling of copper bowl inner surface with individual supply of additional water streams into its nozzle zones is provided due to the improvement of lance tip cooling system.

The prototype for this new design was the previously successfully appraised design of a five-nozzle lance tip with extended Laval nozzles and reduced water passages to 25 mm (Fig. 1, b) which, although it did not allow cooling of the suction areas of the copper bowl (i.e., it had stagnant water areas there (Fig. 2), showed 1.9 times higher resistance compared to the five-nozzle regular lance tip with extended Laval nozzles (Fig. 1, c) with 59 mm distance between the upper and lower bowls.

Location of the Laval nozzles (their angles of inclination to the top lance vertical (α, deg) and in plan between the axes of neighboring nozzles (φ, deg); circumference of their outlet sections location, m) in the lance tip was carried out on the basis of high-temperature top blowing depending on the number of nozzles and taking into account height of the top lances location above the bath level at quiet state (H_f, m), ensuring smooth bleeding of the melt without visible collapse of the oxygen lance and converter mouth by maintaining optimal diameters and depths (D_I, D_{II}, D, L_I, L_{II}) of the formed high-temperature reaction zone (Fig. 3) to prevent accelerated local wear of the converter walls and bottom liner.

The developed construction of lance tip (Fig. 1, a) consists of water feed pipe 1, oxygen feed pipe 2 and water drain pipe 3 (correspondingly central (Ø108 × 8 mm), intermediate (Ø325 × 12 mm) and external (Ø426 × 8 mm)), having lengths from the center of the tip lance of 200 mm, 170 mm 140 mm respectively, upper (steel collector) 4 and lower (copper) 5 bowls, which are held by welding in them by nozzle inserts 6 with Laval nozzles, placed equally spaced around the circle (external Ø252 mm) at an angle 17° to the vertical axis of top lance. Between the bowls 4 and 5 there is an intermediate bowl—splitter of water flows 7, having holes 8, which are made over nozzle zones of the copper bowl 5 for coolant supply into the mentioned zones. In the collector 4 the water feed pipe 1 is positioned in such a way that an annular gap 9 of required size is formed between its lower end-face and the upper surface of the flow divider 7 in order to supply a coolant with the required flow rate to the holes 8. The central hole 10 in flow divider 7 is designed for water supply for cooling the inner surface of copper bowl 5. The size of areas of flow-through sections of holes 8 and ring gap 9 can regulate the amount of water entering the nozzle areas of the copper bowl 5, thus achieving the required intensity of its cooling.

Figure 4 shows the external view of the water flow divider (a), its installation in the lance tip (b), the lance tip from the inside (c) and the outside (d). Water separator 7 (Fig. 1, a; Fig. 4, a) with holes for inserting nozzles 6 and with special cut-outs 8 made in the nozzle areas to supply water to them, is mounted on nozzle inserts 6 (Fig. 1, a; Fig. 4, b), which are previously inserted in the holes made in the copper (bottom) bowl 5.

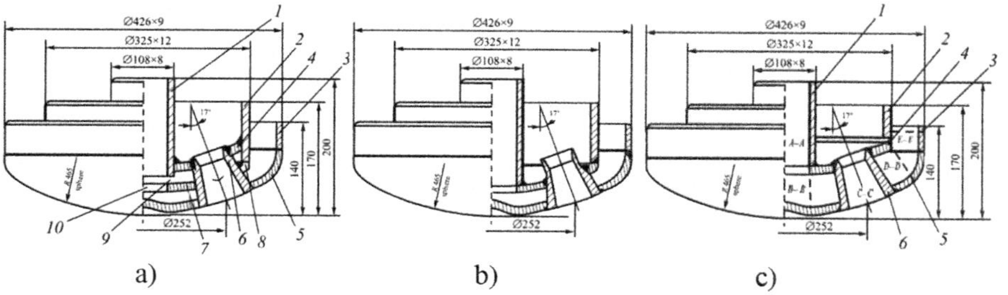

Fig. 1. Designs of five nozzle lance tips (see text for designations): designed (a), with reduced bowl spacing (b) and standard (c).

Then the upper bowl (collector) 4 is put on the nozzles 6 and then the nozzle inserts 6 are welded (Fig. 1, a; Fig. 4, c) to the collector 4 and to the lower bowl 5. In the end of assembly there are installed and welded together central 1, intermediate 2 and outer 3 nozzles so that the required size of the annular gap 9 is created and in the intermediate bowl 7 in the nozzle zones (between the nozzle inserts 6 and intermediate nozzle pipe 2) there are holes 8 for water supply for cooling of the copper lance tip nozzle zones 5.

Lance tip cooling is carried out as follows. Cooling water enters the lance tip through a central nozzle 1, dividing into two groups of flows—main and additional. The main flow group (water consumption is 80% of total cooling water consumption) is supplied through the central opening 10 in the water separator 7 into the outlet line of the lance tip (between the separator 7 and copper bowl 5) to nozzles 6 and, bending around their body, goes into the outlet line of the top lances between the outer 3 and intermediate 2 nozzles. In this case behind the nozzles local stagnant zones of water are formed, into which additional coolant flows are directed (water consumption is 20% of the total) through the holes 8 in the separator 7, to which water is supplied through an annular gap 9 between the bottom end of the central pipe 1 and the separator 7. This way contributes to intensive cooling of the cooper bowl 5 nozzle sections. At the end of the lance tip discharge channel between copper bowl 5 and divider 7, main and additional water flows mix and the coolant is discharged into the discharge line of the top oxygen lance in one continuous flow.

Application of new lance tip design enabled to increase its durability due to more rational distribution of coolant in it and high speed of its circulation, which was achieved

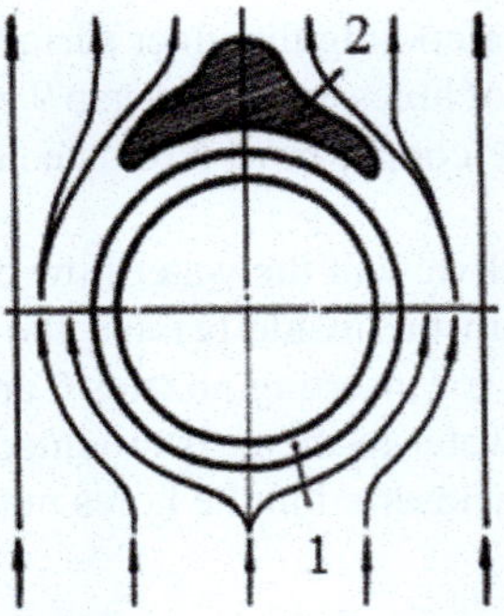

Fig. 2. Flow diagram of the coolant flow around the nozzle (1) and the coolant stagnation zone (2).

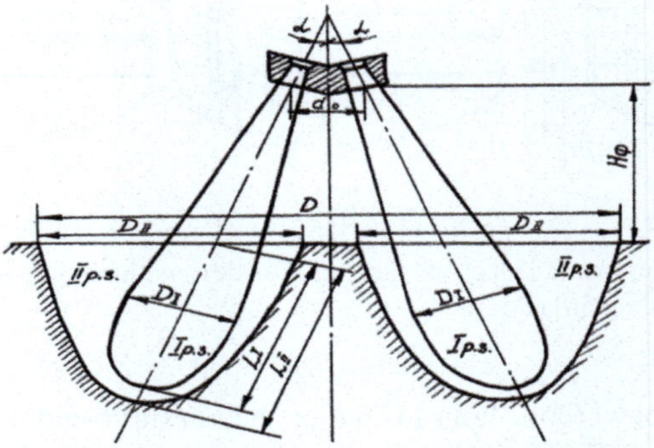

Fig. 3. Schematic diagram of the reaction zone when blowing the converter bath with oxygen from above.

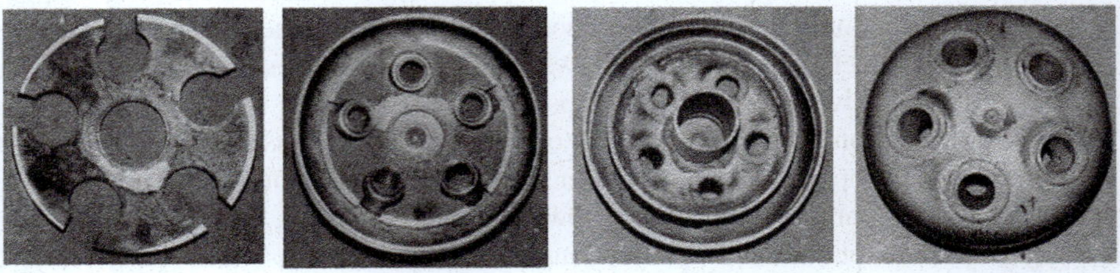

(a) (b) (c) (d)

Fig. 4. Exterior view of the cooler flow divider (a), its installation in the lance tip (b), the lance tip from the inside (c) and the outside (d).

by application of additional intermediate bowl (water flow splitter) 7 in the design of lance tip. This reduced the height of the cooling circuit from 59 mm (standard design) to 23 mm, which kept the flow velocities of cooling water in different sections of the developed lance tip (Fig. 1, a) at the same (increased) level as in the lance tip with reduced to 25 mm distance between the bowls (see Fig. 1, b), while organizing the redirection of water to the nozzle areas of copper bowl 5.

Figure 5 shows calculated velocities of coolant movement (at its flow U_W = 340 m^3/hour) in various sections (A-A, B-B, C-C, D-D, E-E) of water-cooling path in standard (see Fig. 1, c) and designed (see Fig. 1, a) designs of five-nozzle lance tip. In areas of nozzle space of lance tip, the average water velocity was 10.6 m/s at U_W = 340 m^3/hour as it was determined by numerical calculations [18]. Obtained graphic dependence suggests a significant increase in the velocity of water at the inlet into the water-cooled path (section B-B) of the new lance tip design (Fig. 1, a) and between the nozzles (section C-C), somewhat lower—at the outlet of the water-cooled lance tip path (section D-D), but especially high (from 0 to 10.6 m/s)—in the nozzle areas of the lance tip compared with the standard design blowing device (Fig. 1, c).

Presence of special openings 8 in stream splitter 7 behind nozzles 6 enabled feed of down-flow water to nozzle zones of copper bowl 5, thus eliminating congested zones of coolant behind nozzles 6 of lance tip preventing thus overheating of mentioned zones of copper bowl 5 and early failure of blowing device. In areas of nozzles with completely missing stagnation zones of the coolant, the coolant flow speed is kept almost the same as between nozzles 6 (section C-C) owing to additional water supply from above through the holes 8 in flow splitter 7 for cooling of copper bowl 5 areas behind nozzles 6.

4 Results and Discussion

Average service life of the designed lance tips with cooling nozzle zones (Fig. 1, a) was 172 heats (Fig. 6) and in comparison, with five-nozzle constructions without cooling nozzle zones with reduced to 25 mm height (distance between the bowls) of the cooling channel (Fig. 1, b), which showed average service life of 78 heats (Fig. 6), increased by 2.205 times. Compared to standard (shop-floor) five-nozzle lance tips (Fig. 1, c) with average durability of 41 melts (Fig. 6) without any cooling of nozzle zones and with 59 mm high cooling channel, average durability of developed five-nozzle lance tip with cooling of nozzle zones increased by 4.195 times.

Figure 7 shows worked samples of the developed lance tip design with cooling of nozzle zones of the copper bowl.

As would be expected, the copper external welds around the nozzles cannot withstand the high temperatures and thermomechanical stresses generated in the lance tip during operation, which are not adequately addressed by the top lance compensation system, resulting in rapid cracking (Fig. 7, a), followed by coolant leaks, requiring immediate removal of the top lance to replace its lance tip due to the potential for serious failure, as has been repeatedly experienced in many BOF shops. In addition, top lance failure was often caused by a gross violation of process instruction by the distributor, i.e., scrap, which resulted in immediate destruction of the copper lance tip, typically in the central area (Fig. 7, b), where the copper is thinner due to shrinkage in the copper bowl fabrication process.

Work continues to increase the durability of the upper blowing tuyeres by using for their manufacture: arc welding with shielded flux-cored wire [19]; technologies of radial-direct extrusion with compression [20], intense plastic deformation according to

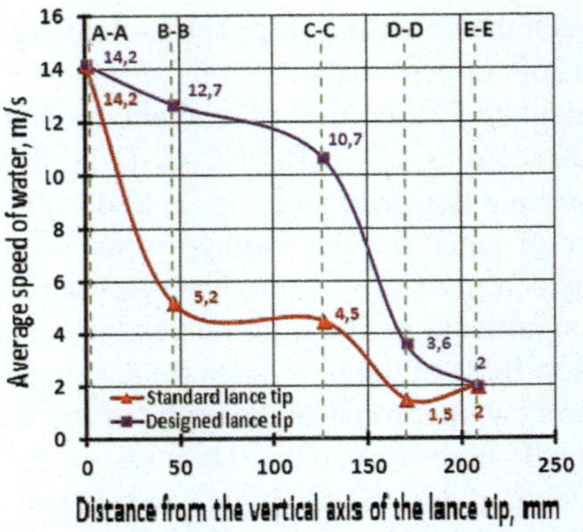

Fig. 5. Dependence of average water velocity ($U_W = 340\ m^3/h$) in different sections of standard and designed lance tip designs.

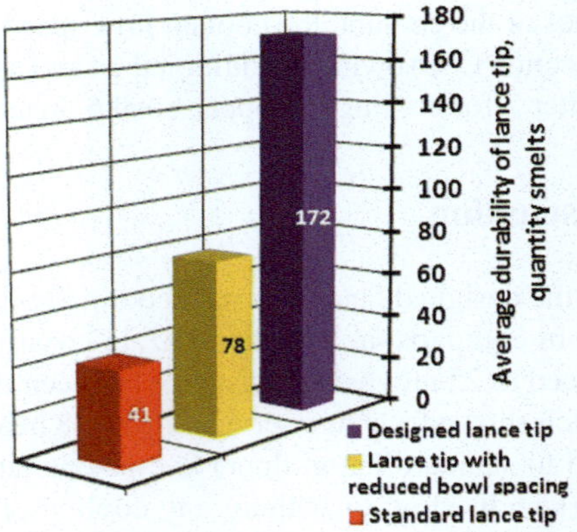

Fig. 6. Average durability of different designs of five-nozzle lance tips with extended Laval nozzles.

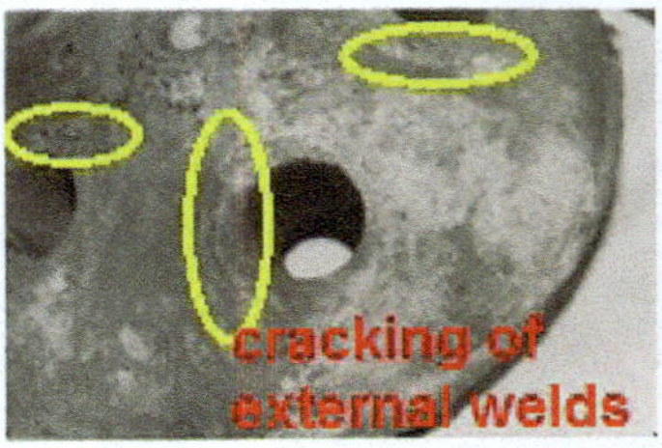

(a)

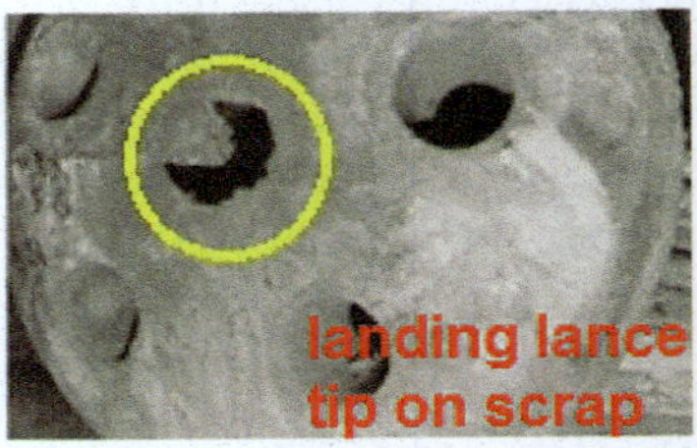

(b)

Fig. 7. Worked samples of the developed lance tip design.

the reverse shear scheme [21], upsetting with combined dies and sequential radial-direct extrusion; new superhard materials – sintered silicon-containing titanium alloys. It is also envisaged to use a system of high-modulus energy sources directly when blowing the melt in the converter.

5 Conclusions

On the basis of the conducted complex of industrial experiments the hypothesis about significant increase of lance tips durability at provision of cooling of inner surface of their copper bowl nozzle zones has been confirmed for the first time. For the first time, a high degree of efficiency of this method is proved—ensuring of cooling of copper bowl sealing zones at intensification of cooling process of all its inner surface is the most effective and necessary measure for increasing the durability of any lance tip constructions.

The application of new elaborated construction of five-nozzle lance tip allows increasing its endurance in 2.205 and 4.195 times in comparison with two similar constructions where the cooling of nozzle-net zones of copper bowl and intensive cooling of all its inner surface are absent.

It was found out that the limiting link that doesn't allow achieving a high endurance of welded structures of lance tips is external weld seams, made of copper around Laval nozzles.

The results of experimental evaluation of efficiency degree of developed lance tip design with extended lavalier nozzles, reflected in the given work, testify to the opportunity of more effective solution of the problem of low durability of lance tips of such type, that is always present at the converter plants in operation.

References

1. Tarasov, O., Vasylieva, L., Altuhov, O., Pavlenko, D., Tkach, D.: Development of integrated CAD/CAE systems based on parameterization of the simulated process. Lect. Notes Netw. Syst. (LNNS) **657**, 679–691 (2023). https://doi.org/10.1007/978-3-031-36201-9_56
2. Shmelov, Y., Bazyk, A., Kitsel, N.: Research of parametric influence of light-emitting diodes on the multicomponent module near-field illuminated zone formation. Lect. Notes Mech. Eng., 801–809 (2022). https://doi.org/10.1007/978-3-030-85230-6_95
3. Tarasov, O.F., Gribkov, E.P., Pavlenko, D.V., Danylenko, M.I., Altukhov, A.V.: Structure and mechanical properties of titanium processed by twist extrusion and subsequent rolling. Adv. Mater. Sci. Eng., 7795273 (2022). https://doi.org/10.1155/2022/7795273
4. Markov, O.E., Aliiev, I.S., Aliieva, L.I., Hrudkina, N.S.: Computerized and physical modeling of upsetting operation by combined dies. J. Chem. Technol. Met. **55**(3), 640–648 (2020)
5. Vlasenko, K., Hrudkina, N., Reutova, I., Chumak, O.: Development of calculation schemes for the combined extrusion to predict the shape formation of axisymmetric parts with a flange. East. Eur. J. Enterp. Technol. **3**(1–93), 51–59 (2018). https://doi.org/10.15587/1729-4061.2018.131766
6. Hrudkina, N.S.: Process modeling of sequential radial-direct extrusion using curved triangular kinematic module. FME Trans. **49**(1), 56–63 (2020). https://doi.org/10.5937/fme2101056H
7. Pavlenko, D.V., Belokon', Y.O., Tkach. D.V.: Resource-Saving technology of manufacturing of semifinished products from intermetallic γ-TiAl alloys intended for aviation engineering. Mater. Sci. **55**(6), 908–914 (2020). https://doi.org/10.1007/s11003-020-00386-1

8. Kostikov, A.A., Perig, A.V., Larichkin, O.V., Stadnik, A.N., Gribkov, E.P.: Research into payload swaying reduction through cable length manipulation during boom crane motion. FME Trans. **47**(3), 464–476 (2019). https://doi.org/10.5937/fmet1903464K
9. Tarasov, A.F., Altukhov, A.V., Gribkov, E.P., Abdulov, A.R.: Development and FEM modeling of a new severe plastic deformation process according to the reverse shear scheme. Model. Simul. Eng. (2019). https://doi.org/10.1155/2019/8563830
10. Hrushko, O.V., Hutsalyuk, O.V., Andreev, I.V., Mel'nychenko, V.V., Studenets', S.F.: Mechanical characteristics of alloys of the W–Ni–Fe system. Mater. Sci. **54**(1), 88–95 (2018). https://doi.org/10.1007/s11003-018-0162-7
11. A model of the assessment of an induction motor condition and operation life, based on the measurement of the external magnetic field. In: 2018 IEEE 3rd International Conference on Intelligent Energy and Power Systems, IEPS 2018—Proceedings (2018). https://doi.org/10.1109/IEPS.2018.8559564
12. Bykov, I.O., Ovchinnikov, A.V., Pavlenko, D.V., Lechovitzer, Z.V.: Composition, structure, and properties of sintered silicon-containing titanium alloys. Powder Metall. Met. Ceram. **58**(9–10), 613–621 (2020). https://doi.org/10.1007/s11106-020-00117-w
13. Savchenko, I., Shapoval, A., Kuziev, I.: Modeling of high module power sources systems safety processes. Mater. Sci. Forum (MSF) **1052**, 399–404 (2022). https://doi.org/10.4028/p-24y9ae
14. Karnaukh, S.G., Markov, O.E., Kukhar, V.V., Shapoval, A.A.: Classification of steels according to their sensitivity to fracture using a synergetic model. Int. J. Adv. Manuf. Technol. **119**(7–8), 5277–5287 (2022). https://doi.org/10.1007/s00170-022-08653-y
15. Jagadeesh, P., Puttegowda, M., Rangappa, S.M., Alexey, K., Gorbatyuk, S., Khan, A., et al.: A comprehensive review on 3D printing advancements in polymer composites: technologies, materials, and applications. Int. J. Adv. Manuf. Technol. **121**(1–2), 127–169 (2022). https://doi.org/10.1007/s00170-022-09406-7
16. Aliiev, I.S., Sivak, R.I., Markov, O.E., Levchenko, V.N.: The evaluation of workpiece deformability for the process of two-stage extrusion of hollow hull. Int. J. Adv. Manuf. Tech. (2023). https://doi.org/10.1007/s00170-023-12353-6
17. Allieva, L., Hrudkina, N., Aliiev, I., Zhbankov, I., Markov, O.: Effect of the tool geometry on the force mode of the combined radial-direct extrusion with compression. East. Eur. J. Enterp. Technol. **2**(1-104), 15–22 (2020). https://doi.org/10.15587/1729-4061.2020.198433
18. Trembach, B., Grin, A., Turchanin, M., Makarenko, N., Markov, O., Trembach, I.: Application of Taguchi method and ANOVA analysis for optimization of process parameters and exothermic addition (CuO-Al) introduction in the core filler during self-shielded flux-cored arc welding. Int. J. Adv. Manuf. Technol. **114**(3–4), 1099–1118 (2021). https://doi.org/10.1007/s00170-021-06869-y
19. Karnaukh, S.G., Markov, O.E., Aliieva, L.I., Kukhar, V.V.: Designing and researching of the equipment for cutting by breaking of rolled stock. Int. J. Adv. Manuf. Technol. **109**(9–12), 2457–2464 (2020). https://doi.org/10.1007/s00170-020-05824-7
20. Zagirnyak, M., Chornyi, O., Nykyforov, V., Sakun, O., Panchenko, K.: Experimental research of electromechanical and biological systems compatibility. Prz. Elektrotech. **92**(1), 128–131 (2016). https://doi.org/10.15199/48.2016.01.31
21. New methods of obtaining materials and structures for light armor protection. In: ICMT 2015—International Conference on Military Technologies (2015). https://doi.org/10.1109/MILTECHS.2015.7153695

Development of the Slag-Splashing Lance for Slag Skull Coating the Converter Lining

Elena Panteikova[1], Sergei Panteikov[2], Oleg Zhulkovskyi[2](✉), Inna Zhulkovska[3], and Iryna Pieieva[4]

[1] Private Joint–Stock Company "KAMET–STAL'", Soborna Str., 18B, Kamianske 51900, Ukraine
hel_pant_in@ukr.net

[2] Dniprovsky State Technical University, Dniprobudivska Str., 2, Kamianske 51918, Ukraine
{ser_pant_in,olalzh}@ukr.net

[3] University of Customs and Finance, Volodymyr Vernadskyi Str., 2/4, Dnipro 49000, Ukraine

[4] Kremenchuk Mykhailo Ostrohradskyi National University, 20, Pershotravneva Street, Kremenchuk 39600, Ukraine

Abstract. The aim of the work is to develop mechanical equipment for resource-saving technology of applying liquid slag to refractory lining of 250-ton converters in order to form a protective layer of slag garnish on its working surface. On the basis of the conducted set of industrial experiments the hypothesis of significant increase of BOF lining durability with organization of liquid slag blow with its primary run-out on the most worn trunnion zones was confirmed for the first time. In order to check this hypothesis, the new design of skull lance (patent protected) with grouped lance nozzles of different cross-section at the face of lance tip was developed, produced and tested. This allowed to form the protective slag layer at the working surface of liner both at the bottom part (walls) of converter and at the mouth of the unit. In the course of the converter's campaigns, this ensured that the more worn areas of the converter's lining would be repaired by forming a thicker slag layer on them, thus obtaining a total lining thickness (masonry + slag layer) on these areas that is almost the same as on the other (less worn) working surfaces of the lining with reduced slag carry-over from the converter and reduced refractory consumption.

Keywords: refractory lining · slag bath splashing · Laval nozzle · lining life

1 Introduction

Ferrous metallurgy is the backbone of mechanical engineering—a third of iron and steel produced goes to the mechanical engineering industry.

BOF is the main steel-making process [1], and accounts for over 60% of global steel production due to the high productivity and high environmental efficiency of modern-day BOF technology [2].

H. Altenbach et al. (Eds.): CAMPE 2023, pp. 143–152, LNME 2025.
https://doi.org/10.1007/978-3-031-82979-6_15

As resource-saving concepts and designs of process and mechanical equipment for converters, including their liners, have improved, the approach to managing and maintaining steelmaking facilities has changed significantly over the past few decades. One of these fundamental approaches in BOF steelmaking that has undergone significant changes is the view of the converter refractory lining. The legacy concept viewed the liner as a consumable non refractory material with an indefinite life span, whilst the modern strategy for BOF steelmaking sees the liner as part of the equipment, which must be maintained continuously during operation in order to achieve maximum life expectancy. Today it is no longer acceptable or economical to simply burn the lining and hope for a long service life. Successful refractory maintenance procedures throughout the life of the converter will ultimately reduce the material cost of lining replacement, downtime of the unit, the amount of waste produced, etc. [3].

To increase the service life of the converter lining, special attention is currently paid to its maintenance by means of semi-dry [4] or flaring concreting and slagging [5]. The use of flare gunning to increase the durability of furnace lining has also become widespread in non-ferrous metallurgy processes [6]. Therefore, steelmakers around the world tried to find cheaper and more effective solutions to the issue of hot lining repairs in converters.

Thus in 1981 a revolutionary way of blowing out high-magnesia end slag on converter walls was invented in Japan, by air spraying of slag on side surfaces of lining with air blow pipe in horizontal position (SBC-slag blow coating), in 1982 with bottom lance, and in 1985 with top oxygen lance. Subsequently, Great Lakes Division (Detroit, Michigan, USA) of National Steel Corporation (Mishawaka, Indiana, USA) developed the technology of end-slag blowing with neutral gas feeding through the top oxygen lance according to recommendations of Japan.

On a large industrial scale this technology was first introduced at the 252-t converters of Indiana Harbor (Chicago, Michigan, USA) by LTV Steel Corporation (Cleveland, Ohio, USA) in 1991 and the results literally stunned and shocked the world converter industry. The record of durability of converter refractories over 10000 heat starts, set by LTV Steel Corporation, meant a radical break from the usual (at that time) duration of converter liner campaigns, which used to be 2000–3000 heat starts at the time in all countries (excluding Japan), when converters were operating without lining slagging, even in case of using new and expensive refractories. After that practically all steel mills in the USA and Canada started urgently and universally introducing technologies of slag garnishing of converters while the majority of steel mills in Japan, China and later in Europe and other regions of the world started sending their engineering staff to the USA to study and master a new perspective method of converters lining life extension.

Nowadays practically all BOF shops use technology of slag garnishing on walls in order to increase lifetime of refractory lining of units. In this case environmentally required slag disposal, which is increasing every year [7], is decreasing significantly, despite popularity of "low-slag" technology of BOF smelting at the present time.

The aim of the work is to develop (based on the developed theoretical principles) mechanical equipment—new designs of skull lance for resource-saving technology of applying liquid slag to refractory lining of 250-ton converters in order to form a protective layer of slag garnish on its working surface. On the basis of the conducted set of industrial experiments the hypothesis of significant increase of BOF lining durability with organization of liquid slag blow with its primary run-out on the most worn trunnion zones was confirmed for the first time. In order to check this hypothesis, the new design of skull lance (patent protected) with grouped lance nozzles of different cross-section at the face of lance tip was developed, produced and tested. This allowed to form the protective slag layer at the working surface of liner both at the bottom part (walls) of converter and at the mouth of the unit. In the course of the converter's campaigns, this ensured that the more worn areas of the converter's lining would be repaired by forming a thicker slag layer on them, thus obtaining a total lining thickness (masonry + slag layer) on these areas that is almost the same as on the other (less worn) working surfaces of the lining with reduced slag carry-over from the converter and reduced refractory consumption.

2 Literature Review

The simplest way of slagging the converter lining is blowing of liquid slag in the cavity of the unit by single-row top blowing lance with circular arrangement in the lance tip of Laval nozzles (Fig. 1a) at replacement of oxidizing blast (oxygen) by neutral gas (nitrogen or argon) [8].

In order to recycle the generated CO_2, it is also possible to use it for blowing the slag melt with the top oxygen lance [9].

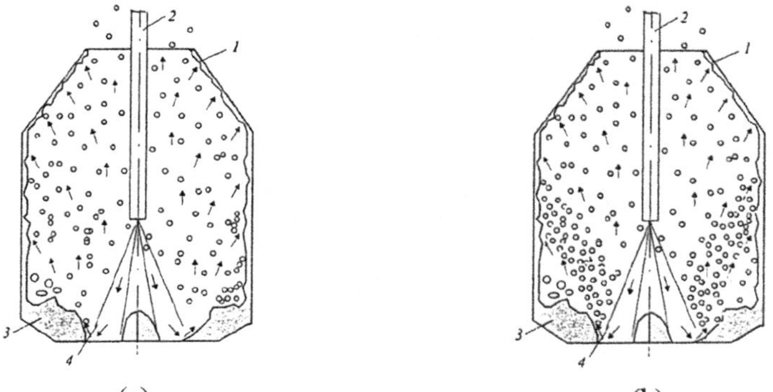

(a) (b)

Fig. 1. Scheme of interaction of gas jets with the slag bath at its blowing in the converter with single-flow top lance with circular uniform (a) and grouped in sectors (b) arrangement of nozzles in the lance tip: 1 - slag garnissage; 2 - skull lance; 3 – slag bath; 4 - neutral gas jet.

However, top blowing lance designs [10] are not suitable for efficient slag application to the walls of the converter by blowing, as

- do not provide the required spray on the working surface of the lining, and in most cases the intensity of nitrogen supply for slag blow-off is less than the consumption of oxygen for purging, i.e., does not correspond to the calculated gas flow rate for a given design of lavalier nozzles, which makes it necessary to lengthen the barrel by 1.0–2.0 m;

- cause more intensive covering of lance barrel with slag-metallic scale which leads to more frequent (1,5–2 times) removal of lance to remove scale with negative consequences (burning of outer pipe, decrease of lance resistance and liquid steel output, increase of labor costs and converter downtime, etc.). The problem of noticing the lance is particularly acute if the slag containing some steel is blown (i.e., incomplete steel output from the converter).

That is, if the top oxygen purging of a deep bath of iron-carbon melt is necessary to completely eliminate or minimize the splashing of metal and slag from the reaction zones, the implementation of technology of blowing with nitrogen jets of the final slag lining, by contrast, it is extremely important to organize a maximum of directional spray particles of liquid slag in the liner from the neutral gas flows to the slag bath, preferably in the zone of high wear.

Thus, a more efficient slag-coating of the converter liner is guaranteed by an optimized design of a skull lance (slagging, blow-out), designed exclusively for the liquid slag blow operation. The following benefits are achieved:

- the dynamic effect of the nitrogen jets on the slag melt is rigid, by designing the lance nozzles strictly to the plant's nitrogen flow rate, so that the slag is sprayed onto the liner more intensively, and the nozzles in the lance tip, having different orientations in space, can be grouped into sectors if necessary to apply more slag garnish at certain points of increased wear on the converter lining, such as in the trunnion areas (Fig. 1, b);
- -significant simplification and lowering of the price of the lance tip and lance itself assembly technology due to the possibility of manufacturing the bottom bowl and nozzle in common carbon steel, increasing the reliability of the construction in comparison with the use of lance tips with copper nozzles and bottom bowl;
- the flow rate of the cooling water in the lance can be significantly reduced or no cooling water at all can be used (when a special gas cooling lance is used), without compromising the durability of the lance tip.

Slag melt blowing is a dynamic phenomenon that depends on the physical [11] and chemical [12] properties of the slag used, which influence the process of formation and crystallization of the skull coating on the converter lining [13]. However, the work on the slag properties research is carried out mainly in the direction of the factors affecting its foaming process [14].

The most important aspect of the slag blowing process is to ensure quick cooling of the slag spray (droplets) directly in contact with the lining, which is caused by preliminary giving the final slag the necessary properties for this purpose in order to form a solid slag garnish on the lining. If the final slag is very liquid and has a low melting point, it has no time to crystallize and, after hitting the converter walls, it runs down the walls, hardly creating a protective layer. Therefore, the defining point of the liquid slag blowing process is the cooling of the slag spray below the melting point of the slag melt directly in the lining with simultaneous acquisition of the required refractory properties by the skull garnish. For this purpose, high-base dense slag containing 8–14% MgO is obtained. As additives, adjusting the composition of slag in the course of blowing, usually use the lump soft-baked dolomite and dolomitized lime in the amount of 15–30 kg / t of steel at the consumption of conventional lime 30–50 kg/t. Recently, as additional magnesia-bearing slag-forming materials ferruginous dolomitized lime (8.9–11.8% MgO; 2.7–2.9% Fe_2O_3), dolomite (31.2–31.8% MgO; 3.3–3 9% Fe2O3) and magnesite (85.7–89.9% MgO; 4,1–4,5% Fe_2O_3), that accelerated formation of slag enriched to 8–14%MgO and ensured its mobility to provide an effective adhesion of slag garnish to the converter lining. To prepare the final slag (reduction of reactivity and thickening) for its blowing, carbonate (lime, dolomite, limestone, etc.) and/or carbonaceous (anthracite, coke, etc.) materials are fed to the converter liner, depending on the characteristics of the final slag resulting from the previous converter melting [15].

As seen from the literature analysis on the subject, there are no works on development of special skull liners designs for effective application of slag garnishing to the places of the biggest wear of lining and no publications on this issue, though the durability of lining is one of the most important problems of oxygen converters, always taking place in all converter shops, which leads to increased consumption of refractories.

3 Research Methodology

The efficiency of the developed design of the skull lance (with grouping of six Laval nozzles of different cross-section in two sectors at the end of the bottom bowl) during the slag garnishing operation on the most worn areas of the trunnion zones of 250-t oxygen converter lining was studied. The object of investigation is refractory lining.

In order to determine real thickness of wear layer, the profile of lining was scanned using LaCamM007 interferometer (Ferro Tron Technologies GmbH, Germany) at different periods of converter's lining before and after slag blow. This laser measuring system provides a surface topography of the lining with an accuracy of ± 5 mm in thickness measurement. The residual thickness of the lining is determined by comparing the previous and current measurement results. The measurement results are displayed on the monitor screen and stored in a database for analysis. The duration of a single scan is 20–25 min. The laser measurement is based on the propagation time of a single light pulse. Here, a semiconductor laser-based LED generates very short wavelength radiation, and the optical pulse is reflected from the surface of the measurement object and hits the light receptor. Simultaneously with the scanning of the lining surface, there is also automatic positioning of the system, based on the detection of reference points installed near the unit. Complete information about the condition of the lining comes out after a series of measurements and processing of the data obtained: converter sweep (identification of areas to be repaired urgently); vertical cuts in the area of both trunnions; vertical cuts in the planes of the drain and loading area; horizontal sections and the bottom.

As slag garnishing is a perspective and cheap way to increase durability of converter refractories and use of top oxygen lance for slag injection is inefficient, the hypothesis was advanced on increase of refractory lining durability by slag injection with priority of slag spraying towards the most worn areas of converter refractory pinch zones.

In order to check this hypothesis, on the base of top oxygen lance for 250-t converters of OJSC "Dneprovsky Metallurgical Combine" (now JSC "KAMET-Stal"", Kamenskoye, Ukraine) the new construction (Patent 117546UA) of the skull lance, which, to increase the information content of its subsequent description, is shown in Fig. 2. The slagging lance tip has 2 diametrically arranged Laval nozzles with critical $d_{kr} = 43$ mm and outlet $d_{op} = 49$ mm diameters and 4 Laval nozzles with $d_{kr} = 37$ mm and $d_{op} =$ 43 mm (located on each side of the first one). The nozzles are circular and arranged in two sectors with angles of 80° between sectors of nozzles and 50° between adjacent nozzles in each sector. Angle of inclination of all nozzles to vertical axis of lance (according to physical process modeling data) is $a = 17°$ (a larger angle as a result of an increase in the momentum of slag droplets leads to the reflection of part of them from the walls and subsequent removal from the converter; a smaller angle due to the low impulse of slag droplets does not allow effective application of the garnissage). Nozzles of larger critical diameter (with a greater gas pulse) are designed to apply the blown slag mainly to the more worn trunnion zones, allowing a more intensive application of slag to obtain a thicker slag layer. Nozzles with smaller critical diameter (and gas pulse) are mainly used to deposit a thinner slag layer on the remaining (less worn) surface of the lining.

At the 250-t converters, a single-layer scheme is used and the thickness of lining is 800 mm in the cylindrical part and 650 mm in the upper conical part. The composition of the refractories used is as follows:

- in the metal area (lower spherical part) - $MgO > 71.0\%$ and $C > 13.0\%$;
- in the trunnion zone - $MgO > 71.0\%$ and $C > 14.0\%$;
- in the loading and discharge zones - $MgO > 68\%$ and $C > 15{,}0\%$;
- in the upper conical part - $MgO > 61.0\%$ and $C > 14.0\%$.

According to the technology adopted at the shop, the final converter slag, preliminary modified with magnesia additives, is blown with nitrogen (flow rate 600700 m^3 /min) for 4–8 min at a step change of lance tip position relative to the converter bottom from 2,5 to 0,5 m.

4 Results and Discussion

In order to determine the actual thickness of the formed skull layer, profile scanning using LaCam M007 interferometer was carried out at the beginning (594 melts) and during the campaign (4131 melts) for 250-t converter lining. For the purpose of fixing by scanning of initial profile of working space before conducting melts N° 593 and N° 4130 (on lining) slag blowing on lining was not carried out. After steel output and slag blowing (its initial mass of 18–20 t) within 5–6 min, the working space was scanned again.

The data of scanning of the converter's workspace geometry before and after bottom ash removal by the designed lance construction (Fig. 2) is shown in Fig. 3. Figure 3, a show the profile of the workspace and distribution of the thickness of skull layer by height (in mm) of the 250-t unit at the beginning of campaign, in Fig. 3b—in the course of the converter's lagging campaign.

The conducted operations under the direct control of slagging operations have allowed establishing the possibility of freezing of the strong slag layer with thickness

25–170 mm on all surface of the lining in the bottom part (on walls) of the converter and 6–10 mm in the area of the converter mouth lining, The wear of the working layer of lining is reliably prevented, with the possibility to form a thicker grainsize layer in the trunnion areas, which helps to have the same total thickness of lining (masonry + grainsize) in these areas as in the other (less worn) areas. When the slag is blown, 30 to 50% of the initial slag bath volume is applied to the lining.

In the course of the campaigns, there was uneven wear of the lining, especially in the area of the more worn trunnion zones, which was detected by interferometer at 783 heat (Fig. 4a); with the five-nozzle blowing lance—already at 595 heat (Fig. 4b).

In general, the proposed technical solution of increasing refractory lining durability of oxygen converters, based on the proposed hypothesis of increasing its durability by directing slag spraying with higher intensity to the most worn areas, which was tested with the developed design of skull lance, allows in the course of 250-t converter refractory campaign to provide repair of gondola zones by forming in them a thicker garnish layer and thus achieve about one.

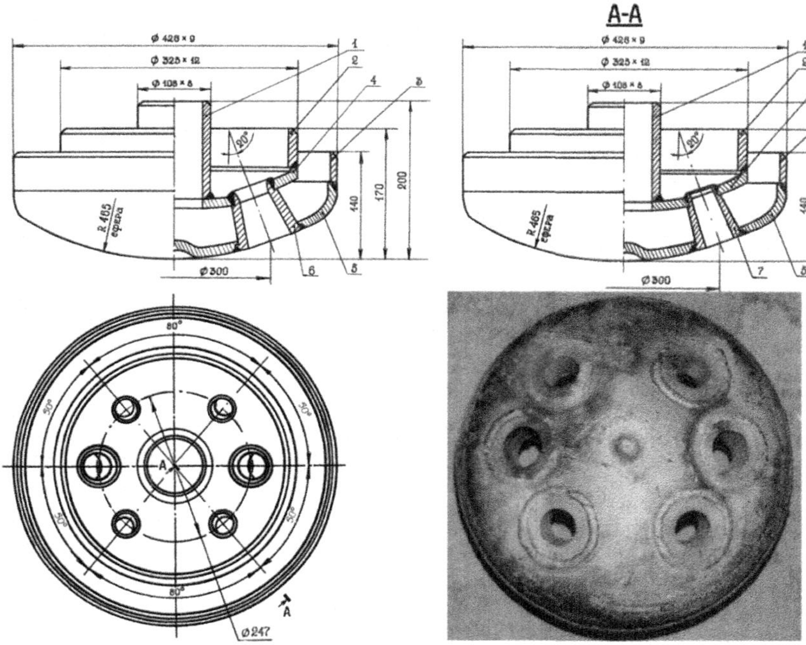

Fig. 2. Construction and appearance of the shower lance tip with grouped in sectors Laval nozzles of different cross-sections: 1 - water inlet, nitrogen inlet and outlet pipes, respectively; 4, 5 - top (steel) and bottom (copper) nozzle, respectively; 6, 7 - large and small section Laval nozzles, respectively.

Further optimization of lance construction with the purpose of improvement of slag spraying process and faster icing of slag on converter walls is continued by means of placing the second tier of blast nozzles on the lance. In perspective development of

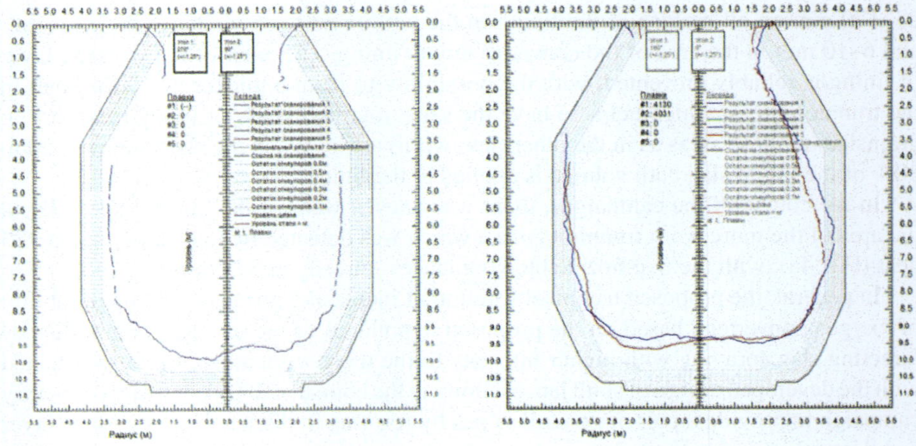

(a) (b)

Fig. 3. Operating space profile and garnish layer thickness distribution at the top of the 250 t convertor (in mm) at the start of (a) and during the liner campaign (b) using the designed skull lance design.

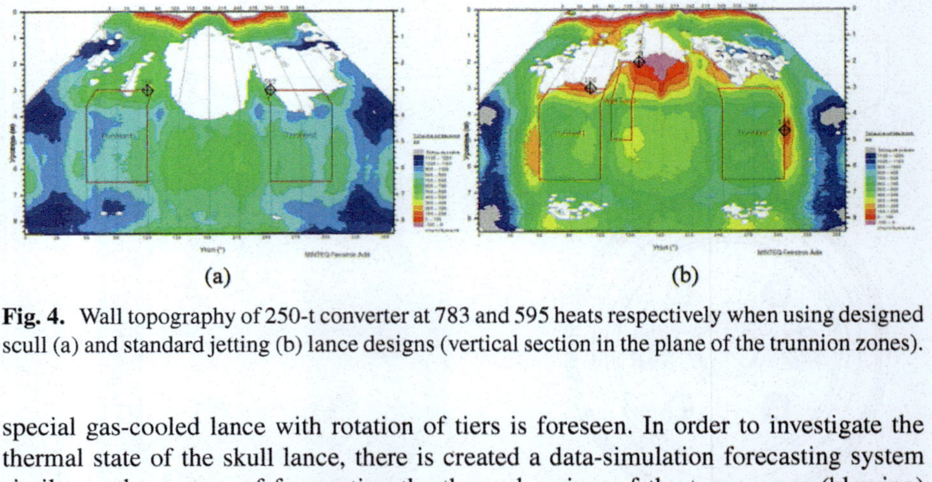

(a) (b)

Fig. 4. Wall topography of 250-t converter at 783 and 595 heats respectively when using designed scull (a) and standard jetting (b) lance designs (vertical section in the plane of the trunnion zones).

special gas-cooled lance with rotation of tiers is foreseen. In order to investigate the thermal state of the skull lance, there is created a data-simulation forecasting system similar to the system of forecasting the thermal regime of the top oxygen (blowing) lance [16].

5 Conclusions

On the basis of performed complex of industrial experiments at 250-t converters of OJSC "Dneprovsky Metallurgical Combine" (at present JSC "KAMET-Stal'", Kamenskoye, Ukraine) the hypothesis of increasing refractory lining resistance by means of liquid slag blow-out with preferentially splashing out on the most worn-out sections of its gland areas is confirmed. In order to check this hypothesis a new (patent-protected) construction of the top blowing lance was developed, produced and tested [10].

The results of melts obtained using the developed design of skull lance with grouping in two sectors of six nozzles of different cross-section Laval indicate the possibility of formation on the working surface of the liner of skull slag layer with thickness of 25–170 mm in the lower part (on the walls) of the unit and 6–10 mm in the area of its throat.

Thus, the developed designs of the skull lance are more effective in the final repair of the lining in accordance with the conventional blowing lances declared for slagging, since it allows to provide hot repair of trunnion zones due to formation of thicker skull layer in their areas, reaching in these zones the total lining thickness (masonry + skull liner) almost the same size as in its other (less worn out) zones. This results in reduced slag carry-over from the converter and reduced consumption of refractory materials, which provides an economic effect of at least \$0.19 per ton of steel produced.

Further work is underway to increase the durability of slag-splashing (skull) lances by using for the manufacture of their structures: arc welding with shielded flux-cored wire [17], produced by rolling on a metal substrate [18]; technologies for forging without upsetting the ingot (casting) [19] and upsetting with combined dies [20]; new superhard materials [21]—titanium in various structural states, including with a submicrocrystalline structure obtained by screw extrusion, and sintered silicon-containing titanium alloys. It is also envisaged to use a system of high-modulus energy sources directly during the process of slagging the converter lining.

References

1. Zhul'kovskii, O.A., Panteikov, S.P., Zhul'kovskaya, I.I.: Information-modeling forecasting system for thermal mode of top converter lance. Steel Transl. **52**, 495–502 (2022). https://doi.org/10.3103/S0967091222050138
2. Savchenko, I., Gurenko, A., Naumenko, O.: Cutting-edge industrial technology of mining tool manufacturing. Mining of Mineral Deposits **10**(4), 105–110 (2016). https://doi.org/10.15407/mining10.04.105
3. Computer Simulation of Safety Processes of Composite Structures Rheological Properties. Proceedings of the 2022 IEEE 4th International Conference on Modern Electrical and Energy System, MEES 2022. 2022 https://doi.org/10.1109/MEES58014.2022.10005747
4. Control Systems for the Temperature Field During Drawing, Taking into Account the Dynamic Modes of the Technological Installation. Proceedings of the 2022 IEEE 4th International Conference on Modern Electrical and Energy System, MEES 2022. 2022. https://doi.org/10.1109/MEES58014.2022.10005724
5. Pavlenko, D.V., Tkach, D.V., Danilova-Tretyak, S.M., Evseeva, L.E.: Thermophysical properties and temperature of the start of titanium recrystallization in different structural states. J. Eng. Phys. Thermophys. **90**(3), 685–696 (2017)
6. Bykov, I.O., Ovchinnikov, A.V., Pavlenko, D.V., Lechovitzer, Z.V.: Composition, structure, and properties of sintered silicon-containing titanium alloys. Powder Metall. Met. Ceram. **58**(9–10), 613–621 (2020)
7. Olshanetskii, V.E., Stepanova, L.P., Greshta, V.L., Pavlenko, D.V., Tkach, D.V.: Deformation behavior of titanium VT1-0 with submicrocrystalline structure formed by the method of helical extrusion. Met. Sci. Heat Treat. **55**(11–12), 603–607 (2014)

8. Savchenko, I., Shapoval, A., Kuziev, I.: Modeling of high module power sources systems safety processes. Mater Sci Forum 2022;1052 MSF:399–404
9. Jagadeesh, P., Puttegowda, M., Rangappa, S.M., Alexey, K., Gorbatyuk, S., Khan, A., et al.: A comprehensive review on 3D printing advancements in polymer composites: technologies, materials, and applications. Int. J. Adv. Manuf. Technol. **121**(1–2), 127–169 (2022)
10. Karnaukh, S.G., Markov, O.E., Kukhar, V.V., Shapoval, A.A.: Classification of steels according to their sensitivity to fracture using a synergetic model. Int. J. Adv. Man. Tech. **119**(7–8), 5277–5287 (2022). https://doi.org/10.1007/s00170-022-08653-y
11. Allieva, L., Hrudkina, N., Aliiev, I., Zhbankov, I., Markov, O.: Effect of the tool geometry on the force mode of the combined radial-direct extrusion with compression. East Eur. J Enterp Technol **2**(1–104), 15–22 (2020)
12. Drahobetsky, V.V., Shapoval, A.A., Shchetynin, V.T., et al.: New solution for plastic deformation process intensification. Metallurgist **65**(9–10), 1108–1116 (2022). https://doi.org/10.1007/s11015-022-01253-x
13. Markov, O.E., Khvashchynskyi, A.S., Musorin, A.V., et al.: Investigation of new method of large ingots forging based on upsetting of workpieces with ledges. Int. J. Adv. Man. Tech. https://doi.org/10.1007/s00170-022-09989-1
14. Kulynych, V., Shapoval, A., Dragobetskii, V.: Hard alloys recycling as a promising direction of technological equipment for machine-building production. Mater. Sci. Forum; 1052 MSF:423–428 (2022)
15. Shmelov, Y., Bazyk, A., Kitsel, N.: Research of parametric influence of light-emitting diodes on the multicomponent module near-field illuminated zone formation. Lect. Notes. Mech. Eng. 801–809 (2022)
16. Mathematical Model of Rheological Processes of Composite Materials Deformation. Proceedings of the 2022 IEEE 4th International Conference on Modern Electrical and Energy System, MEES 2022. (2022). https://doi.org/10.1109/MEES58014.2022.10005658
17. Markov, O.E., Aliiev, I.S., Aliieva, L.I., Hrudkina, N.S.: Computerized and physical modeling of upsetting operation by combined dies. J. Chem. Technol. Met. **55**(3), 640–648 (2020)
18. Vlasenko, K., Hrudkina, N., Reutova, I., Chumak, O.: Development of calculation schemes for the combined extrusion to predict the shape formation of axisymmetric parts with a flange. East. Eur. J. Enterp. Technol. **3**(1–93), 51–59 (2018)
19. Hrudkina, N.S.: Process Modeling of Sequential Radial-Direct Extrusion Using Curved Triangular Kinematic Module. FME Trans. **49**(1), 56–63 (2020)
20. Potapkin, V.F., Levkin, A.N., Satonin, A.V., Romanov, S.M., Vorob'yev, Y.A., Gribkov, E.P.: Stressed state and kinematics in the rolling of powder materials on a metal substrate. Powder Metall. Met. Ceram. **39**(1–2), 11–17 (2000)
21. Kostikov, A.A., Perig, A.V., Larichkin, O.V., Stadnik, A.N., Gribkov, E.P.: Research into payload swaying reduction through cable length manipulation during boom crane motion. FME Trans **47**(3), 464–476 (2019)

Techno-Economic Method for the Rational Choice of Air Source Heat Pumps for Bivalent Heating Systems

Mikhail Kuznetsov(✉), Victoria Tarasova, and Andrii Kostikov

Anatolii Pidhornyi Institute of Mechanical Engineering Problems of the National Academy of Sciences of Ukraine, Pozharskogo 2/10, Kharkiv 61046, Ukraine
childeric1975@gmail.com

Abstract. The actual problem of modern energy, both in Ukraine and around the world, is the transition to energy-saving technologies, such as heat pumps using renewable energy sources. The purpose of this study is to develop a techno-economic method for the rational choice of air source heat pumps offered on the market for use in bivalent heating systems, taking into account the technical characteristics specified by the manufacturers and local climatic and economic conditions. The scientific novelty is the optimization of the objective function of resulting costs according to energy, climatic and economic parameters for the subsequent choosing of the most rational heat pump from the model range. The practical value of this study lies in the development of a software package that makes it possible to obtain an unambiguous analytical solution for choosing the most rational heat pump option from a model range based on the data given in the manufacturer's catalogs and climate tables for a specific region, without carrying out labor-intensive traditional technical and economic calculations. Using this method, a rational model of a heat pump was chosen for the bivalent heating system of a building in Kharkiv. The result of the calculation is the construction of a graph of the objective function of the resulting costs for the analyzed number of heating seasons and the accepted electricity tariff. The rational model of the heat pump corresponds to the minimum of the resulting costs.

Keywords: Ambient Temperature · Bivalence Point · Energy Efficiency · Low-Grade Heat · Resource Use Efficiency

1 Introduction

Due to the rapid growth of world prices for scarce fossil fuels and a significant decrease in its reserves, the actual problem of modern energy, both in Ukraine and around the world, is the transition to energy-saving technologies, such as heat pumps using renewable energy sources. The heat pump allows you to significantly save fuel and reduce thermal pollution of the environment. This is the reason for the widespread use of heat pump technology in heating systems of residential, public and industrial buildings in most developed countries of the world.

H. Altenbach et al. (Eds.): CAMPE 2023, pp. 153–163, LNME 2025.
https://doi.org/10.1007/978-3-031-82979-6_16

Ukraine is just starting to develop heat pump technology and does not have its own production of heat pumps. The introduction of imported heat pumps in Ukraine takes place according to integral characteristics and often, in order to guarantee the coverage of the required heat load, firms recommend to the customer equipment with excess capacity, ignoring the issues of competitiveness of its use in the economic conditions of Ukraine. Direct replication of expensive foreign projects in Ukraine can not only lead to the irrational use of heat pumps, but in some cases generally discredit the very idea of introducing heat pump technology.

An actual task is to carry out research on the development of methods for the rational implementation of heat pumps in order to adapt their technical characteristics to the parameters of specific technological processes, climatic conditions and consumer requirements. This will help reduce the resulting costs for the purchase and operation of heat pumps, reduce their payback period and increase competitiveness in comparison with traditional heat generators, which will lead to the creation of conditions for the large-scale introduction of heat pump technology in Ukraine.

The purpose of this research can be formulated as the development of a method for the rational choice of heat pump equipment offered on the market for heating systems of buildings, taking into account the technical characteristics specified by the manufacturers and local climatic and economic conditions.

2 Literature Review and Problem Statement

Rational design of a heat pump heating system using heat pump equipment offered on the market with technical characteristics specified by manufacturers is based on the correct choice of its heating capacity and the source of low-grade heat used by the heat pump [1].

An indicator of the energy efficiency of energy conversion in a heat pump is coefficient of performance (COP) [2], which is the ratio of thermal energy generated in a heat pump and given to the consumer per unit of time to the electric power supplied from outside to implement the thermodynamic cycle of the heat pump. The COP value depends on a large number of factors and, first of all, on the temperatures of the low-grade heat source and the heat carrier directed to the consumer [3].

If the operating temperatures in technological processes are maintained at a certain constant level and the source of low-grade heat for heat pumps is, for example, ground or groundwater, the temperature of which depends little on fluctuations in environmental parameters [4], then the COP value is stable during heat pump operation. However, water heat pumps are much more expensive than air heat pumps, and additional capital investments in the subsystem for extracting heat from the ground can exceed the cost of the heat pump itself [5].

The most accessible source of low-grade heat for a heat pump is free ambient air. However, the heating capacity and COP of air source heat pumps significantly depend on the ambient temperature [6]. With a decrease in the ambient temperature, just when, under conditions of comfortable heating, the need for heat in a heated building increases, the heating capacity and COP of air source heat pumps decrease [7].

If the ambient temperature level is high enough, the required heat load can be fully covered by the heat pump. It is obvious that it is irrational from an economic point of view to choose a heat pump to cover the object peak heat load, because the number of hours with minimum ambient temperature during the heating season is relatively small [8]. Therefore, if at low ambient temperatures the heat pump heating capacity is not enough to cover the required heat load of the object, then a bivalent heating system is used, which includes a heat pump and an additional heater – a heat booster, for example, an electric or gas boiler [9]. The heat booster is used to cover the object peak heat loads and can operate in parallel with a heat pump or separately [10].

The problem arises of expedient distribution of the heating capacity between the heat pump and the heat booster [11]. The key task here is the calculation of the bivalent temperature T_{bw} [12], the optimal value of which determines the rational choice of the heating capacity of the heat pump.

The existing methods used in the design and chosen of rational modes of heat pumps operation are based mainly on traditional techno-economic calculations, which are based on the energy balance method.

Paper [13] presents a techno-economic feasibility analysis related to a heat pump installation in the poly-generative smart grid located at the University of Genoa Campus, to convert the overproduction electricity into thermal power. The results show that the introduction of a heat pump reduces gas consumption by 15% (24 tons per year). The optimal size of a high-temperature heat pumps for waste heat recovery and the contribution of each component are discussed in detail from the techno-economic perspective in [14], highlighting the greatest impact of the evaporator. In [15], the techno-economic aspects of using high-temperature heat pumps for waste heat recovery are also investigated. The result is a compromise between high efficiency and high cost to identify the most cost-effective heat pump configuration at each temperature level.

In these studies, for the chosen equipment and a given mode of operation, a techno-economic calculation is carried out, ending with the determination of the value of the resulting costs. After that, one or more defining parameters are changed and the calculation is repeated. The option that provides the lowest resulting costs is considered rational.

Thus, the use of traditional techno-economic methods implies an individual approach to each specific heat pump and does not provide for obtaining an unambiguous analytical solution for choosing the most rational heat pump option from the model range of a particular manufacturer. Therefore, the research tasks are:

1. Development of a techno-economic method for the rational choice of air source heat pumps offered on the market for use in bivalent heating systems, taking into account the heat pumps cost and the local electricity tariff and changes in the characteristics of heat pumps and the heat load of the object during the heating season.
2. Choice from the manufacturer's model range for the heating system of a typical residential building of an air-to-water heat pump model that is rational in terms of minimizing the resulting costs for the analyzed number of heating seasons and the accepted electricity tariff.

3 Research Methodology

A techno-economic method was developed for the rational choice of air source heat pumps offered on the market for use in bivalent heating systems, taking into account the technical characteristics specified by the manufacturers and local climatic and economic conditions.

This method is based on the description of the amount of generated heat and consumed electricity for the billing period, as well as the cost of heat pumps from the considered model range by interpolation functions from the time coordinate τ_{bv}, corresponding to the bivalent temperature T_{bv}. Under T_{bv} is meant the ambient temperature T_{am}, corresponding to the equality of the heat load of the object and the heating capacity of the heat pump (bivalence point). The time coordinate τ_{bv} is calculated for each heat pump from the considered model range. Resulting costs from time τ are taken as the objective function. The minimum of the resulting costs corresponds to τ_{opt}. Based on the τ_{bv} value closest to τ_{opt}, a rational model of the heat pump is chosen.

As a presentation of this method for a typical residential building with a design heat load $Q^d{}_h = 58$ kW in Kharkiv, a rational model of an air-to-water heat pump was chosen from the Clivet WSAN-EE 82–162 model range previously offered on the market. An electric boiler is used as a heat booster. The room air temperature T_r is taken equal to 20 °C, and the design ambient temperature $T^d{}_{am}$ for Kharkiv is −23 °C.

Data on the dependence of heating capacity and consumed power of the Clivet WSAN-EE 82–162 heat pumps on ambient temperature, which are given in the manufacturer's catalog, are shown in Table 1.

Table 1. Dependence of heating capacity Q_{hpi} and consumed power N_{hpi} of the Clivet WSAN-EE 82–162 heat pumps on ambient temperature T_{am}.

Civet heat pump model	$Qhpi$ N_{hpi}	T_{am}, °C					
		− 5	0	5	7	10	15
WSAN-EE 82	Q_{hp82}	19.2	21.8	24.8	26.5	28.2	32.6
	N_{hp82}	6.59	6.76	6.87	6.91	6.93	6.94
WSAN-EE 102	Q_{hp102}	22.4	25.9	29.3	31.0	32.7	36.7
	N_{hp102}	7.87	7.93	8.00	8.03	8.05	8.11
WSAN-EE 122	Q_{hp122}	25.6	29.4	33.1	35.0	36.8	41.1
	N_{hp122}	9.19	9.19	9.23	9.26	9.30	9.42
WSAN-EE 142	Q_{hp142}	29.1	33.8	38.4	40.7	43.0	48.5
	N_{hp142}	10.5	10.8	10.9	11.0	11.0	11.1
WSAN-EE 162	Q_{hp162}	34.7	38.9	43.5	45.9	48.3	54.3
	N_{hp162}	12.2	12.3	12.4	12.5	12.5	12.6

The object heat load is related to the ambient temperature T_{am} by linear dependence

$$Q_h = Q_h^d \frac{T_r - T_{am}}{T_r - T_{am}^d}$$

and was considered at ambient temperatures from $T^0{}_{am} = -5$°C to $T^{end}{}_{am} = 8$ °C – at the most efficient operation range of Clivet air-to-water heat pumps of WSAN-EE models during the heating season (for Kharkiv, the heating season lasts at ambient temperatures $T_{am} \leq 8$ °C). It is assumed that at ambient temperatures below −5 °C only an electric boiler operates, the installed heating capacity of which is selected for covering the peak heat load of the object. At -5 °C $\leq T_{am} < T_{bv}$, the heat pump and electric boiler operate in parallel. After the ambient air reaches the temperature T_{bv}, the electric boiler switches off and only the heat pump operates.

Often, to reduce capital costs, a heat booster with a lower installed heating capacity is chosen, resigned to the fact that during short periods of peak loads, the temperature at the facility will be lower than comfortable. In any case, the value of the installed heating capacity of electric boiler and, consequently, its cost are not taken into account further, since they are considered unchanged and does not influence on the rational choice of the heat pump.

Based on the calculation results the line graph of the dependence of the object heat load on an ambient temperature $Q_h = f\ (T_{am})$ was constructed (Fig. 1). The number of hours with one or another ambient temperature for the heating season is given in the tables of climate data for each specific region. Based on these data, for the Kharkiv region, a graph of the dependence of the object heat load on the number of hours with various T_{am} during the heating season $Q_h = f\ (\tau)$ was also constructed (Fig. 1).

The dependences of the heating capacity of the Clivet heat pumps of the WSAN-EE 82–162 models, selected for the analysis, on the ambient temperature $Q_{hpi} = f\ (T_{am})$ (see Table 1) and on the number of hours with various ambient temperatures T_{am} during the heating season $Q_{hpi} = f\ (\tau)$ are also shown in Fig. 1. The figure from bottom to top shows the heating capacity curves of the following models of heat pumps from Clivet: WSAN-EE 82, WSAN-EE 102, WSAN-EE 122, WSAN-EE 142 and WSAN-EE 162.

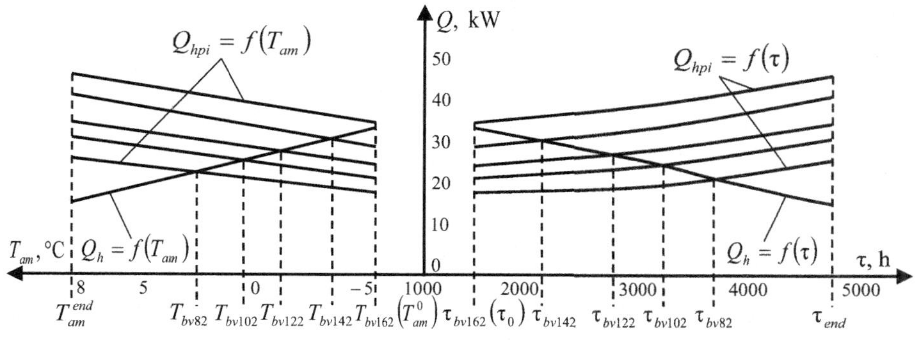

Fig. 1. Dependence of the object heat load and the Clivet WSAN-EE 82–162 heating capacity on the T_{am} change and the number of hours with various T_{am} values during the heating season τ.

The graphs $Q_{hpi} = f\ (T_{am})$ intersect with the straight line $Q_h = f\ (T_{am})$ at the points whose projections on the axis T_{am} correspond to the temperatures T_{bvi} for each heat pump from the considered model range.

Each temperature T_{bvi} corresponds to the certain time coordinates τ_{bvi} – the projections of the intersection of the curves $Q_{hpi} = f\ (\tau)$ and $Q_h = f\ (\tau)$ on the τ axis. To determine the τ_{bvi} values for each heat pump from the considered model range the curves $Q_{hpi} = f\ (\tau)$ and $Q_h = f\ (\tau)$ should be described by interpolation functions as polynomials according to the known values of the function and the argument at the node points:

$$
\begin{aligned}
Q_h(\tau) &= -9.999 \cdot 10^{-7}\tau^2 + 3.674 \cdot 10^{-4}\tau + 35.216;\\
Q_{hp82}(\tau) &= 6.299 \cdot 10^{-7}\tau^2 - 1.237 \cdot 10^{-3}\tau + 19.686;\\
Q_{hp102}(\tau) &= 5.404 \cdot 10^{-7}\tau^2 - 2.867 \cdot 10^{-4}\tau + 21.718;\\
Q_{hp122}(\tau) &= 5.891 \cdot 10^{-7}\tau^2 - 3.223 \cdot 10^{-4}\tau + 24.870;\\
Q_{hp142}(\tau) &= 7.354 \cdot 10^{-7}\tau^2 - 4.289 \cdot 10^{-4}\tau + 28.226;\\
Q_{hp162}(\tau) &= 8.543 \cdot 10^{-7}\tau^2 - 1.266 \cdot 10^{-3}\tau + 34.775. \qquad (1)
\end{aligned}
$$

Equating the function $Q_h = f\ (\tau)$ to each of the $Q_{hpi} = f\ (\tau)$ functions allows to get the row of equations whose solution determines the value of the time coordinate τ_{bvi} for each heat pump from the considered model range. The obtained values of τ_{bvi} are shown in Table 2. Shown also in the same place the values of temperature T_{bvi} (to which τ_{bvi} correspond) were determined from Fig. 1, are informative in nature and are not used in further calculations. In Table 2 also shows the cost of each heat pump from the considered model range, which is given in the manufacturer's price list.

Table 2. The time coordinates τ_{bvi} and the cost of heat pumps Clivet WSAN-EE 82–162.

Clivet heat pump model	Temperature T_{bvi}, °C	Time coordinate τ_{bvi}, h, corresponding to T_{bvi}	Heat pump cost C_{hpi}, USD
WSAN-EE 82	2.65	3618	9864
WSAN-EE 102	0.52	3180	10924
WSAN-EE 122	−1.15	2778	12416
WSAN-EE 142	−2.98	2249	13774
WSAN-EE 162	−5	1420	15566

According to the data given in Table 1, the coefficients of performance $COP_{hpi} = Q_{hpi}/N_{hpi}$ of the considered models of heat pumps depending on the ambient temperatures were calculated and, according to the time coordinates corresponding to these temperatures, the interpolation functions $COP_{hpi} = f\ (\tau)$ were constructed:

$$COP_{hp82}(\tau) = 8.758 \cdot 10^{-8}\tau^2 - 2.025 \cdot 10^{-4}\tau + 3.024;$$

$$COP_{hp102}(\tau) = 6.065 \cdot 10^{-8}\tau^2 - 1.571 \cdot 10^{-5}\tau + 2.746;$$
$$COP_{hp122}(\tau) = 5.694 \cdot 10^{-8}\tau^2 - 2.953 \cdot 10^{-6}\tau + 2.675;$$
$$COP_{hp142}(\tau) = 6.811 \cdot 10^{-8}\tau^2 - 8.673 \cdot 10^{-5}\tau + 2.757;$$
$$COP_{hp162}(\tau) = 6.093 \cdot 10^{-8}\tau^2 - 7.887 \cdot 10^{-5}\tau + 2.833. \quad (2)$$

Based on the data given in Table 2, a graph of the dependence of the cost of the considered heat pump models on the time coordinates τ_{bvi}, lying on the time axis τ was constructed, that is $C_{hp} = f(\tau)$ (Fig. 2), which is described by the following interpolation function:

$$C_{hp}(\tau) = 1.153 \cdot 10^{-9}\tau^4 - 1.163 \cdot 10^{-5}\tau^3 + 0.0422\tau^2 - 67.389\tau + 54854.333. \quad (3)$$

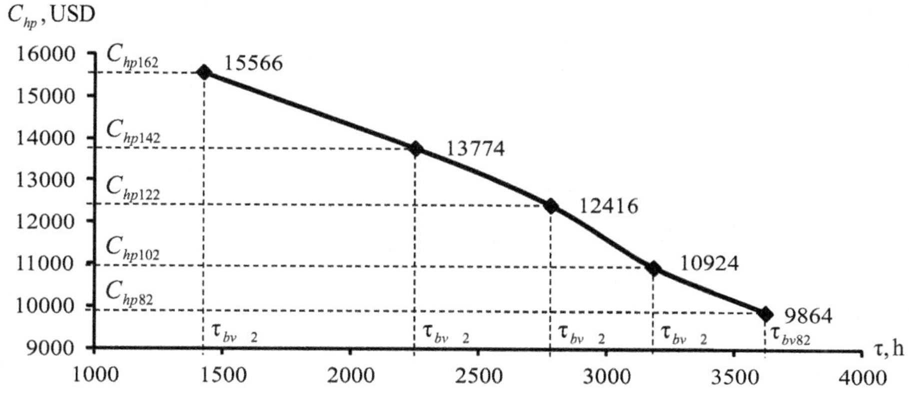

Fig. 2. Dependence of the cost of Clivet WSAN-EE 82–162 heat pumps on the τ_{bvi} values.

During the parallel operation of the heat pump and the electric boiler as a heat booster, which provides coverage of the object heat load at ambient temperatures below T_{bvi}, the amount of electricity consumed by each heat pump E_{hpi} and electric boiler E_{ebi} to cover the given heat load of the object during the heating season is calculated using the equations:

$$E_{hpi} = \int_{\tau_0}^{\tau_{bvi}} \frac{Q_{hpi}(\tau)}{COP_{hpi}(\tau)} d\tau + \int_{\tau_{bvi}}^{\tau_{end}} \frac{Q_h(\tau)}{COP_{hpi}(\tau)} d\tau;$$
$$E_{ebi} = \int_{\tau_0}^{\tau_{bvi}} \left(Q_h(\tau) - Q_{hpi}(\tau)\right) d\tau, \quad (4)$$

where τ_0 and τ_{end} are the time coordinates corresponding to the lowest and highest values of T_{am} at which the heat pump operates.

Equations (4) take into account the time periods ($\tau_0 - \tau_{bvi}$), when the heat pump operates in parallel with the electric boiler, and ($\tau_{bvi} - \tau_{end}$), when the heating capacity of the heat pump is sufficient to cover the entire heat load of the object. The coefficient of performance of the electric boiler can be taken as 1. In this case, the amount of heat generated by the electric boiler is equal to the amount of electricity consumed by this electric boiler to cover the given heat load of the object during the heating season. Therefore, when the heat pump and electric boiler operate in parallel, the E_{ebi} value is equal to the area bounded by the curves $Q_h = f(\tau)$ and $Q_{hpi} = f(\tau)$, as well as by the time coordinates τ_0 and τ_{bvi} (see Fig. 1).

Substitution in (4) of Eqs. (1) and (2), as well as the corresponding values of the time coordinates τ_{bvi} (see Table 1), and integration of (4) taking into account the fact that for the Kharkiv region $\tau_0 = 1420$ h and $\tau_{end} = 4550$ h, allow us to determine the numerical values of E_{hpi} and E_{ebi} for each of the considered variants of the heating capacity ratio of the heat pump and the electric boiler.

Based on the results of the calculations, the graphs of the dependence of the amount of electricity consumed by the Clivet WSAN-EE 82–162 heat pumps and the electric boiler on the time coordinates τ_{bvi} lying on the time axis τ were constructed, that is $E_{hp} = f(\tau)$ and $E_{eb} = f(\tau)$ (Fig. 3).

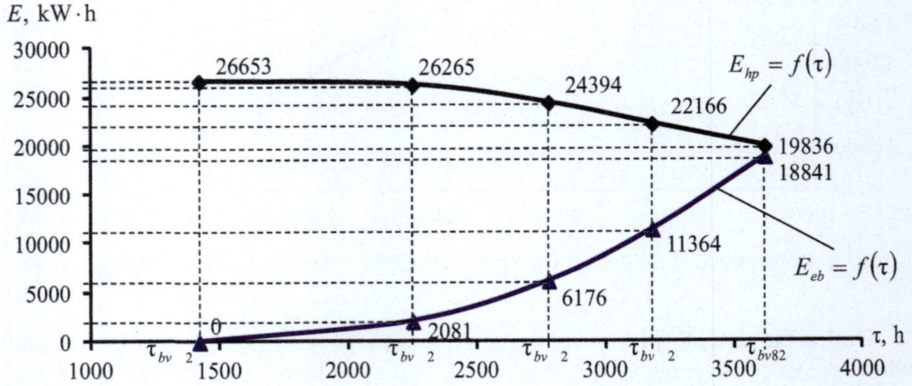

Fig. 3. Dependence of the amount of electricity consumed by the electric boiler and heat pumps Clivet WSAN-EE 82–162 on the time coordinates τ_{bvi}.

The graphs shown in Fig. 3 are described by the following interpolation functions:

$$E_{hp}(\tau) = 7.641 \cdot 10^{-10}\tau^4 - 7.287 \cdot 10^{-6}\tau^3 + 0.023\tau^2 - 30.543\tau + 41031.327;$$
$$E_{eb}(\tau) = -6.060 \cdot 10^{-10}\tau^4 + 6.780 \cdot 10^{-6}\tau^3 - 0.023\tau^2 + 32.181\tau - 16724.197. \quad (5)$$

The resulting costs, taken as the objective function, were determined by the equation

$$RC(\tau) = \left(E_{hp}(\tau) + E_{eb}(\tau)\right)c_{EE}n_{hs} + C_{hp}(\tau), \quad (6)$$

where c_{EE} is the electricity tariff, USD/(kW·h); n_{hs} is the amount of heating seasons.

4 Results and Discussion

After substituting in (6) the Equations for E_{hp} (τ) (5), E_{eb} (τ) (5) and C_{hp} (τ) (3) at the electricity tariff $c_{EE} = 0.07$ USD/(k•Wh) and the amount of heating seasons $n_{hs} = 7$, the dependence of the objective function of the resulting costs on time τ was obtained:

$$RC(\tau) = 1.231 \cdot 10^{-9}\tau^4 - 1.187 \cdot 10^{-5}\tau^3 + 0.042\tau^2 - 66.586\tau + 66764.826,$$

shown in Fig. 4. By the minimum of the resulting costs, the time coordinate τ_{opt} is determined and the rational model of the heat pump is chosen.

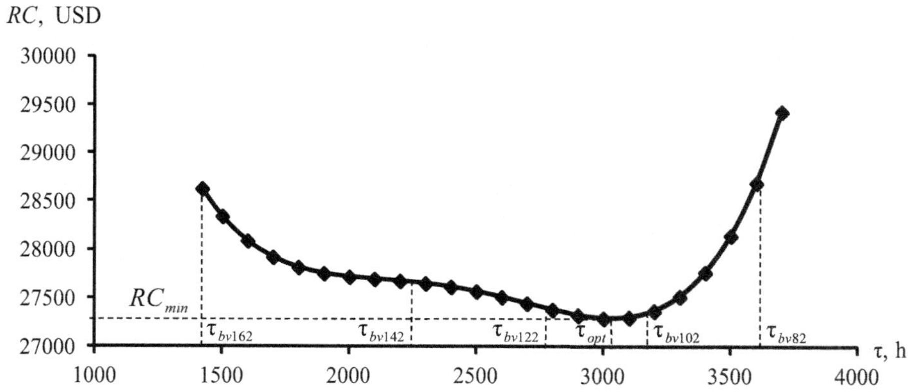

Fig. 4. Dependence of the resulting costs on time τ.

Figure 4 shows that the resulting costs reach a minimum at the time coordinate τ_{opt} = 3033 h. This made it possible to choose the most rational heat pump from the model range presented in Table 1 in terms of minimizing the resulting costs. This is a Clivet WSAN-EE 102 heat pump, whose time coordinate $\tau_{bv102} = 3180$ h (see Table 2) is closest in value to τ_{opt}.

Thus, the scientific novelty of this study is the optimization of the objective function of resulting costs according to energy, climatic and economic parameters for the subsequent choosing of the most rational heat pump from the model range. The practical value of this study lies in the development of a software package that makes it possible to obtain an unambiguous analytical solution for choosing the most rational heat pump option from a model range based on the data given in the manufacturer's catalogs and climate tables for a specific region, without carrying out labor-intensive traditional technical and economic calculations presented in [4, 6, 13–15].

The limitations of the presented method include the disadvantage inherent in the techno-economic approach associated with the impossibility to predict the explosive growth in prices for equipment and electricity due to political crises.

The results obtained will help reduce the resulting costs for the purchase and operation of air source heat pumps, reduce their payback period and increase competitiveness in comparison with traditional heat generators, which will help create conditions for the large-scale implementation of energy-saving and environmentally friendly heat pump technology in Ukraine.

5 Conclusions

1. A techno-economic method was developed that allows making a rational choice of air-to-water or air-to-air heat pump equipment offered on the market for use in bivalent heating systems, taking into account the technical characteristics of the heat pumps specified by the manufacturers and local climatic and economic conditions.
2. Using this method, a rational model of a heat pump was chosen for the bivalent heating system of a typical residential building with a design heat load of 58 kW in Kharkiv. The result of the calculation is the construction of a graph of the dependence of objective function of the resulting costs on the values of the time coordinate τ for the analyzed number of heating seasons and the accepted electricity tariff. By the minimum of the resulting costs, the optimal time coordinate τ_{opt} is determined and the rational model of the heat pump closest to it is chosen.

References

1. Kijo-Kleczkowska, A., Bruś, P., Więciorkowski, G.: Economic analysis of heat pump exploitation—a case study. Energy **280**, 128184 (2023)
2. Kharlampidi, D., Tarasova, V., Kuznetsov, M.: Experimental study of water heating modes by a heat pump. In: 2020 IEEE KhPI Week on Advanced Technology (KhPIWeek 2020), pp. 119–123. IEEE, NYC (2020). https://doi.org/10.1109/khpiweek51551.2020.9250148
3. Verdnik, M., Rieberer, R.: Influence of operating parameters on the COP of an R600 high-temperature heat pump. Int. J. Refrig. **140**, 103–111 (2022)
4. Nolting, L., Praktiknjo, A.: Techno-economic analysis of flexible heat pump controls. Appl. Energy **238**, 1417–1433 (2019). https://doi.org/10.1016/j.apenergy.2019.01.177
5. Zhou, K., Mao, J., Zhang, H., Li, Y., Yu, X., Chen, F., Li, M.: Design strategy and techno-economic optimization for hybrid ground heat exchangers of ground source heat pump system. Sustain. Energy Technol. Assess. **52B**, 102140 (2022)
6. Yu, M., Li, S., Zhang, X., Zhao, Y.: Techno-economic analysis of air source heat pump combined with latent thermal energy storage applied for space heating in China. Appl. Therm. Eng. **185**, 116434 (2021). https://doi.org/10.1016/j.applthermaleng.2020.116434
7. Chae, S., Bae, S., Nam, Y.: Performance improvement of air-source heat pump via optimum control based on artificial neural network. Energy Rep. **10**, 460–472 (2023)
8. Congedo, P.M., Baglivo, C., D'Agostino, D., Mazzeo, D.: The impact of climate change on air source heat pumps. Energy Convers. Manag. **276**, 116554 (2023)
9. Beccali, M., Bonomolo, M., Martorana, F., Catrini, P., Buscemi, A.: Electrical hybrid heat pumps assisted by natural gas boilers: a review. Appl. Energy **322**, 119466 (2022)
10. Dongellini, M., Naldi, C., Morini, G.L.: Influence of sizing strategy and control rules on the energy saving potential of heat pump hybrid systems in a residential building. Energy Convers. Manag. **235**, 114022 (2021). https://doi.org/10.1016/j.enconman.2021.114022
11. Blackman, C., Gluesenkamp, K.R., Malhotra, M., Yang, Z.: Study of optimal sizing for residential sorption heat pump system. Appl. Therm. Eng. **150**, 421–432 (2019)

12. Li, H., Yang, Q., Xu, Z., Shao, S., Wang, Z., Sun, X., Wang, Y., Xu, C., Zhao, W.: Impact of water volume on the energy saving potential of air source heat pump systems. Int. J. Refrig. **130**, 128–139 (2021). https://doi.org/10.1016/j.ijrefrig.2021.06.025.
13. Barberis, S., Rivarolo, M., Bellotti, D., Magistri L.: Heat pump integration in a real poly-generative energy district: a techno-economic analysis. Energy Convers. Manag. **X**(15), 100238 (2022). https://doi.org/10.1016/j.ecmx.2022.100238
14. Vannoni, A., Sorce, A., Traverso, A., Massardo, A.F.: Techno-economic optimization of high-temperature heat pumps for waste heat recovery. Energy Convers. Manag. **290**, 117194 (2023). https://doi.org/10.1016/j.enconman.2023.117194
15. Kosmadakis, G., Arpagaus, C., Neofytou, P., Bertsch, S.: Techno-economic analysis of high-temperature heat pumps with low-global warming potential refrigerants for upgrading waste heat up to 150 °C. Energy Convers. Manag. **226**, 113488 (2020)

Dynamics and Strength of Power Equipment

Multi-scale Finite-Element Methodology for Predicting the Thermodeformed State of Welded Large-Size Structures

Alexey Milenin[1](✉), Elena Velikoivanenko[1], Galina Rozynka[1], Nina Pivtorak[1], and Serhii Volkov[2]

[1] E.O. Paton Electric Welding Institute of the NAS of Ukraine, 11, K. Malevich Str, Kyiv 03150, Ukraine
asmilenin@ukr.net
[2] Kyiv Academic University, 36, Akad. Vernadskoho Blvd, Kyiv 03142, Ukraine

Abstract. The determination of thermodeformed state of large-size structures is a crucial step in optimizing assembly welding and assessing structural reliability. Conducting extensive laboratory tests for this purpose can be challenging, which high-lights the need for modern modeling methods to predict stress and strain kinetics and its impact on the structural integrity under different external loads. However, the resource-intensive nature of computer simulations, particularly in finite-element analysis, imposes limitations on numerical analyses. To overcome these limitations and reduce computational time and resource requirements, a multi-scale modeling methodology has been developed. This methodology involves two levels of analysis: the level of welding stresses and strains, and the level of macroscopic deformation of the entire structure. The connection between these levels is achieved through the averaging of properties and numerical parameters of specific finite elements. The developed methodology has been successfully applied to a typical case study involving the assembly welding of a large aluminum pressure vessel. The study also demonstrates the limitations of the proposed approach, highlighting its applicability.

Keywords: large-size structures · assembly welding · state of stresses and strains · multi-scale model · finite-element model

1 Introduction

Optimization of assembly welding of large-size structures and analysis of their reliability is often related to the typical problem of the prediction of their state of stresses and strains (SSS). This problem is related to the different scales of the temperature kinetics and SSS in the area of assembly welding and the structure as a whole. Carrying out relevant laboratory studies on full-scale samples is too time-consuming and expensive, while computer simulation methods require significant computational resources.

Modern principles of discrete description of non-stationary multiphysical processes kinetics (e.g., finite element methods) provide wide opportunities for solving fundamental and applied problems [1–3]. The adequacy and accuracy of the obtained numerical

H. Altenbach et al. (Eds.): CAMPE 2023, pp. 167–175, LNME 2025.
https://doi.org/10.1007/978-3-031-82979-6_17

results depends on the mesh-fineness, i.e., the size of the finite element (FE) [4]. But with regard to the numerical description of the large-sized structures in welding, this means excessive resource-intensive problems, because the sufficient gradient of temperatures, stresses and strains in the welding area requires a small spatial division, which cannot be used throughout the structure, on the other hand, a big difference in FE dimensions leads to instability of solving the corresponding boundary value problems. This makes it reasonable the development of simplified models that allow gaining the necessary accuracy of prediction of physical-mechanical processes in welding and related processes.

In particular, the inherent strain methods have become widespread for prediction the residual distortion of large-size structures as a result of assembly welding [5–7]. This class of methods consists in the assumption that the residual plastic strains caused by welding can be set as the initial state of a specific structure, which is especially convenient in the case of a large number of identical welds, e.g., in ship shells, stringer panels, etc. [8]. The disadvantage of these methods is that they do not allow assessment the current SSS of structures during welding, but only the residual one. In addition, in the absence of data on the kinetics of the temperature field and the strain state in the welding process, there are objective difficulties in determining the effective area for setting the initial integrated or distributed inherent strains in the weld metal and the heat-affected zone.

The so-called 2D-X models [9] allow simulation of the current state of large-size structures, but their applicability is quite limited. Therefore, it is practically important to further develop numerical methods for prediction the current SSS of large-size structures in the assembly welding process, in particular, on the example of typical cylindrical pressure vessels, which is the main aim of this work.

2 Multi-Scale Finite-Element Procedure for Prediction of State of Stresses and Strains of Large-Size Structures in Welding

Peculiarity of the state of large-sized structures during welding is the scales of physical and mechanical processes in the welding area and on the periphery. This complicates the numerical implementation of the corresponding mathematical models using conventional methods, but makes it rational to use multi-scale models. Thus, the development of temperatures and SSS during welding has a local character and can be described by joint models of heat-conducting and elastic-plastic deformation. For a number of cases, the local current and residual stress and strain distributions can be described by simplified two-dimensional models, which allows using small FEs for their description without significantly increasing the calculation time. The transfer of a certain calculation data into 3D model of a structure with its own FE mesh (that allows the appropriate calculations in real time) with the subsequent calculation of the total SSS formalizes the algorithm of multi-scale relationship between calculation levels.

In this work two scale levels were considered, that is typical for the problems of predicting the kinetics of temperature and stress-strain states of large structures. They include, the meso-level (for description the welding processes, with a characteristic spatial scale of about 1 mm), and the macro-level (for problems of distortion of large-size structures, with a spatial scale of 1 cm or more). For the numerical description of non-stationary temperature and stress-strain states in assembly welding, the FE solution of interrelated problems was used. These algorithms are the adaptation of approaches developed by the authors [10–12] for multi-scale problems.

Thus, the temperature field kinetics was predicted by numerically solving the non-stationary heat conduction equation in 2D approximation of a rapidly moving heating source [13, 14]. This made it possible not only consider the temperature dependences of the thermal conductivity and heat capacity of the material, but also reducing the spatial mesh of the calculation area without significant increasing the calculation time.

On the basis of calculated temperature fields in welding, the boundary value problem of non-stationary thermoplasticity was formulated with the corresponding FE implementation based on eight-node FEs (Fig. 1). Thus, the growth of the strain tensor can be represented in accordance with the following expression [10]:

$$d\varepsilon_{ij} = d\varepsilon_{ij}^{e} + d\varepsilon_{ij}^{p} + \delta_{ij} \cdot d\varepsilon_{T} \tag{1}$$

where $d\varepsilon_{ij}^{e}$, $d\varepsilon_{ij}^{p}$, $\delta_{ij} \cdot d\varepsilon_{T}$ are the components of the increments strain tensor due to the elastic mechanism, plasticity and the temperature field kinetics, respectively.

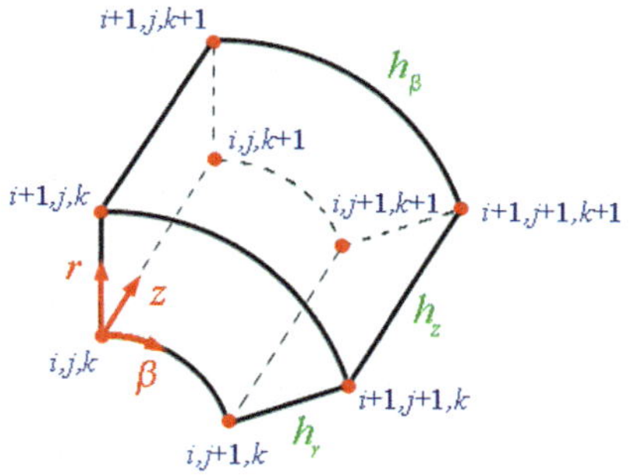

Fig. 1. Diagram of an eight-node finite element in a cylindrical coordinate system.

The growth of strains caused by the temperature field is calculated as $d\varepsilon_T = \alpha dT$, where α is coefficient of thermal expansion.

The increments of the strain tensor can be represented in the form of a superposition of the increments of the corresponding components of stress tensor [11]:

$$\Delta\varepsilon_{ij} = \Psi \cdot \left(\sigma_{ij} - \delta_{ij} \cdot \sigma\right) + \delta_{ij} \cdot (K \cdot \sigma + \Delta\varepsilon_T) - \frac{1}{2 \cdot G} \cdot \left(\sigma_{ij} - \delta_{ij} \cdot \sigma\right)^* - (K \cdot \sigma)^*, \tag{2}$$

where the symbol "*" refers to the corresponding variable to the previous tracing step, Ψ is the function of the material state that determines the condition of plastic flow according to the von Mises criterion [15], σ is the average value of the normal components of the stress tensor σ_{ij}, i.e. $\sigma = (\sigma_{rr} + \sigma_{\beta\beta} + \sigma_{zz})/3$, $K = (1\text{-}2\nu)/E$, $G = 0.5\ E/(1 + \nu)$, E is Young's modulus, ν is Poison ratio.

The determination of the function Ψ is carried out by iteration at each step of the numerical tracing by time or by load increment. At each step of the iteration along Ψ, the stresses σ_{ij} are calculated according to the following algorithm (summation takes place according to repeated indices) [12]:

$$\sigma_{ij} = \frac{1}{\Psi} \cdot \left(\Delta\varepsilon_{ij} + \delta_{ij} \cdot \frac{\Psi - K}{K} \cdot \Delta\varepsilon \right) + J_{ij}, \tag{3}$$

where $\Delta\varepsilon = \frac{\Delta\varepsilon_{ii}}{3}$, $J_{ij} = \frac{1}{\Psi} \cdot \left[(b_{ij} - \delta_{ij} \cdot b) + \delta_{ij} \cdot \left(K \cdot \sigma^* - \frac{\Delta\varepsilon_T}{K} \right) \right]$, $b = \frac{b_{ii}}{3}$,

$$b_{ii} = \left(\frac{\sigma_{ij}}{2 \cdot G} \right)^* + \delta_{ij} \cdot \left[\sigma^* \cdot \left(K - \frac{1}{2 \cdot G} \right)^* - \Delta\varphi \right].$$

The components of the stress tensor satisfy the static equation for internal FEs and the boundary conditions for surface ones. The solvable system of equations in the variables of the vector of movement increments ΔU_i at the FE nodes at each step of tracing and iterations by Ψ is determined by the minimization of the following functional:

$$L_1 = -\frac{1}{2} \cdot \iiint\limits_{V_b} (\sigma_{ij} + J_{ij}) \cdot \Delta\varepsilon_{ij} \cdot \mathrm{d}V_b + \iint\limits_{S_p} F_i \cdot \Delta U_i \cdot \mathrm{d}S_p, \tag{4}$$

where F_i is surface load vector, V_b and S_p are volume and surface of the object, i.e., the following system of equations allows to obtain a solution in the components of displacement increments at each step of tracing and iterations by Ψ:

$$\frac{\partial L_1}{\partial \Delta U_i} = 0. \tag{5}$$

The 2D plane stress state model can be used for cylindrical structures with a girth weld. The stress tensor σ_{ij} $(i, j = r, \beta, z)$ with this description includes four nonzero components $\sigma_{rr}, \sigma_{zz}, \sigma_{\beta\beta}, \sigma_{rz}$. The strain increment tensor $\Delta\varepsilon_{ij}$ contains similar nonzero components. The components of this tensor are related to the components of displacement increments ΔU_r and ΔU_z as follows:

$$\Delta\varepsilon_{rr} = \frac{\partial \Delta U_r}{\partial r}, \Delta\varepsilon_{zz} = \frac{\partial \Delta U_z}{\partial z}, \Delta\varepsilon_{rz} = \frac{1}{2} \cdot \left(\frac{\partial \Delta U_r}{\partial z} + \frac{\partial \Delta U_z}{\partial r} \right), \Delta\varepsilon_{\beta\beta} = \frac{\Delta U_r}{r}. \tag{6}$$

The relationship between σ_{ij} and $\Delta\varepsilon_{ij}$ are as follows:

$$\begin{aligned} \sigma_{rr} =& A_1 \cdot \Delta\varepsilon_{rr} + A_2 \cdot (\Delta\varepsilon_{\beta\beta} + \Delta\varepsilon_{zz}) + Y_{rr}, \sigma_{\beta\beta} = A_1 \cdot \Delta\varepsilon_{\beta\beta} + A_2 \cdot (\Delta\varepsilon_{rr} + \Delta\varepsilon_{zz}) \\ &+ Y_{\beta\beta}, \sigma_{zz} = A_1 \cdot \Delta\varepsilon_{zz} + A_2 \cdot (\Delta\varepsilon_{rr} + \Delta\varepsilon_{\beta\beta}) + Y_{zz}, \sigma_{rz} = A_3 \cdot \Delta\varepsilon_{rz} + Y_{rz} \end{aligned} \tag{7}$$

where $A_1 = \frac{\Psi + 2 \cdot K}{3 \cdot K \cdot \Psi}$, $A_3 = \frac{1}{\Psi}$, $A_2 = A_1 - A_3 = \frac{\Psi - K}{3 \cdot K \cdot \Psi}$, $Y_{ij} = \frac{1}{\Psi} \cdot \left(\frac{\sigma_{ij} - \delta_{ij} \cdot \sigma}{2 \cdot G} \right)^* + \delta_{ij} \cdot \frac{(K \cdot \sigma)^* - \Delta\varphi}{K}$.

Equations (6) form a system of linear algebraic equations, which is the basis of the FE implementation.

As was mentioned above, the use of a similar approach to predict the temperature state, current and residual SSS of a large-sized structure based on a small spatial mesh is complicated or impossible due to the excessive resource-intensiveness of such problem. But increasing the steps of spatial mesh in the macro-scale approach can lead to a loss of calculation model accuracy. Therefore, the results of meso-scale calculation, but taking into account the integral interpretation, were used as input data for the macro-scale calculation. It is necessary using the averaged state of several FEs of a meso-problem during assembly welding to trace the state of one FE of a macro-scopic problem. The result of such averaging should be the integral values of the mechanical properties of the material and the parameters of its current deformed state. Thus, a simplified approach (similar to the rule of mixtures) can be used to average the mechanical properties of the material, namely Young's modulus, coefficient of thermal expansion and yield strength. This is justified by characteristically small gradients of properties at distances of the order of FE, which is necessary for a stable solution of the boundary value problem (1)-(7). Therefore, if the state of the element *mn* in the macro-problem is described by several FEs *ij* in the meso-scale formulation, then the corresponding characteristics can be estimated as follows:

$$E'_{mn} = \frac{\sum_i \sum_j E(T_{ij}) \cdot s_{ij}}{\sum_i \sum_j s_{ij}}; \ \alpha'_{mn} = \frac{\sum_i \sum_j \alpha(T_{ij}) \cdot s_{ij}}{\sum_i \sum_j s_{ij}}; \ \sigma'_{Tmn} = \frac{\sum_i \sum_j \sigma_T(T_{ij}) \cdot s_{ij}}{\sum_i \sum_j s_{ij}} \quad (8)$$

where s_{ij} is the area of the *ij*-th FE, T_{ij} is the temperature at point *ij*, "′" relates the parameter to the macro-problem.

It should be noted that relation (8) is formulated in 2D approach, but averaging from a 2D problem to a 3D one can be implemented in a similar way.

The development of SSS in a macroscopic level requires taking into account the force factor, namely the non-uniform deformation of the material. To calculate the distribution of current and residual SSS, which satisfies the equilibrium condition (5) and does not receive significant distortion on coarse spatial mesh, the following spatial averaging of the numerical components of the matrix (7) and the function of the material state Ψ was proposed:

$$Y_{mn'} = \frac{\sum_i \sum_j Y_{ij} \cdot s_{ij}}{\sum_i \sum_j s_{ij}}; \ \Psi_{mn'} = \frac{\sum_i \sum_j \Psi_{ij} \cdot s_{ij}}{\sum_i \sum_j s_{ij}}; \ A_{kmn'} = \frac{\sum_i \sum_j A_{k\,mn} \cdot s_{ij}}{\sum_i \sum_j s_{ij}}. \quad (9)$$

The solution of the system of linear algebraic Eqs. (7) within the formulation of the macroscopic SSS problem allows obtaining the equilibrium distributions of the stresses and strains at each step of the tracing.

3 Results and Discussion

The applicability and features of the developed multi-scale approach were studied on a typical example of assembly welding of girth welds of a cylindrical structure made of aluminum alloy AMg6 with a diameter of $D = 3900$ mm and a wall thickness of $t = 10$ mm, welding parameters are as follows: voltage $U = 20$ V, welding current $I =$

380 A, welding rate $v_w = 2.23$ mm/s. It should be noted that the choice of structure for research corresponds to typical design solutions for space purposes, in particular, fuel tanks of launch vehicles [16], although the versatility of the proposed approach allows considering a wide range of materials, structural elements and welding technologies.

The meso-scale formulation was implemented as joint solution of the temperature kinetics problem and the problem of thermoplasticity in welding in a 2D formulation, which made it possible to use a fine and regular spatial FE mesh with a element size in the radial and axial directions $h_r = h_z = 1$ mm. For 3D formulation, the linear size of the FEs in the radial and axial directions was multiplied by 2, 4 and 8, and it was made a comparison of the stresses and strains obtained within the framework of the multi-scale problem and verification 3D modeling with a spatial step $h_r = h_z = h_\beta = 1$ mm.

As shown in Fig. 2, a comparison of the calculated distributions of circumferential $\sigma_{\beta\beta}$ and axial σ_{zz} stresses for different macro-scale spatial meshes confirms the high degree of correspondence between the results of multi-scale modeling and verification calculations. Significant differences appear when the characteristic spatial distribution of local stresses becomes smaller than one step of the macro-scale problem, where appropriate averaging occurs (in particular, when h_z is increased to 8 mm). At the same time, in other parts of the structure, where the gradient of residual stresses is less significant the error is less than 2%. Such accuracy of the multi-scale approach is related to the physical essence of the mathematical formulation (9), namely, the transfer of the averaged internal energy of the deformed material. This makes it possible to use the proposed method for the analysis of the stress state of large-sized welded structures under the influence of a complex operational load, as well as for the analysis of brittle strength.

Similar calculations of the strain kinetics showed that the influence of the multi-scale approach on the error of the prediction results of the residual strain state is significantly higher (Fig. 3). Thus, an increase in h_r within the selected limits slightly affects the calculated magnitude of circumferential strains $\varepsilon_{\beta\beta}$ (the error is about 1.7%), while for longitudinal strains ε_{zz} a satisfactory result of multi-scale modeling is observed for a small increase in spatial division (2 times up to the size 2 mm). A similar conclusion about the relatively small possibilities of increasing of h_z: the coarsening the mesh up to 4 times allows obtaining a satisfactory value of the error of multi-scale prediction of $\varepsilon_{\beta\beta}$ (error less than 7%), while for correct results ε_{zz} the increase in h_z should not be more than 2 times up to the size 2 mm.

These analysis results are explained by the fact that one of the main assumptions of FE modeling is the homogeneity of the properties of each FEs. A significant increase in the FE size leads to incorrect formulation of the problem and discrepancy of strain field with the true solution on fine mesh. However, this instability concerns the strains state in the axial direction, where the maximum gradient of strains is observed, while for $\varepsilon_{\beta\beta}$ and the stress state the possibilities of optimizing the calculations of large-sized structures are much higher. As the calculations showed, the simultaneous increase of the spatial steps in the radial and axial directions does not significantly change these conclusions, because the solutions instability cased with mesh coarsening in the axial direction prevails.

In comparison with alternative methodologies of simplified FE calculation of SSS of large scale structures in welding, proposed approach shows similar demands for

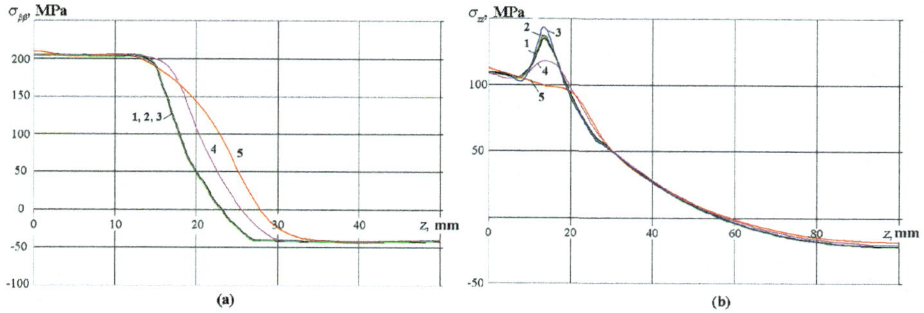

Fig. 2. Numerical distributions of residual post-welding circumferential $\sigma_{\beta\beta}$ (a) and axial σ_{zz} (b) stresses. Here and in Fig. 3: 1 – verification 3D calculation at $h_r = h_z = h_\beta = 1$ mm; 2 – multi-scale calculation at $h_r = 4$ mm, $h_z = h_\beta = 1$ mm; 3 – multi-scale calculation at $h_r = 8$ mm, $h_z = h_\beta = 1$ mm; 4 – multi-scale calculation at $h_z = 4$ mm, $h_r = h_\beta = 1$ mm; 5 – multi-scale calculation at $h_z = 8$ mm, $h_r = h_\beta = 1$ mm. Note: in Fig. 2 and curves 1, 2 and 3 actually coincide

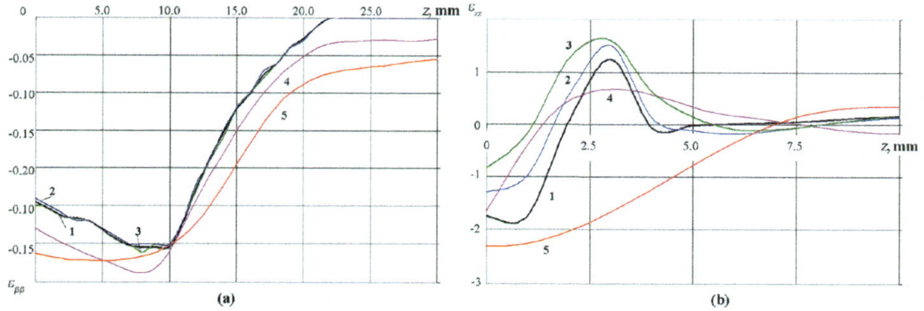

Fig. 3. Numerical distributions of residual post-welding circumferential $\varepsilon_{\beta\beta}$ (a) and axial ε_{zz} strains (b)

obtaining adequate. Thus, the use of inherent strain method in [17] for prediction of residual deformation of welded thin-sheet structures FE of 2 mm size were used, while their enlarging leads to distortion of calculation results. But the approach developed within the framework of this work allows calculating the kinetics of SSS in the welding process, but not only the residual state. It also allows implementation a multiphysics formulation of the welding problem without additional dimensionality assumptions.

4 Conclusions

1. The multi-scale approach for numerical prediction of temperature kinetics and stress-strain state is proposed as an effective method of analysis of welding the large-sized structures without significantly demands to the resource intensity of the calculations. This technique is based on the FE solution of non-stationary thermoplasticity problem at the meso-scale level with a fine mesh of the structure and the subsequent transfer of a certain amount of calculation data into a macroscopic model with a coarse mesh.

For this, algorithms for appropriate averaging of the properties and parameters of the stress-strain state are proposed. This allows the analysis of the state of large-size structures in welding without the need to involve significant computing recourses.
2. The example of assembly welding of girth weld of a large-sized cylindrical pressure vessel made of aluminum alloy AMg6 shows a high correspondence between the results of FE calculations of the stress state within the framework of the developed model and verification calculations using standard approaches. It is shown that a significant increase in spatial FE mesh is possible until the characteristic spatial distribution of local stresses becomes smaller than one step of the macro-scale problem, where appropriate averaging takes place.
3. It is shown that the possibilities of the proposed approach for predicting the strain state of large-size structures in welding are limited. This is caused by high strain gradients, especially in the axial direction of cylindrical structure, in which the spatial averaging of the material properties can lead to the convergence of FE problem to an incorrect solution.

Acknowledgements. The authors of this work are grateful to the Ukrainian armed forces and the world community, which is now helping Ukraine in the struggle for freedom and independence. Thanks to the unity and unbroken spirit of the Ukrainian nation during this period of our history, it is possible carrying out scientific researches for our peaceful and successful future.

References

1. Wang, B., Cai, Y., Li, Z., Ding, C., Yang, T., Cui, X.: Stochastic stable node-based smoothed finite element method for uncertainty and reliability analysis of thermo-mechanical problems. Eng. Analy. Boundary Elem. **114**, 23–44 (2020)
2. Zhentian, H., Dong, L., Zi, H., Ji, L.: Boundary moving least squares method for 3D elasticity problems. Eng. Anal. Boundary Elem. **121**, 255–266 (2020)
3. Fascetti, A., Feo, L., Abbaszadeh, H.: A critical review of numerical methods for the simulation of pultruded fiber-reinforced structural elements. Compos. Struct. **273**, 114284 (2021)
4. Carrera, E., Elishakoff, I., Petrolo, M.: Who needs refined structural theories? Compos. Struct. **264**, 113671 (2021)
5. Wu, C., Kim, J.-W.: Numerical prediction of deformation in thin-plate welded joints using equivalent thermal strain method. Thin-Wall. Struct. **157**, 107033 (2020)
6. Honaryar, A., Iranmanesh, M., Liu, P., Honaryar, A.: Numerical and experimental investigations of outside corner joints welding deformation of an aluminum autonomous catamaran vehicle by inherent strain/deformation FE analysis. Ocean Eng. **200**, 106976 (2020)
7. Li, L., Luo, C., Shen, J., Zhang, Y.: Numerical prediction of welding deformation in ship block subassemblies via the inhomogeneous inherent strain method. J. Manuf. Process. **80**, 860–873 (2022)
8. Kang, S., Kim, J., Jang, Y., Lee, K.: Welding deformation analysis, using an inherent strain method for friction stir welded electric vehicle aluminum battery housing. Consider. Product. Appl. Sci. **9**, 3848 (2019)

9. Makhnenko, V.I., Milenin, A.S., Semyonov, A.P.: Mathematical modeling of thermodeformational and physicochemical processes during welding of lap and butt dissimilar titanium-aluminum joints. In: Computer Technology in Welding and Manufacturing, 16th Int. Conf. and Mathematical Modelling and Information Technologies in Welding and Related Processes, 3rd Int. Conf. Proc. of Joint Int. Conf, Kyiv, 234–239 (2006)
10. Velikoivanenko, EA., Milenin, A.S., Popov, A.V., Sidoruk, V.A., Khimich, A.N.: Methods and technologies of parallel computing for mathematical modeling of stress-strain state of constructions taking into account ductile fracture. J. Autom. Inf. Sci. **46**(11), 23–35 (2014)
11. Makhnenko, V.: Problems of examination of modern critical welded structures. Paton Weld. J. **5**, 21–28 (2013)
12. Milenin, A.S., Velikoivanenko, EA., Rozynka, G.P., Pivtorak, N.I.: Residual strength and reliability of corroded pipelines–monte-carlo approach for consideration of spatially nonuniform material properties. Struct. Integ. **8**, 321–326 (2019)
13. Karkhin, V.A.: Thermal Processes in Welding. Springer Singapore (2019)
14. Akhonin, S.V., Milenin, A.S., Pikulin, A.N.: Simulation of evaporation processes of alloying elements during electron beam surface melting of ingots-slabs from titanium-base alloys. Problemy Spetsial'noj Electrometallugii **3**, 29–32 (2005)
15. Cazacu, O., Revil-Baudard, B.: Plasticity of Metallic Materials. Modeling and Applications to Forming. Elsevier (2021)
16. Bin Rayhan, S., Pu, X., Xi, H.: Modeling of fuel in aircraft crashworthiness study with auxiliary fuel tank. Int. J. Impact Eng **173**, 104449 (2023)
17. Makhnenko, O.V.: Increase of efficiency of thermal straightening of welded thin-sheet structures on basis of mathematical modeling. Paton Weld. J. **9**, 6–9 (2008)

Creep and Damage Processes in Cyclically Loaded Model of Turbine Rotor

Holm Altenbach[1], Dmytro Breslavsky[1,2](✉), Alyona Senko[2], and Oksana Tatarinova[2]

[1] Otto-Von-Guericke-University Magdeburg, Magdeburg, Germany
Dmytro.Breslavsky@khpi.edu.ua
[2] National Technical University "Kharkiv Polytechnic Institute", Kharkiv, Ukraine

Abstract. A novel approach to solving the creep-damage problem of cyclically loaded elements of power engineering structures is presented. Cycles of short duration caused by vibration as well as long cycles caused by turbine shutdowns are considered. The total hidden damage of the material is estimated taking into account the damage due to creep and high cycle fatigue. A formulation of the boundary – initial value problem based on the methods of many time scales and averaging over cycle periods is presented. The new results of solving the creep-damage problem in the model of the rotor are given for different cases of static and cyclic loading. It is shown that taking into account the accumulation of high cycle fatigue damage significantly reduces the time until the completion of the hidden damage accumulation and leads to the fracture development in rotor.

Keywords: Creep · Damage · Fatigue · Cyclic Loading · Fracture · Elevated Temperatures · Damage Evolution Equation · Turbine Rotor

1 Introduction

The problems of long-term strength of turbines have always attracted the attention of engineers [1]. In many cases irreversible creep strains occur in the turbine material. Creep processes are accompanied by the accumulation of hidden damage, which is caused both by creep and fatigue processes [1,2]. The stress-strain state in structures from a mathematical point of view is described by a boundary - initial value problem. For its adequate solution, it is necessary to formulate the constitutive equations, which reflect the main effects of deformation and damage accumulation [1].

Steam and gas turbines operate in conditions of cyclical changes in temperatures and loads, which varying due to the turbine's operating modes or due to its removal from operation (long-term cycles). Due to the impossibility of complete control of vibrations, there are cycles with a short period, which can lead to high cycle fatigue damage [1, 2].

Calculation of the stress-strain state of turbine elements in cyclic conditions is a more complex task and it is in the focus of researchers' attention. The constitutive equations include hardening and softening to describe the creep of a turbine rotor were developed [3]. Further, Finite Element Method (FEM) calculations were performed for the rotor

H. Altenbach et al. (Eds.): CAMPE 2023, pp. 176–185, LNME 2025.
https://doi.org/10.1007/978-3-031-82979-6_18

model for cases of cold and hot start-ups. The purpose of [4] was the analysis of local changes in the stress-strain state under heating conditions. The constitutive model include so-called "mechanism-based" evolution equations for various damage parameters. Steam temperatures and load profiles are taken into account, which correspond to the idealized sequence of starting, holding and stopping of the turbine.

The stress-strain state in the rotor of a steam turbine under conditions of cyclic creep and fatigue is analysed in [5]. The applied material model is based on the Chaboche model [1]. Both temperature and centrifugal force were considered, but the degree of their influence was different at different locations.

The results of the analysis of the local change in the stress-strain state and damage behaviour in the turbine component made of heat-resistant steel 9–12% Cr at high temperature are described in [6]. The steam temperature and load profiles correspond to an idealized sequence of starting, holding and stopping. Thirteen cycles were analyzed.

The study of damage from the interaction of creep and fatigue for a steam turbine rotor made of steel with 9–12% Cr under the action of cyclic thermomechanical loads is presented in [7]. Various factors affecting the creep and fatigue damage of a steam turbine rotor are analyzed, including the dwell period and the rotor speed.

The importance of analyzing the stress state, which is caused by fatigue and creep damage in the area of blade and rotor connection, is presented in [8]. Elements from nickel alloys were studied. A resource estimation method based on the damage parameter evaluation is presented. This damage parameter is included in a complex material model, taking into account the static, cyclic and viscoplastic behavior of the materials.

The FEM modeling of the high temperature behavior in the gas turbine rotor is described in [9]. The improved Chaboche – Lemaitre model [2] was used. The influence of thermal processes on the development of residual stresses was studied.

The method for creep and long-term strength evaluation in cyclically loaded elements of turbomachines is presented in [10]. The creep-damage and fracture analysis of the turbine locking joints are presented. It was shown that the influence of cyclic loading on stress relaxation, deformation and fracture of the lock joint is significant.

The typical behavior of a cooled turbine blade was analyzed in [11]. The authors used an original approach in which the complex three-dimensional design of the blade is represented by simplified schemes. This allows to identify the main factors affecting such nonlinear deformation processes as plasticity, ratcheting, creep, etc., and to formulate the corresponding laws of high temperature behavior.

A similar approach was used in this paper. The rotor is represented by a simplified model, in which the main factors characterizing the thermal and stressed behavior of a typical heat-resistant alloy, are considered. An approach based on the method of many time scales and averaging over the cyclic load period [12–14] was used.

2 Problem Statement and Constitutive Equations

Let us limit here by 2D problem statement, which is wide spread for rotor analysis (see for example [3]). From the other hand, it allows us to present and analyze all typical kinds of deformation and hidden damage accumulations in a rotor material. In order to solve the creep problem with consideration of damage accumulation, caused by creep and high cycle fatigue processes in material of structural element, the following statement will be done. Let us regard the solid with volume V which is fixed in a surface part S_1. In the coordinate system $\boldsymbol{x} = (x_1,x_2)$ the motion of the continuum of material points will be described by the displacement vector $\boldsymbol{u} = \boldsymbol{u}(x_i,t)$, tensors of stresses $\boldsymbol{\sigma}$ and strains $\boldsymbol{\varepsilon}$ with components $\sigma_{ij}, \varepsilon_{ji}(i,j = 1,2)$, which are functions of coordinates and time t. Total strain consists of elastic e_{ij}, temperature ε_{ij}^T and creep c_{ij} components. The tensor connection for them with the stress tensor and time is determined by the accepted constitutive equations. The small strains are considered.

The solid is loaded by volume forces f_i, the traction p_i acts on the surface part S_2. The case of constant in time volume centrifugal forces is regarded: $f = f(x_i)$, $x_i \in V$, $i = 1,2$. Traction has a time-independent value $p_i^{(0)} = p_i^{(0)}(x_i)$ and can vary in a time through the cycle $\Phi_i(t)$: $p_i = p_i^{(0)} + \Phi_i(t)$, $x \in S_2$,

$$\Phi_i(t) = p_i^{\max} \sum_{k=1}^{\infty} Z_k \sin(\Omega_k t + \beta_k) + p_i^a \sin(\Omega t) \tag{1}$$

where p_i^{max}, p_i^a are the amplitudes of appropriate components of traction; $Z_k = \sqrt{a_k^2 + b_k^2}$; $\beta_k = \mathrm{arctg}(a_k/b_k)$; a_k, b_k are the Fourier series coefficients, $\Omega_k = 2\pi k/T$; $\Omega = 2\pi/T_2$. The problem statement for creep problem has the following form [12, 14]:

$$\begin{aligned} &\varepsilon_{ij} = e_{ij} + c_{ij} + \varepsilon_{ij}^T, \sigma_{ij,j} + f_i = \rho \frac{\partial^2 u_i}{\partial t^2}, \varepsilon_{ij} = \frac{1}{2}(u_{i,j} + u_{j,i}), x_i \in V; \\ &\sigma_{ij} = D_{ijkl}(\varepsilon_{kl} - c_{kl} - \varepsilon_{kl}^T); \sigma_{ij} n_j = p_i^{(0)} + \Phi_i(t), x_i \in S_2; \\ &u_i = \overline{u}_i, x_i \in S_1; u_i(x_i, 0) = c_{ij}(x_i, 0) = 0 \end{aligned} \tag{2}$$

where n_j are the components of the unit vector which is normal to solid's surface, $j = 1,2$; D_{ijkl} are the components of the tensor of elastic material properties; u_i are known values of the displacements on the surface S_1, which are constant in a time; ρ is the material density.

To solve the system (2) we use the asymptotic expansions method with subsequent averaging over a period of cyclic loading [12]. The use of these methods allows us to divide the system of Eqs. (2) into two. First system (3) arises from expressions which result from averaging and describes the processes which are run in slow basic motion; the second (4) is obtained by subtraction from (2) the first system. These systems are not independent, they are connected by constitutive equations [12, 13].

So, in order to determine the stress-strain state of creeping solid subjected to cyclic loading, we have to solve the following first system of equations, where all unknowns vary in slow time scale:

$$\begin{aligned} &\sigma_{ij,j} + f_i = 0, x_i \in V; \sigma_{ij} n_j = p_i, x_i \in S_2; \varepsilon_{ij} = C_{ijkl}\sigma_{kl} + c_{ij} + \varepsilon_{ij}^T, \\ &\varepsilon_{ij} = \frac{1}{2}\left(u_{i,j} + u_{j,i}\right), u_i = \overline{u}_i, x_i \in S_1; u_i(x_i, 0) = c_{ij}(x_i, 0) = 0 \end{aligned} \tag{3}$$

Constitutive equations have to be added. Components of amplitude stress tensors which are included in this constitutive model are determined from systems (4) and (5), formulated for fast varying time scale ($k = 1,2,\ldots$) [12, 13]:

$$
\begin{aligned}
&\sigma_{ij,j}^{ak} = -\rho(\Omega_k)^2 u_i^{ak},\ x_i \in V;\ \sigma_{ij}^{ak} n_j = p_i^{\max} Z_k,\ x_i \in S_2,\\
&\varepsilon_{ij}^{ak} = \frac{1}{2}\left(u_{i,j}^{ak} + u_{j,i}^{ak}\right) = C_{ijmn}\sigma_{mn}^{ak},\ x_i \in V;\ u_i^{ak} = 0,\ x_i \in S_1;
\end{aligned} \tag{4}
$$

$$
\begin{aligned}
&\sigma_{ij,j}^{a} = -\rho(\Omega)^2 u_i^{a},\ x_i \in V;\ \sigma_{ij}^{a} n_j = p_i^{a},\ x_i \in S_2;\\
&\varepsilon_{ij}^{a} = \frac{1}{2}\left(u_{i,j}^{a} + u_{j,i}^{a}\right) = C_{ijmn}\sigma_{mn}^{a};\ x_i \in V;\ u_i^{ak} = 0,\ x_i \in S_1.
\end{aligned} \tag{5}
$$

By solving systems (4) for each harmonic components k and (5) the values of amplitude stress tensors are determined. They are used in constitutive model (6–7).

Thereby the solution of creep-damage problem in the case of cyclic loading is divided into two systems. System of slow motion (3) has to be solved jointly with Eqs. (6–7) where amplitude stresses are determined from auxiliary systems (4) and (5).

Using the classical approach for the system of constitutive equations formulated for strain rates, the final form of the equations for the cyclic stress varying at the complex stress state was obtained and used [12, 13]:

$$
\dot{c}_{ij} = \frac{3}{2} B g_n \frac{\left(1 + H_n(A_n^i)\nu_f(A_n^i)\right)\sigma_v^{n-1}}{(1-\omega)^l} S_{ij},\ c_{ij}(0) = 0, \tag{6}
$$

$$
\dot{\omega} = D g_m \alpha_f\left(A_m^e\right)\frac{\sigma_e^m}{(1-\omega)^l} + \beta_f(A_m^e) F \frac{\chi\left(\sigma_e^a, b g_m^{1/m}\sigma_e\right)^p}{(1-\omega)^q},\ \omega(0) = \omega_0,\ \omega(t_*) = \omega_*, \tag{7}
$$

$$
\alpha_f(A) = 1 + K(A_m^e)\nu_f(A_m^e)\ \beta_f(A_m^e) = 1 - \exp\left(-\pi\frac{A_m^{e2}}{A_{cr}^2}\right);\ \nu_f(A_m^e) = \exp\left(-\pi\frac{A_m^{e2}}{A_{cr}^2}\right).
$$

$$
H_n(A_n^i) = \int_0^1 \left(1 + A_n^i \sin(2\pi\xi)\right)^n d\xi - 1\ K_m(A_m^e) = \int_0^1 \left(1 + A_m^e \sin(2\pi\xi)\right)^m d\xi - 1, \tag{8}
$$

$$
\begin{aligned}
&\sigma_i = \sqrt{\frac{3}{2}S_{ij}S_{ij}},\ A_n^i = \frac{A^i}{g_n^{1/n}}\ A^i = \frac{\sigma_v^a}{\sigma_v},\ A_m^e = \frac{A^e}{g_m^{1/m}},\ A^e = \frac{\sigma_e^a}{\sigma_e},\ \sigma_v^a = \sqrt{\frac{3}{2}S_{ij}^a S_{ij}^a};\\
&\sigma_e = (1-\lambda)\sqrt{\frac{3}{2}S_{ij}S_{ij}} + \lambda\sigma_I;\ \sigma_v^{ak} = \sqrt{\frac{3}{2}S_{ij}^{ak}S_{ij}^{ak}},\ \sigma_e^{ak} = (1-\lambda)\sqrt{\frac{3}{2}S_{ij}^{ak}S_{ij}^{ak}} + \lambda\sigma_I^{ak},
\end{aligned}
$$

where ω is scalar damage parameter, t_* is the time value of the hidden damage accumulation finishing; $\chi\left(\sigma_e^a, b\sigma_e\right)$ is the equivalent stress describing the fatigue damage at complex stress state [1]; σ_I is maximum principal stress, S_{ij} are stress deviator components; S_{ij}^a, S_{ij}^{ak} are the deviatoric components of amplitude stress tensors σ_{ij}^a, σ_{ij}^{ak}. Influence functions $H_n(A_n^i)$ and $K_m(A_m^e)$ reflect the effect of creep-damage acceleration due to developed dynamic creep, which occurs due to vibration [12, 15].

Let us determine the expression for the functions g_n and g_m, which reflect the influence of cycles of long duration, as proposed in [13]. Let us consider the process of turbine operation as cyclic with a full period of $L = 300$ h (Fig. 1). In the first approximation, we neglect the components describing the parameters increasing before the normal mode obtaining as well as the gradual stops. We model the process with a rectangular cycle (Fig. 1). Due to the integral form of consideration of cyclicity, additional sections are not critical, they are taken into account in the rectangular representation of the cycle. Operating time $t_1 = 240$ h, the last 60 h the turbine is not operated. For this cycle, we present the load function, which use the Fourier series expansions:

$$P = p\left(a_0 + \sum_{k=1}^{\infty}\left(a_k \cos\left(\frac{2\pi k}{L}t\right) + b_k \sin\left(\frac{2\pi k}{L}t\right)\right)\right), \tag{9}$$

Using the method of two time scales and averaging over the period of cyclic action L [13], we obtain an expression for the functions of the long cycles influence on creep g_n and on the damage accumulation g_m:

$$g_n = \int_0^1 \left(a_0 + \sum_{i=1}^{\infty} (a_i \cos(2\pi i\xi) + b_i \sin(2\pi i\xi))\right)^n d\xi. \tag{10}$$

$$g_m = \int_0^1 \left(a_0 + \sum_{k=1}^{\infty} (a_i \cos(2\pi k\xi) + b_i \sin(2\pi k\xi))\right)^m d\xi.$$

where $a_0 = \frac{t_1}{L} = 0.8$, $a_k = \frac{1}{\pi k}\sin\left(\frac{2\pi k}{L}t_1\right)$, $b_k = \frac{1}{\pi k}\left(1 - \cos\left(\frac{2\pi k}{L}t_1\right)\right)$.

The influence of the vibration component of the load transmitted to the rotor as a result of the oscillations of the discs with the blades is represented in the first approximation by the harmonically load varying with the amplitude p_a and the frequency Ω in expression (1).

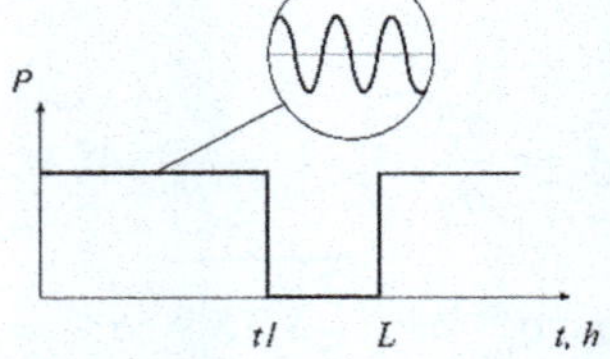

Fig. 1. Schematic representation of the turbine operation cycle

The constitutive Eqs. (6–7) generally include 11 constants that are determined experimentally. Values of *B, n, D, m, l* are determined using creep and long-term strength curves. These constants are usual for the Bayley-Norton-Rabotnov-Kachanov equations [1]. Constants *F, p, q* are obtained by use of diagrams of long-term strength at high cycle fatigue; constant *b* reflects the effect of static loading on the accumulation of fatigue damage [1]. A_{cr} is so-called critical value of the stress cycle asymmetry parameter, which separates the processes of dynamic creep acceleration and pure high cycle fatigue superimposed on static stress, $(0.8 < A_{cr} < 1)$ [13]. Constant λ $(0 < \lambda < 1)$ is the parameter of material sensitivity for fracture type [1].

The presented problem statement with the proposed constitutive equations is implemented in the calculation method built on the basis of a combination of FEM and time integration methods [12, 14]. The system (3) derives into the system (11). Here we consider only quasistatic processes, when the frequencies of oscillations are located far from the eigen ones. Then the systems of types (4) and (5) takes the form (12):

$$[K]\{\dot{u}\} = \{\dot{F}\} + \left\{\dot{F}^C\right\}; \left\{\dot{F}^C\right\} = \sum_{N_\beta} \int_{V_\beta} [\overline{B}]^T [C]\{\dot{c}\}\, dV;$$

$$\{\dot{F}\} = \sum_{N_\beta} \int_{V_\beta} [N^p]^T \{\dot{f}\}\, dV + \sum_{N_\beta} \int_{S_2^\beta} [N^p]^T \{\dot{p}\}\, dS + \sum_{N_\beta} \int_{V_\beta} [\overline{B}]^T [C]\left\{\dot{\varepsilon}^T\right\} dV; \quad (11)$$

$$[K]\{u^a\} = \{F^a\}; \{F^a\} = \sum_{N_\beta} \int_{S_2^\beta} [N^p]^T \{p^a\}\, dS; \quad (12)$$

Here $\boldsymbol{K}$ is the system stiffness matrix; S_2^β is the surface area of the finite element, which is under the traction action. $\overline{\boldsymbol{B}}$ is the matrix differentiation operator; $\boldsymbol{u}$ is the global vector of nodal displacements; $\boldsymbol{F}$ is the vector of instantaneous nodal loads caused by traction, volume forces and thermal strains; $\boldsymbol{F^c}$ is nodal vector of loading caused by creep strains; $\boldsymbol{C}$ is the matrix of elastic constants; $\boldsymbol{N}$ is shape functions matrix; β is a finite element number; V_β is a volume of the finite element; $\sum N_\beta$ is the summation operator over all finite elements. Index a shows the amplitude values. Systems (11), (12) use to simulate the creep and damage accumulation in the turbine rotor model.

3 Numerical Modelling Results

The formulated FE problem statements (11), (12) for the plain strain approach will be used here in order to provide numerical modelling of turbine rotor. As the modelling tool we use the *FEM Creep Damage Fracture* [16] software developed at NTU "KhPi" basing on the presented methods [12–14].

As the rotor model, let us consider a long body of revolution (tube) with a circular cross-section, the inner diameter of which is 0.4 m, the outer diameter is 1.4 m. The rotor material is a nickel alloy with 13% Cr. We assume the temperature field with logarithmic law along the radius. The temperature on inner surface is 550 °C, on the outer surface is 350 °C.

After processing the experimental data from [15], constants describing the strain rate and the accumulation of creep damage in the material of the rotor were obtained. Creep and long-term strength curves were used. To determine the values of the constants that are included in the second part of Eq. (7) and correspond to high cycle fatigue damage, data from [1] and [17] were used. A recalculation from the dependence on the number of cycles to the dependence on time for harmonic loading was done.

So, the values of obtained constants are: $n = m = 14.24$, $B = 1.53 \times 10^{-41}$ MPa^{-n}/h, $D = 1.09 \times 10^{-41}$ MPa^{-m}/h, $k = l = 8$, $F = 4.01 \times 10^{-23}$ MPa^{-p}/h, $p = q = 7.16$. The critical value of the stress cycle asymmetry parameter $A_{cr} = 1$, the von Mises criterion of long-term fracture [1] is used, $\lambda = 0$. Young's modulus is assumed to be equal $E = 1.7 \times 10^5$ MPa, Poisson ratio $\nu = 0.3$, coefficient of thermal expansion $\alpha = 1.5\ 10^{-5}$ (° C)$^{-1}$.

Centrifugal forces (rotation velocity 50 rev/s) act on the rotor, there is a non-uniform distribution of temperature stresses. As a result of the vibrations of the blades and disks, there is a time-varying load represented by the harmonic law (the second term in (1)). The cyclical operation of the turbine is modeled by relations (9).

Due to the symmetry of the model, the calculations were carried out using the plane strain approach, as the model we consider one fourth part of the structural element. The symmetry boundary conditions were applied: zero displacements in the direction x_2 for the nodes in the axis x_1 as well as zero displacements in the direction x_1 for the nodes in the axis x_2. After convergence studies, a scheme with 1500 finite elements was used.

First, we will provide the data of the classical creep-damage calculation under static loading (all influence functions $g_n = g_m = 1$, $H_n = K_m = 0$, Fig. 2). Figure 2a) contains the cross-sectional distribution of the von Mises stresses, Fig. 2b) – the distribution of the damage parameter at the moment before the completion of hidden damage accumulation, $t_* = 41013$ h. As can be seen from Fig. 2b), the distribution of the damage parameter corresponds to the classical understanding about the location of the fracture site, which corresponds to the maximum values in the initial stress distribution (Fig. 2a). The fracture occurs approximately in the middle of the arc of the inner circle of the model. For up to 100,000 h, the completion of damage accumulation does not occur at any additional point of the model. The von Mises strains at $t_* = 41013$ h are not greater 0.25%.

Let us present new results of cyclic creep-damage investigations in turbine rotor models. Calculations were carried out due to the cyclical operation of the turbine (scheme in Fig. 1). The values of the influence functions were: for the creep rates $g_n = 0.74$ and for damage accumulation $g_m = 0.75$. Due to this, the time until the completion of damage accumulation finishing turned out to be much longer, $t_* = 62567$ h, the distribution of the damage parameter in this case is similar to that presented in Fig. 2b.

It is known [15], that this alloy exhibits significant properties of dynamic creep. Taking into account the acceleration of creep and damage accumulation which leads to a significant intensification of processes, the time until the completion of hidden fracture is reduced to $t_* = 15495$ h. The distribution of the damage parameter is also similar to that obtained in the static calculation, but the deformation is more intense, with a maximum strain level of 0.5%. The distribution of the von Mises equivalent strains at the moment $t_* = 15495$ h is shown in Fig. 3.

As a result of dynamic creep, there is a significant stress relaxation. It is especially intense in the zones of maximum stresses. Here the stress cycle asymmetry coefficients increase, surpassing the critical value of the parameter $A_{cr} = 1$. This means that high cycle fatigue damage begins to develop in the material, which is taken into account by the second term of the evolution Eq. (7). The data of complex numerical modeling, which takes into account the cyclic nature of turbine operation with vibrational influence due to dynamic creep and high cycle fatigue, indicate that the completion of hidden damage accumulation occurs faster, at $t_* = 13930$ h, and after that the intensive development of the macroscopic defect (crack) begins. Its current configuration with the distribution of the damage parameter along the cross-section of the model at $t_* = 18472$h, is presented in Fig. 4. At the same time, the dimensions of the macrodefect along the radius of the rotor reach one-tenth of its part, which can be considered dangerous.

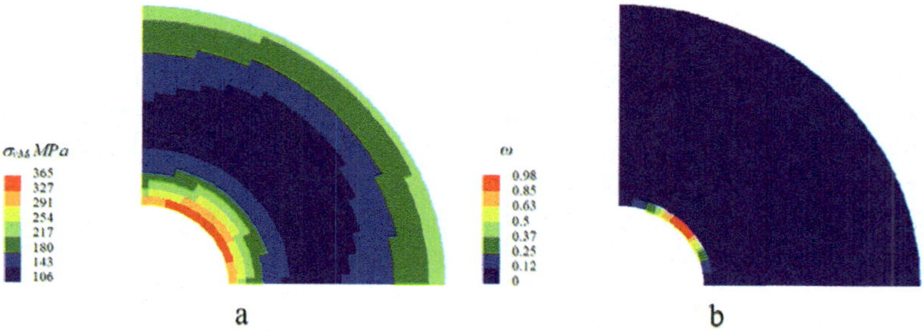

Fig. 2. Static loading. Distribution of von Mises stresses, $t = 0$ a) and damage parameter, $t_* = 41013$ h, b)

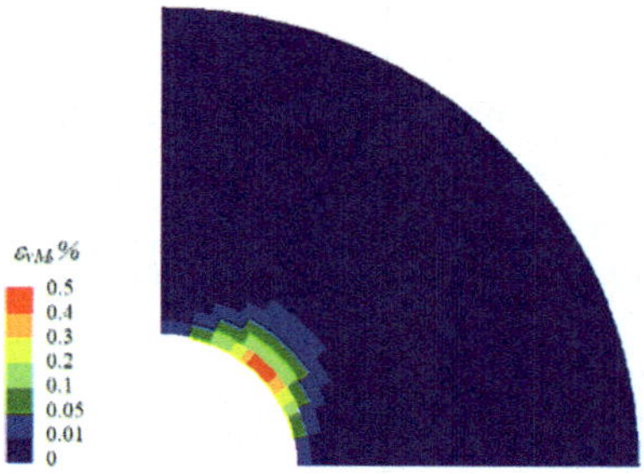

Fig. 3. Cyclic loading with two different periods. Distribution of von Mises strains, $t_* = 15495$ h

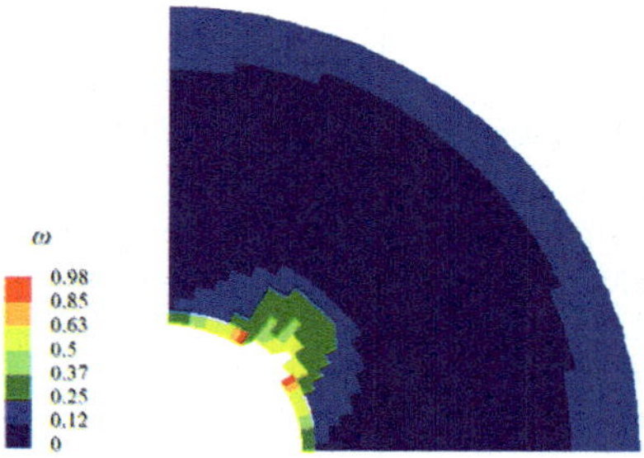

Fig. 4. Cyclic loading with two different periods. Distribution of damage parameter and form of the macroscopic defect, $t_* = 18473$h

4 Conclusions

The main aim of the paper is the presentation of the calculation method, which can be used for the simulation of cyclic creep processes in turbine structures with consideration of the total damage accumulation due to interaction between creep and high cycle fatigue damage up to development of macroscopic defect [13, 14]. The calculational effectiveness of the method and software was obtained due to use the asymptotic methods with averaging through a cycle, which allow us to exclude the necessity of the analysis of the inner-cyclic behavior. The method includes the possibility of the analysis the macroscopic defect (crack) development by use of the information about current initial and boundary conditions of the structural element with fractured area [14]. With the help of the method and developed software, an example of the solution of the long-term high temperature cyclic deformation problem for a steam turbine rotor model were obtained. The new results of the modelling for the real cyclic program with two periods of short and long durations were presented. It was shown that consideration of only creep process leads to the overestimation of the time to fracture. It was firstly shown, that due to important role of the high cycle fatigue damage, which during stress relaxation interacts and intensifies the total rate of damage accumulation, the time to the appearance of a macrodefect is reduced significantly, more than three times, compared to the data obtained for the case of static loading. The obtained data show that due to the analysis of the macrodefect development the possibility of the analysis of times of its safe operating arising. It is established that the dangerous size of macrodefect is obtained in twice a little time as it takes for hidden fracture to be completed when only the static load is taken into account. The time values like obtained in considered example value of approximately 4500 h can be used by engineers to evaluate the real dangerous state of the structural element and detect the time when dimensions of the defect (crack) cannot be considered in the analysis of general long term strength as acceptable.

Acknowledgements. Dmytro Breslavsky acknowledges the support by the Volkswagen Foundation within the program "Visiting research program for refugee Ukrainian scientists" (Az. 9C184).

References

1. Lemaitre, J., Chaboche, J.L.: Mechanics of solid materials. Cambridge University Press, New York (1994)
2. Breeze, P.A.: Power generation technologies. Newnes, Oxford (2014)
3. Eisenträger, J., Naumenko, K., Kostenko, Y., Altenbach, H.: Analysis of a power plant rotor made of tempered martensitic steel based on a composite model of inelastic deformation. In: Naumenko, K., Krüger, M. (eds) Advances in Mechanics of High-Temperature Materials. Advanced Structured Materials, vol. 117, pp. 1–34. Springer, Cham (2020). https://doi.org/10.1007/978-3-030-23869-8_1
4. Naumenko, K., Kutschke, A., Kostenko, Ye., Rudolf, T.: Multi-axial thermo-mechanical analysis of power plant components from 9–12% Cr steels at high temperature. Eng. Fract. Mech. **78**(8), 1657–1668 (2011). https://doi.org/10.1016/j.engfracmech.2010.12.002

5. Liu, Y., Wang, W.: Evolution of principal stress of a turbine rotor under cyclic thermo-mechanical loading. Eng. Fail. Anal. **109**, 104242 (2020). ISSN 1350-6307. https://doi.org/10.1016/j.engfailanal.2019.104242
6. Wu, S., Kang, M., Cai, Z., Wang, W.: Multi-axial creep-fatigue analysis of a cracked groove structure subjected to cyclic thermal loading. Eng. Fail. Anal. **131**, 105826 (2022). https://doi.org/10.1016/j.engfailanal.2021.105826
7. Zhu, X., Chen, H., Xuan, F., Chen, X.: On the creep fatigue and creep rupture behaviours of 9–12% Cr steam turbine rotor. Eur. J. Mech. A. Solids **76**, 263–278 (2019). https://doi.org/10.1016/j.euromechsol.2019.04.017
8. Rauch, M., Roos, E.: Life assessment of multiaxially cyclic loaded turbine components. Fatigue Fract. Eng. Mater. Struct. **31**(6), 441–451 (2008)
9. Benaarbia, A., Rae, Y., Sun, W.: Unified viscoplasticity modelling and its application to fatigue-creep behaviour of gas turbine rotor. Int. J. Mech. Sci. **136**, 36–49 (2018)
10. Anishchenko, G.O., Breslavskii, D.V., Morachkovskii, O.K.: Creep and long-term strength of herringbone lock joints of gas-turbine engines under the combined action of static and cyclic loads. Strength Mater. **30**(1), 25–30 (1998)
11. Skamniotis, C., Cocks, A.C.F.: Analytical shakedown, ratchetting and creep solutions for idealized twin-wall blade components subjected to cyclic thermal and centrifugal loading. Eur. J. Mech. A/Solids **95**, 104652 (2022). https://doi.org/10.1016/j.euromechsol.2022.104652
12. Breslavskii, D.V., Morachkovskii, O.K.: Nonlinear creep and the collapse of flat bodies subjected to high-frequency cyclic loads. Int. Appl. Mech. **34**(3), 287–292 (1998)
13. Breslavsky, D.V., Morachkovsky, O., Tatarinova, O.: Creep and damage in shells of revolution under cyclic loading and heating. Int. J. Nonlinear Mech. **66**, 87–95 (2014)
14. Breslavsky, D., Senko, A., Tatarinova, O.: Creep damage and fracture of notched specimens under static and fast periodic loading. Int. J. Damage Mech. **30**(6), 964–983 (2021)
15. Taira, S., Koterazawa, R.: Dynamic Creep and Fatigue of 13 Chromium Steel at Elevated Temperature. Bull. JSME **4**(15), 460–465 (1961)
16. Breslavsky, D.V., Senko, A.V., Tatarinova, O.A.: Computer program "Solution of two-dimensional problems of creep, damage and fracture" ("FEM Creep-Damage-Fracture"). Certificate of author right on the production, #82758, Ukraine. Published 12 Nov 2018. (In Ukrainian)
17. Zhang, W., Zhang, T., Wang, X., Ji, Y., Gong, J.: The correlation between prior fatigue damage and remaining creep properties in a high chromium steel welded joint. Mater. Sci. Eng., A **790**, 139717 (2020). https://doi.org/10.1016/j.msea.2020.139717

Structure Formation Features of White Chrome Cast Irons Inside Working Layer of Massive Mill Rolls

Oksana Klochko[1(✉)], Volodymyr Volchuk[2], Mariia Bilinska[3], Evgeniya Deryabkina[1], and Oleksandr Voronov[1]

[1] State Biotechnological University, 44, Alchevskih Street, Kharkiv 61002, Ukraine
klochko.hntysh@gmail.com

[2] Prydniprovska State Academy of Civil Engineering and Architecture, 24a, Chernyshevsky Street, Dnipro 49005, Ukraine

[3] Jagiellonian University, 171, Orla Street, 30-244, Kraków, Poland

Abstract. Our present study is aimed at increasing the operational durability of massive two-layer mill rolls with a working layer of white chrome cast iron (15.9–16.4%Cr) by decreasing the retained austenite fraction. The latter is achieved by additional alloying with molybdenum and tungsten, as well as by applying cyclic annealing with double heating at subcritical temperatures (450 or 550 °C, depending on the chemical composition of studied cast iron). The scientific novelty includes establishing the differing influence of molybdenum or tungsten alloying on the properties of the studied alloys. Via the fractal analysis of the microstructure it is discovered that the microhardness level is sensitive to the fractal dimension of the structural components. The difference in microhardness value is especially noticeable in a metal matrix: its levels are 1.5 times higher if the matrix is only alloyed with Mo. Such difference is associated with an increase in the fraction of ferritic component due to precipitation hardening which happens in the as-cast condition in alloys containing W. The practical value of our study lies in decrease of the residual austenite fraction, and, accordingly, in the reduction of tendency to cracking during the operation of products which is possible due to both dispersion hardening in the process of additional alloying with carbide-forming elements and cyclic annealing. The obtained results confirm possibility of using the fractal dimension to assess the microhardness of precipitation hardening products in cases when measuring via standard methods is difficult due to the small size of the objects of study.

Keywords: White chromium cast irons · precipitation hardening · retained austenite · microstructure · fractal analysis

1 Introduction

At present, two-layer centrifugally cast rolls with working layers of white chromium cast iron [1] and cores made of less brittle graphitized cast irons are widely used as hot-rolled sheet rolls in finishing stands [2]. White chromium cast irons have high wear resistance

H. Altenbach et al. (Eds.): CAMPE 2023, pp. 186–196, LNME 2025.
https://doi.org/10.1007/978-3-031-82979-6_19

and corrosion resistance. However, when it comes to the production of massive (6–30 tons) two-layer rolls for hot-rolled sheet mills, one of the significant disadvantages of such materials is the excessive proportion of retained austenite. If the rolls are being operated [3] under high unit pressures and thermal cyclic loads [4], heated to critical temperatures with subsequent immediate cooling down to 50 °C, the decay of residual austenite is accompanied by an increase of local stresses and structural instability. These effects contribute to a significant decrease in the performance of the working layer material and operational resistance of the entire product. Basing on our own research [5] and the literature ([6–8]), the purpose of our study is to find the additional ways of improving the performance of high-alloy chromium cast irons by reducing the fraction of retained austenite through the use of alloying additives and heat treatment. We pay special attention to the ways molybdenum, tungsten and niobium influence the structure formation, as well as to the relationship between the mechanical properties and the fractal structure formed during the manufacture of products.

2 Literature Review

The given problem can be solved by reducing the fraction of residual austenite at the stage of manufacturing such products by additionally introducing alloying additives [6, 9, 10] into the working layer material [11] and by identifying heat treatment parameters [12] which cause precipitation hardening [13]. Such measures ensure both the decay of residual austenite and reduce of the stresses [7]. Additional alloying with carbide-forming elements would help to reduce the carbon concentration in residual austenite [8]. On one hand, it would also reduce austenite stability and on the other, stabilize the properties of the material under operation. From a geometric point of view it is difficult to mathematically assess the structure of such materials due to their heterogeneity [14]. Approximation of its elements by integer Euclidean figures (scoring included) can result in introducing an error into the worked out structure-property prediction models. Implementing the fractal approach requires some additional information about the evolution of the metal structure through the most differentiated assessment of the elements, due to their fractal (non-integer) dimension. The prospects for its application are wide due to the possibility of comparing the spectrum of fractal dimensions of the structural elements [15] with the spectrum of properties of the studied metal [16]. The use of fractal formalism made it possible to most fully study the interaction, decay and stability characteristics of the mesoscopic structures [17, 18]. With such an approach, the choice of the most sensitive quality indicators and indicators of fractal dimensions makes it possible to establish previously unexplored relationships between structure and properties. We thus believe the discussed approach will help improve the performance of white chromium cast irons.

3 Research Methodology

In our present paper we study white chromium cast irons of the working layer of two-layer centrifugally cast rolls (850 × 2000 mm) of hot sheet rolling in finishing stands. The decomposition of retained austenite under the influence of dispersion hardening has been studied by additional alloying with carbide-forming elements and subsequent heat treatment on the samples taken from the roller body (No. 1–3). To ensure precipitation hardening we investigate the impact of various additives: Mo (0.3–1.4%), W (0.7–2.6%), and Nb (~0.14%). The Cr/C ratio in these samples varies from 5.68 to 6.14. The chemical composition of the studied samples is given in Table. 1. Thermal treatment has been carried out in laboratory conditions. The microstructure has been examined using a scanning electron microscope (SEM) JEOL JSM-6390LV equipped with microprobe analyser Link AN10/85S EDS system. The chemical composition of the carbides is determined using energy dispersive X-ray spectroscopy (EDS). We have used picric acid reagent for chemical etching. X-ray diffraction (XRD) analysis is carried out using a DRON-3M X-ray diffractometer with a Cu Kα lamp.

Table 1. Chemical composition of the studied cast irons

Alloy No.	Element content,%										
	C	Si	Mn	Cr	Ni	Mo	V	Cu	W	Nb	Mg
1	2.67	0.99	1.1	16.4	1.3	1.33	0.44	0.14	-	-	0.04
2	2.85	1.1	1.12	16.2	1.41	0.91	0.24	0.09	0.86	0.14	0.04
3	2.62	1.02	0.95	15.9	1.19	0.21	0.22	0.23	2.6	-	0.03

The local inhomogeneity of the phase components is evaluated by measuring the microhardness level. We carry out the fractal analysis using images from Optika IM metallographic microscope. The hardness of the microstructural constituents has been measured using via the Vickers microhardness tester 402MVD Instron Wolpert Wilson Instruments with a load of 0.49 N (HV50). The hardness of the alloys has been evaluated via Shore hardness test with hardness unit HSD. The fractal dimension is estimated according to the method from [19], which is based on the search for its convergence by cell [20] and point [21] methods. The latter makes it possible to refine its values. The cell dimension D is calculated via the Hausdorff Eq. (1) [22]. The fractal dimension D is calculated from the average value of the cells covering the object (2):

$$D = -\lim_{\delta \to 0} \frac{\ln N(l)}{\ln l}, \quad (1)$$

$$N(L) = \sum_{m=1}^{K} (1/m) P(m, L) \sim L^{-D} \quad (2)$$

where N is the number of square cells of size l, which cover the object of study (a phase or another structural element). The point dimension is based on calculating the sum of probabilities P(m, L), where m is the number of points of the object under study in a square cell of size L.

4 Results and Discussion

We have identified both carbide phases and metal matrix domains (light and dark) of different degree of etchability in the considered as-cast alloys under study (Table 2).

Table 2. Results of EDS analysis

Analysed structure	N*	Element content%								HV50**
		Fe	Cr	Ni/Nb	V	Mo	Si	W	Mn/Ti	
As-cast condition										
Carbides Me_7C_3	1	47.57	47.83	0.06	2.16	2.27	–			–
	2	51.3	44.24	–	0.78	0.85	0.112	1.02	1.45	–
	3	47.5–50	45–49	0.2–0.55	0.77–0.79	0.17–0.3	0.02	1.92–3.81		–
Carbides MeC	1	32.4	12.88	0.40	0.79	53.3	0.34			
	2–3	34.23–47.0	7.75–8.7	1.09	0.37–0.45	26.06	–	30.77–38.0	0.75	–
Light matrix	1	78.11	15.99	1.08	0.54	3.01	1.30			623
	2	89.8	4.3–5.6	1.65	0.03	0.56–0.9	1.09–1.22	0.6–1.2	0.73–1.17	423
	3	87.8–89.4	5.76	2.09	0.07	0.36	1.28	1.04		404
Dark matrix	1	78.4	15.7	1.32	0.59	2.93	1.06			336
	2	80.9	12.8	1.56	0.03	0.60	0.71	1.66	1.74	304
	3	85.6	9.99	1.74	0.17	0.27	1.08	–		329

(continued)

Table 2. *(continued)*

Analysed structure	**N***	**Element content%**								**HV50****
		Fe	Cr	Ni/Nb	V	Mo	Si	W	Mn/Ti	
After heat treatment (cyclic annealing with two heatings: alloys 1 and 2: up to 550°C; alloy 3: up to 450°C)										
Carbides Me_7C_3	1	51.39	46.07	–	0.92	1.15	0.48			1003
	2	48.95	44.53	/0.09	2.35	2.51	0.062	0.71	0.81	–
	3	48.90	47.00	0.32	1.05	0.09	–	2.65		–
Carbides MoC	2	14.27	10.68	0.11	1.04	70.0	0.26	–	0.49	–
Carbides NbC		9.92	9.07	/68.8	2.38	8.34	0.07	0.02	0.93/ 0.54	–
Carbides MeC	1–3	37.26–42.01	8.71–10.03	0.78–1.09	0.26–0.37	52.52–7.89	0.47–0.64	0.00–38.05		
Light matrix	1	88.28	10.06	–	0.35	0.20	1.10			373
	2	74.84–77.4	17.0–20.0	1.29–1.87	0.14–0.41	1.81–0.70	0.95–1.1	0.41–0.33	1.04–0.75	331
	3	85.60	10.16	1.58	0.03	0.32	1.31	0.99		469
Dark matrix	1	91.78	5.61	–	–	1.05	1.56			291
	2	84.93	10.43	1.23	0.30	1.21	0.87	0.11	0.83	283
	3	86.65	9.04	1.89	0.15	–	1.26	1.00		341
Matrix on the border with carbides	1	92.03	5.17	–	0.13	1.16	1.49			373
	2–3	87.9–89.4	6.5–7.38	1.38–1.61	0.11	0.76	0.82–1.01	0.47–0.23	0.63–0.99	–

Notes: 1. N* is alloy number; 2. ** measuring the microhardness in fine carbides occurred to be impossible

Precipitation hardening (i.e., formation of secondary carbides from austenite) occurs in the process of long-term cooling during casting crystallization, which takes up to three days [23]. In alloy No. 1 with Mo, it is found in the form of separate oriented inclusions 0.1–5 μm (the size depends on the cross section of the crystal, Fig. 1a). The secondary carbides are way smaller (up to 0.2 μm) and their total amount is greater in alloys No. 2 and 3 alloyed with W and Nb, which is especially noticeable in alloy No. 3 (Fig. 1b,c).

It has been determined that alloy No. 1 with molybdenum (hardness: 77 HSD) consists residual austenite (up to 80%), dark matrix (including a ferrite-carbide mixture) and bainite, were revealed [24]. In alloys No. 2 and 3 with tungsten and niobium (74 HSD and 66 HSD, respectively). The values of hardness are lower in case of alloy No. 3 (2.6% W) due to a larger amount of ferrite-carbide mixture. There is a decrease in the content of residual austenite (down to 60%) in the considered structure while the size of the dark matrix increases (see Fig. 1 b,c).

According to SEM-EDS analysis, the components content in such domains differs significantly (see Table 2), especially in the amount of chromium and iron.

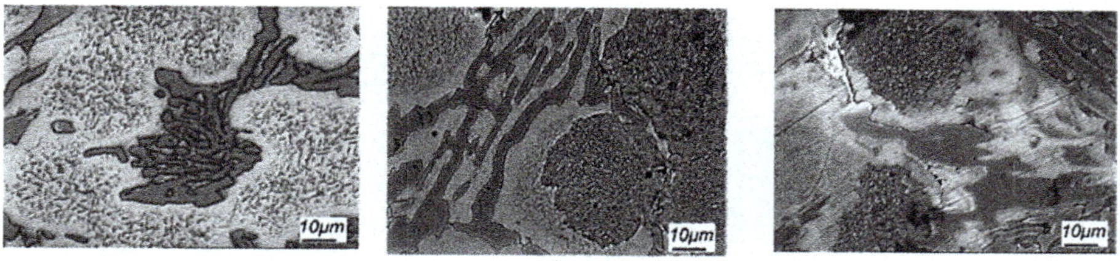

a b c

Fig. 1. SEM image of the studied microstructures of as-cast cast irons, × 1000 in alloys: (a) No. 1; (b) No. 2, (c) No. 3.

The light matrix of alloy No. 1 contains three times more chromium (up to 16.0%) than alloys No. 2 and No. 3 with tungsten do (up to 5.76%). We do not observe such a sharp difference in chromium concentration in the dark matrix. It has been found that as the tungsten concentration increases from 0.86% to 2.6% in the alloy, its content in the dark and light matrices slightly decreases (down to 1.0%). This effect is associated with an increase in the proportion of tungsten containing inclusions. In case of the cast iron exclusively alloyed with molybdenum, its metal matrix is more homogeneous by the content of its components.

The studies have been previously conducted (see [11]) dedicated to examining the effects of tungsten alloying on increasing the wear resistance of high-chromium cast iron by increasing the ductility and indentation modulus of $(Fe, Cr)_7C_3$ carbides. In [12], the MatCalc modeling of the influence of the distribution of the eutectic carbide phase and elements on the level of properties of white chromium cast irons in the as-cast condition has been carried out. In [8], the authors investigate the influence of molybdenum additive on the level of macrohardness due to dispersion hardening. These studies, however, do not pay ample attention to the distribution of the carbide-forming elements in the context of reducing the proportion of retained austenite, as well as to predicting the influence of the resulting local phase heterogeneity on the level of the working layer application performance during the production and operation of mill rolls.

Chromium carbides of the Me_7C_3 type are found in all the studied alloys. On average, they contain 47–51% of Fe and 44–49% of Cr. Alloy No. 1 contains up to 2.2% vanadium and molybdenum uniformly distributed between the carbide phase and the matrix. There is around 0.78% V and 0.25–0.85% Mo in the Me_7C_3 carbide phase. During our study, the MeC type carbides have been discovered (see Table 2): (Fe,Mo,Cr)C in alloy No. 1 and (Me,W)C in samples containing tungsten. In the alloy No. 1, such phases contain 53.25% Mo, 32.35% Fe and 12.88% Cr. The carbides of this type contain 7.75 – 8.7% Cr; 1.09% Ni; 0.37 – 0.45% B; up to 26.06% Mo; 30 – 38% W; 34–47% Fe in alloys of alloys No. 2 and 3. A noticeable decrease (~4–5%) in the Cr concentration was established compared as to the phase of alloy No. 1.

In [15–18] the authors discuss the application of fractal formalism in the study of the metal alloys structure evolution. According to [18], this method makes it possible to successfully rank the criteria for chromium-nickel static grooved cast iron rolls (2.96% C, 1.5% Cr, 0.5% Mo, 4.5% Ni) with a ratio error of 5 – 8%.

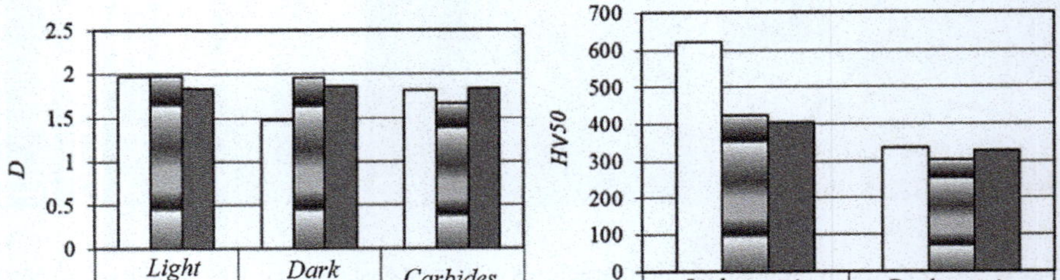

	Light matrix	Dark matrix	Carbides
Alloy1	1.972	1.482	1.811
Alloy2	1.968	1.965	1.658
Alloy3	1.830	1.859	1.836

a

	Light matrix	Dark matrix
Alloy1	623	336
Alloy2	423	304
Alloy3	404	329

b

Fig. 2. Changes in the fractal dimensions of the elements of the microstructure of the studied cast irons (panel a) and their microhardness HV50 (panel b), where the row number corresponds to the number of the studied alloy, according to Table 1

It seems appropriate to use this approach to evaluate the effect of dispersion hardening on the microstructure properties of the heterogeneous alloys under study. The calculations of the fractal dimension of the structural elements of the three studied as-cast alloys (see Table 1) have been carried out according to the proposed method [19]. A change in the fractal dimension of the light and dark domains, as well as that of the Me_7C_3 and MeC carbide types has been identified in the images. The existing relationship between the HV50 microhardness indices of the light and dark matrices (Fig. 2b) and their fractal dimensions (Fig. 2a) indicate that the microhardness properties are sensitive to changes in the configuration of the shape of the structure elements, expressed through fractals (see Fig. 2b). A decrease in the HV50 microhardness indices of the light matrix from 623 to 404 has been recorded together with a decrease in its fractal dimension from

1.972 to 1.830. A decrease in the fractal dimension of the dark matrix from 1.965 to 1.482 corresponds to an increase in its microhardness from 336 to 304.

In order to reduce the amount of residual austenite and obtain a more equilibrium structure, the studied alloys have been subjected to cyclic annealing with double heating: for 8 h up to 550 °C (alloys No. 1 and 2); for 3 h up to 450 °C (alloy No. 3). The carried out heat treatment has enhanced the effect of precipitation hardening (Fig. 3).

Analysis of the alloys microstructure in the heat-treated state has revealed existence of Me_7C_3 and MeC carbides, and dark and light matrices. With approximately the same nickel content in the alloys, its amount is large in the carbide phase of alloy No. 3 with 2.6% W. It has also occurred that after heat treatment Me_7C_3 carbide contains niobium (up to 0.086%) in alloy No. 2 containing 0.86% W and 0.14% Nb. We have also registered niobium carbide with around 68.85% of Nb. Molybdenum is a part of MoC carbide. Significant changes in the concentration of tungsten in the carbide phases have not been revealed (see Table 2). Vanadium mostly dissolves in the carbide phase.

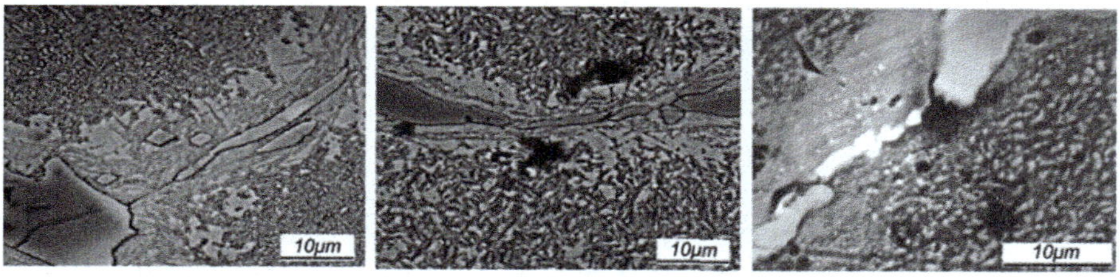

Fig. 3. SEM image of the microstructures of the studied cast irons after two annealing cycles: (a) alloy No. 1 and (b) alloy No. 2, × 2000, at 550 °C; (c) alloy No. 3, × 3000, at 450 °C. Tungsten carbide glow is visible.

Molybdenum in alloy No. 1 (without tungsten) is uniformly distributed between Me_7C_3 and the matrix (2.2–3.0%). Its amount depends on the concentration in the alloy. Tungsten is distributed in a different way: in addition to MeC carbides, it is mainly concentrated in carbides of the Me_7C_3 type. Its share in the latter reaches 3.81%, and in the matrix its amount varies from 0 to 1.04%.

The matrix mainly consists of a ferrite-carbide mixture. XRD analysis established a sharp decrease of the fraction of residual austenite with the introduction of tungsten (around 11% in alloy No. 3). The fractal analysis of the studied cast iron structures has been carried out. The results obtained (Fig. 4) indicate changes in the fractal dimensions of light and dark metal matrices after heat treatment of the three studied alloys.

We record a decrease in the fractal dimension of the light matrix from 1.920 to 1.897 after heat treatment and an increase in the size of the dark matrix from 1.869 to 1.913 in the case of alloy No. 2 (Fig. 4a). For alloy No. 1, the fractal dimension of the light matrix after heat treatment increased from 1.847 to 1.867, while that of the dark matrix decreased from 1.887 to 1.848. The change in the fractal dimension of the structure after heat treatment (Fig. 4a, b) is due to the additional decomposition of residual austenite (light matrix) with the formation of a ferrite-carbide mixture and the appearance of secondary dispersed carbides precipitated along the boundaries of decomposed grains.

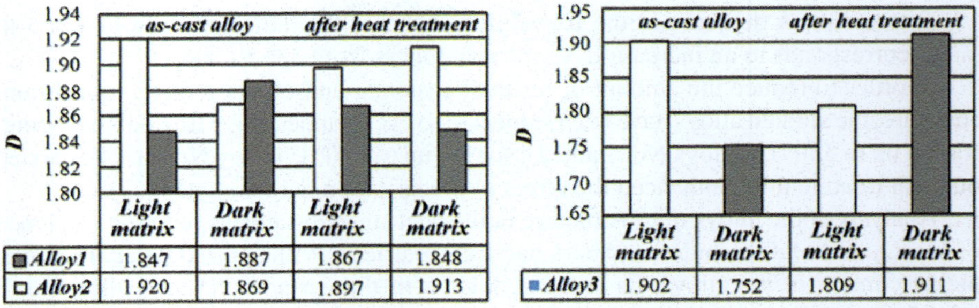

a b

Fig. 4. Changes in the fractal dimensions of the microstructure elements of the studied cast irons after two annealing cycles: (a) at 550 °C; (b) at 450 °C. The row number corresponds to the number of the studied alloy (see Table 1).

Hardness of the studied alloys has changed as a result of heat treatment. Precipitation hardening contributes to an increase in hardness only in alloy No. 3 (2.6% W) from 66 HSD to 73 HSD. In the other cases, hardness decreases only slightly. In alloys No. 1 (without tungsten) and No. 2 (0.86% W), the hardness drops from 77 HSD to 69 HSD and to 62 HSD, respectively.

Alloying of white chrome cast iron with molybdenum or tungsten affects the level cast iron properties in different ways, which is associated with the distribution of these components between phases in cast iron (see Table 2). The difference is especially noticeable in the matrix microhardness level, which is 1.5 times higher with W additions than with only Mo alloying (291 HV50 and 341 HV50, respectively). Such a difference in microhardness occurs due to an increase in the proportion of the ferrite component in the as-cast cast irons without tungsten. The difference becomes less pronounced after heat treatment since there occurs an additional decomposition of residual austenite. According to [6] and [25], Ni and Si promote the graphitization of cast iron. In the cast alloy, Ni is contained in austenite; it can presumably form superfine inclusions of the intermetallic type after the heat treatment. However, due to their dispersion, it has proven to be impossible to identify them and assess the content of components.

5 Conclusion

Carbide-forming elements that are present in cast iron in the considered amounts contribute to the effect of precipitation hardening and the formation of a ferrite-carbide mixture. Apart from MeC, tungsten is mainly concentrated in carbides of Me_7C_3 type, its amount reaching 3.81%. Mo is part of MoC carbide (up to 70%). The differences in the influence of Mo and W on the properties of the alloys under study have been revealed, which presents the scientific novelty of our research.

Alloying with these elements reduces the proportion of retained austenite from 80% to 60% at 2.6% W at the as-cast condition. The obtained result reduces the tendency to cracking of a mill roll surface layer during operation which is of undoubted practical relevance. Molybdenum can be replaced with tungsten and niobium. In this case, these elements should be introduced using foundry alloys, which would reduce the cost of mill rolls manufacture.

Since Ni and Si contribute to the occurrence of spotted graphite in white chromium cast iron, the further studies of the influence of these elements on the wear resistance in such cast iron inside the working layer of two-layer mill rolls are undoubtedly of great interest.

It is recommended to subject the two-layer mill rolls manufactured from white chromium cast irons to heat treatment in order to improve the performance of these alloys, according to the modes that provide the greatest decay of residual austenite. Heat treatment under 450–550 °C enhances the effect of precipitation hardening.

The results of the fractal analysis of the microstructure before and after heat treatment indicate that the levels of microhardness are sensitive to the fractal dimension of the light and dark matrices and the carbides identified in the photographs. The obtained results confirm the possibility of using the fractal dimension as an additional method for recording changes in the structure inhomogeneity and microhardness of cast iron due to the precipitation hardening.

References

1. Kiss, I.: Cast iron rolls – an overview on the proper hardness assured by the manufacturing process. Technical Journal **13**(2), 92–99 (2019). https://doi.org/10.31803/tg-20180516131304
2. Cofiño-Villar, A., Alvarez-Antolin, F., Alvarez-Perez, C.H.: Improvement in the Resistance to wear of work-rolls used in finishing stands of the hot strip mills. Metals **11**(11), 1873 (2021). https://doi.org/10.3390/met11111873
3. Palit, P., Patel, S., et al.: Analysis of a progressive failure of a work roll in hot strip mill. J. Fail. Anal. and Preven. **19**, 1297–1303 (2019). https://doi.org/10.1007/s11668-019-00688-w
4. Trishevskij, O., Kaliuzhnyi, O., et al.: Establishing patterns in the temperature distribution within a deformation zone during thin strip rolling. East.-Eur. J. Enterp. Technol. **2**(5–104), 21–28 (2020). https://doi.org/10.15587/1729-4061.2020.198296
5. Skoblo, T.S., Klochko, O.Yu., et al.: Features of the carbide phase degradation under heating and deformation. Lett. Mater. **11**(1), 22–27 (2021). https://doi.org/10.22226/2410-3535-2021-1-22-27
6. Avtukhov, A.K., Bantkovsky, V.A., et al.: Influence of technological methods of production on the level of residual austenite in the working layer of products from chromium-nickel cast iron. Naukovi notatky **73**, 80–85 (2022). https://doi.org/10.36910/775.24153966.2022.73.12
7. Ngqase, M., Pan, X.: Microstructural Investigation on Heat Treatment of Hypoeutectic High Chromium White Cast Irons. J. Phys. Conf. Ser. **1495**, 012024 (2020). https://doi.org/10.1088/1742-6596/1495/1/012024
8. Guerra, F.V., Bedolla-Jacuinde, A., Zuno-Silva, J., et al.: Effect of the simultaneous Ti and W addition on the microstructure and wear behavior of a high chromium white cast iron. Met. Res. Technol. **116**, 602 (2019). https://doi.org/10.1051/metal/2019031
9. Chen, C., Wang, J., Ge, Y., Ma, L.: Investigations on microstructures and properties of (Fe, Cr, $_W)_7C3$ carbides by first principles and experiments. Coatings **12**(9), 1363 (2022). https://doi.org/10.3390/coatings12091363
10. Nayak, U.P., Guitar, M.A., Mücklich, F.A.: A comparative study on the influence of chromium on the phase fraction and elemental distribution in as-cast high chromium cast irons: simulation vs. experimentation. Metals **10**, 30 (2020). https://doi.org/10.3390/met10010030

11. Kutsova, V.Z., Kovzel, M.A., Shvets, P.U., et al.: Structure, phase composition of supercooled austenite, and kinetics of its decomposition in perlite temperature range of chromium–manganese cast iron. Metallofiz. Noveishie Tekhnol. **40**(4), 551–560 (2018). https://doi.org/10.15407/mfint.40.04.0551
12. Wiengmoon, A., Pearce, J. T. H., Nusen, S., et al.: Electron microscopy study of carbides precipitated during destabilization and tempering heat treatments of 25 wt.%Cr-0.7 wt.%Mo high chromium cast irons. Micron **143**, 103025 (2021). https://doi.org/10.1016/j.micron.2021.103025
13. Guitar, M.A., Suárez, S., Prat, O., et al.: High chromium cast irons: destabilized-subcritical secondary carbide precipitation and its effect on hardness and wear properties. J. of Materi Eng and Perform. **27**, 3877–3885 (2018). https://doi.org/10.1007/s11665-018-3347-1
14. Mandelbrot, Benoit B.: The fractal geometry of nature. Echo Point Books & Media, LLC (2021). ISBN 978-1648370403
15. Fernández, R., González-Doncel, G., Garcés, G.: Fractal analysis of strain-induced microstructures in metals. Fractal Analysis - Selected Examples. IntechOpen (2020). https://doi.org/10.5772/intechopen.91456
16. Romaniuk, S.P., Volchuk, V.M., Taran, A.V., et al.: Characterization of ARC-PVD ZrN nanostructured coatings by using the fractals theory. Problems of Atomic Science and Technology **142**(6), Series: Plasma Physics, 123–128 (2022). https://doi.org/10.46813/2022-142-123
17. Suarez-Dominguez, E.J., Perez-Rivao, A., Sanchez-Medrano, M.T., et al.: Mesoscopic model for the surface fractal dimension estimation of solid-solid and gas-solid dispersed systems. Surfaces and Interfaces **18**, 100407 (2020). https://doi.org/10.1016/j.surfin.2019.100407
18. Bolshakov, V.I., et al.: Regularization of one conditionally ill-posed problem of extractive metallurgy. Metallofiz. Noveishie Tekhnol. **40**(9), 1165–1171 (2018). https://doi.org/10.15407/mfint.40.09.1165
19. Vakhrusheva, V.S., Volchuk, V.M., et al.: Fractal model of estimating quality of cold worked fuel cladding tubes // Problems of Atomic Science and Technology. Ser.: Phys. Radiat. Eff. Radiat. Mater. Sci. **135**(5), 57–63 (2021). https://doi.org/10.46813/2021-135-057
20. Zhang, H. Zhu, J. Hong, R., et al.: Convergence-improved congruent matching cells (CMC) method for firing pin impression comparison J. Forensic Sci., **66**(2), **571**(2021). https://doi.org/10.1111/1556-4029.14634
21. Okuno, T.: Local convergence of primal-dual interior point methods for nonlinear semi-definite optimization using the family of Monteiro-Tsuchiya directions. *arXiv:* Optimization and Control (2020). https://doi.org/10.48550/arXiv.2009.03020
22. F. Hausdorff: Dimension und äusseres Mass. Math. Ann. **79**, 157–179 (1919). ISSN: 0025-5831; 1432–1807/e
23. Wiengmoon, A., Pearce, J.T.H., Nusen, S., et al.: Electron microscopy study of carbides precipitated during destabilization and tempering heat treatments of 25 wt.%Cr-0.7 wt.%Mo high chromium cast irons. Micron **143**, 103025 (2021). https://doi.org/10.1016/j.micron.2021.103025
24. Ngqase, M., Pan, X.: An overview on types of white cast irons and high chromium white cast irons. J. Phys. Conf. Ser. **1495**, 012023 (2020). https://doi.org/10.1088/1742-6596/1495/1/012023
25. Skoblo, T.S., Klochko, O., Belkin, E.L., et al.: Structure formation of high-chromium cast irons in the temperature range of the magnetic transformation of carbide phases. Lett. Mater. **10**(2), 129–134 (2020). https://doi.org/10.22226/2410-3535-2020-2-129-134

Buckling of Cylindrical Sandwich Panels with Imperfections in Honeycomb Cores Manufactured by Fused Deposition Modeling

Maryna Chernobryvko[1], Konstantin Avramov[1,2,3](✉), and Christophe Pierre[3]

[1] Anatolii Pidhornyi Institute of Power Machines and Systems, National Academy of Science of Ukraine, St. Komunalnykiv 2/10, Kharkiv 61046, Ukraine
kvavramov@gmail.com

[2] Department of Aircraft Strength, National Aerospace University N.Ye. Zhukovsky "KhAI", St. Chkalov 17, Kharkiv 61070, Ukraine

[3] Department of Mechanical Engineering, Stevens Institute of Technology, Castle Point Terrace, Hoboken, NJ 07030, USA

Abstract. The sandwich panels are elements of launch vehicles and autonomous aerospace vehicles. The buckling of cylindrical sandwich panels with imperfections is considered. The core layer of such panel is a honeycomb structure made from polylactide (PLA) material and manufactured by fused deposition modeling (FDM) additive technology. The inner and outer panel layers are made of carbon fiber-reinforced materials. The purpose of this work is determination the effect of cell breakings on the buckling cylindrical sandwich panels with imperfections under longitudinal compression and radial pressures. This will make possible to use some sandwich panels with imperfections in the production. The scientific novelty of this study is the following. The influence of honeycomb imperfections in the form of cell breakings on the critical buckling modes is studied. Eight cases of imperfection patterns, in terms of numbers and locations of the cell breakings, are considered. Numerical computations show that the sandwich panels under longitudinal compression are moderately sensitive to imperfections, with decreases in critical load ranging from 5.3% for a single cell breaking in the center of the honeycomb to 8.6% for multiple cell breakings located along different directions. Also, the corresponding buckling mode is global. In contrast, the buckling behavior of sandwich panels subject to radial pressure is highly sensitive to imperfections, such that the presence of a single cell breaking results in a critical load more than three times smaller than that of the perfect panel. The results on cell breakings influence on the shell buckling can be used for thin-walled structure design.

Keywords: Cell Breaking · Additive Technologies · Multilayer Shell

1 Introduction

The use of thin-walled sandwich structures with honeycombs manufactured by addi-tive technologies has several advantages over structures made from more traditional materials: large strength and rigidity together with small weight. It is being of great importance

H. Altenbach et al. (Eds.): CAMPE 2023, pp. 197–206, LNME 2025.
https://doi.org/10.1007/978-3-031-82979-6_20

for aerospace applications [1, 2]. Also additive manufacturing simplifies significantly the production of honeycombs. Thin-walled structures featuring cylindrical shells and panels are commonly used in launch vehicles and autonomous vehicles, and buckling is one of the primary ways in which these components lose their strength [3, 4]. Therefore, in the design phase, thin-walled structural components must be analyzed to prevent buckling. In addition, imperfections in the form of the cell breakings can originate during 3D printing of honeycombs, which reduce the strength and rigidity of the honeycomb and the sandwich structure [5]. The purpose of this work is to determination the effect of the cell breakings on the buckling cylindrical sandwich panels under longitudinal compression and radial pressures. This makes possible to use sandwich panels with imperfections for the production. The scientific novelty of this study is the following. The influence of honeycomb imperfections in the form of cell breakings on the critical buckling modes is studied.

2 Literature Review

Many past research efforts focused on the analysis of thin-walled sandwich structures, due to their advantages over single-layer orthotropic structures. A few representative recent studies are as follows. The dynamic of a cylindrical sandwich panel with auxetic honeycomb under the action of blast loads was analyzed in [1]. The buckling of a composite lattice panel obtained by additive technology was analyzed experimentally in [2]. Carlsson et al. [3] obtained the elastic-plastic response of a honeycomb under uniaxial compression and uniaxial tension in a state of finite strain. Tankasala et al. [4] calculated the exact elastic-plastic response of honeycombs with small strain using beam theory. Shear tests were used to study the response of a honeycomb core in the case of large deformations, and the experimental results were validated by a numerical analysis based on a hyperelastic orthotropic model. Experimental studies of specimens made with additive manufacturing, along with the numerical analysis of honeycomb panels with geometrical imperfections, were reported in [5]. Finally, formulas for the calculation of honeycomb mechanical properties were presented in [6]. In addition, several recent research papers were devoted to buckling of composite shells, as follows. Franzoni et al. analyzed cylindrical composite shells in [8], using validated finite element models to predict the buckling loads. They presented a new approach for calculating the axial buckling load of orthotropic skin-dominated cylindrical shells in [8], and obtained the relationship between the compressive force and natural frequencies for imperfect cylindrical shells in [9]. Salloomi et al. [10] studied the influence of geometrical parameters of a cylindrical shell with elliptical cut-outs on the buckling loads. As follows from [11], the materials of components made using additive manufacturing technology are anisotropic. Finally, an experimental approach for the determination of the mechanical characteristics of components printed by Fused Deposition Modeling (FDM) technology was suggested in [12]. It follows from this brief literature review that the buckling of cylindrical sandwich panels with cell breakings in honeycombs manufactured by additive technologies has not yet been treated. In this paper, the buckling of sandwich panels with imperfect honeycombs manufactured by FDM additive technology is analyzed. First, the case of global buckling under longitudinal compression is considered. Second, the local buckling of honeycomb cells is studied when a sandwich panel is subject to radial pressure. The effect of different types of honeycomb imperfections on the buckling is analyzed.

3 Research Methodology

The buckling of cylindrical sandwich panels with a honeycomb core is considered, where the honeycomb is manufactured by FDM additive technology. Figure 1 shows an example honeycomb with single cell and multiple cell breakings. Such imperfections are typically observed in the structure after its operation, and the goal of this study is to analyze their influence on the buckling of sandwich panels. Both cases of longitudinal compression of the panel and buckling under radial pressure are considered. The influence of different boundary conditions on the structure's buckling loads is analyzed numerically.

It should be noted that longitudinal compression can cause the global buckling of the whole sandwich structure. In contrast, compression exerted by applying radial pressure results in local buckling of the honeycomb core as a plate-like structure. These two types of buckling are analyzed in this paper.

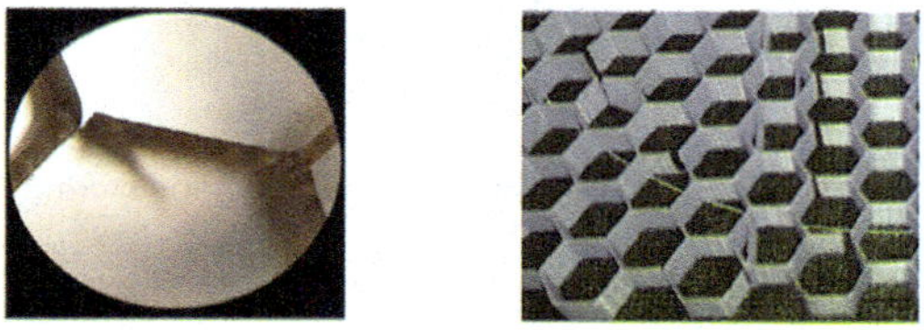

a. *b.*

Fig. 1. Imperfections in honeycomb: *a.* Single cell; *b.* multiple cells

The structure studied is shown in Fig. 2 and its geometry is as follows. The cylindrical sandwich panel has length $L = 184$ mm, arc length $l = 92$ mm, internal panel radius $R = 280$ mm, and thickness $h = 14$ mm. The thicknesses of the inner and outer layers are both equal to 2 mm. The thickness of hexagonal honeycomb core layer is $h_c = 10$ mm. The cell length of a honeycomb is $h_a = 6.3$ mm, and the cell width is $h_b = 0.4$ mm.

The software ANSYS is used to perform finite element computations. The stress state of the structure is modeled in the cylindrical coordinate system (x, φ, z), as shown in Fig. 2. The validity of the finite element model is ensured by refining the finite element mesh until the buckling load solution is reasonably converged. Finite element models for the buckling of sandwich panels with eight different cell breaking locations were developed.

Figure 3 shows the sketches of imperfection locations where cell breakings occur. The first structure in Fig. 3 features a single cell breaking. The second and third structures in Fig. 3 have three and five successive such cell breakings, respectively, located on lines which are parallel to the upper edge of the shell. The fourth structure contains three cell breakings, which are located on a line parallel to the lateral edges of the panel. The fifth structure contains four cell breakings, which are located on an oblique line. The sixth structure contains ten cell breakings located on two parallel horizontal lines, with five breakings on each line. The seventh structure has six cell breakings located on two vertical lines. Finally, the eighth structure contains seven cell breakings, four of which on a horizontal line and three on a vertical line.

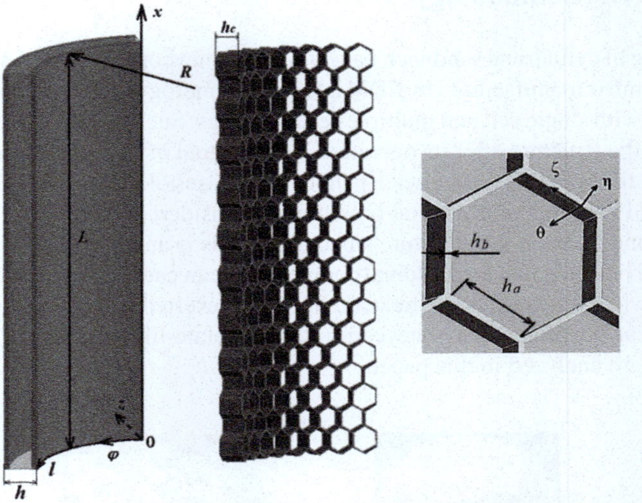

Fig. 2. Geometry of the sandwich cylindrical panel, the honeycomb core and the cell

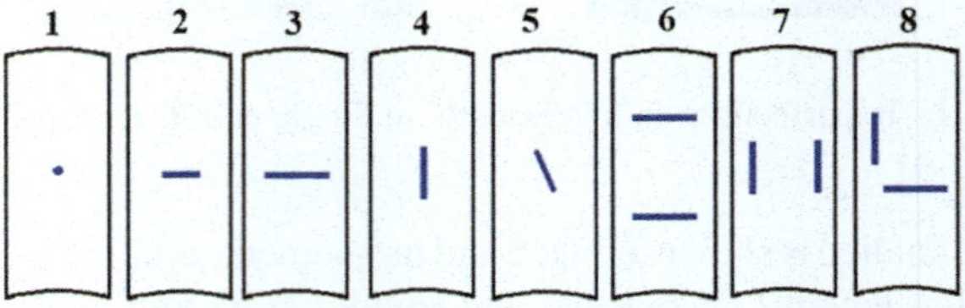

Fig. 3. Sketch of the locations of cell imperfections in the honeycomb core

The honeycomb core is manufactured from polylactide (PLA). The mechanical properties of this material are reported in [12, 13]. The mechanical axes of this orthotropic honeycomb coincide with the directions of 3D printing. As indicated in Fig. 2, the axis ζ is directed along the cell walls, which coincides with the printing direction. The axis η is perpendicular to ζ and is parallel to the tangent to the exterior layers. The axis θ is chosen in such a way that the coordinate system is right-handed.

The mechanical properties of the honeycomb orthotropic material are the following: Young's moduli $E_{\zeta\zeta} = 3.58$ GPa, $E_{\eta\eta} = 3.00$ GPa, $E_{\theta\theta} = 3.81$ GPa; shear moduli $G_{\zeta\eta} = 1.07$ GPa, $G_{\eta\theta} = 1.41$ GPa, $G_{\zeta\theta} = 1.40$ GPa; Poisson ratios $\nu_{\zeta\eta} = 0.298$, $\nu_{\eta\theta} = = 0.224$, $\nu_{\zeta\theta} = 0.207$; and density $\rho = 1240$ kg/m^3. The material satisfies Hooke's law:

$$\begin{bmatrix} \sigma_{\zeta\zeta} \\ \sigma_{\eta\eta} \end{bmatrix} = \begin{bmatrix} \overline{\overline{C}}_{11} & \overline{\overline{C}}_{12} \\ \overline{\overline{C}}_{12} & \overline{\overline{C}}_{22} \end{bmatrix} \begin{bmatrix} \varepsilon_{\zeta\zeta} \\ \varepsilon_{\zeta\zeta} \end{bmatrix};$$

$$\sigma_{\zeta\eta} = 2\overline{\overline{C}}_{66}\varepsilon_{\zeta\eta}, \sigma_{\zeta\theta} = 2\overline{\overline{C}}_{55}\varepsilon_{\zeta\theta}, \sigma_{\eta\theta} = 2\overline{\overline{C}}_{44}\varepsilon_{\eta\theta}, \tag{1}$$

where $\sigma_{\zeta\zeta}, \sigma_{\eta\eta}, \sigma_{\zeta\eta}, \sigma_{\zeta\theta}, \sigma_{\eta\theta}, \varepsilon_{\zeta\zeta}, \varepsilon_{\eta\eta}, \varepsilon_{\zeta\eta}, \varepsilon_{\zeta\theta}, \varepsilon_{\eta\theta}$ are the elements of the stress and strain tensors.

The exterior layers of the cylindrical panels are manufactured from carbon fiber-reinforced composite material, which is a brittle orthotropic material. The mechanical properties of this material in the panel cylindrical coordinate system (x, φ, z) are the following: Young's moduli $E_{xx} = 35$ GPa, $E_{\varphi\varphi} = 35$ GPa, $E_{zz} = 8$ GPa; shear moduli $G_{x\varphi} = 6$ GPa, $G_{\varphi z} = 30$ GPa, $G_{xz} = 30$ GPa; Poisson ratios $\nu_{x\varphi} = 0.01$, $\nu_{\varphi z} = 0.09$, $\nu_{xz} = 0.09$; and density $\rho = 1477$ kg/m^3. The exterior layers satisfy Hooke's law, as:

$$\begin{bmatrix} \sigma_{xx}^{(j)} \\ \sigma_{\varphi\varphi}^{(j)} \end{bmatrix} = \begin{bmatrix} \overline{C}_{11} & \overline{C}_{12} \\ \overline{C}_{12} & \overline{C}_{22} \end{bmatrix} \begin{bmatrix} \varepsilon_{xx}^{(j)} \\ \varepsilon_{\varphi\varphi}^{(j)} \end{bmatrix};$$

$$\sigma_{x\varphi}^{(j)} = 2\overline{C}_{66}\varepsilon_{x\varphi}^{(j)}; \sigma_{xz}^{(j)} = 2\overline{C}_{55}\varepsilon_{xz}^{(j)}; \sigma_{\varphi z}^{(j)} = 2\overline{C}_{44}\varepsilon_{\varphi z}^{(j)}; j = b, t \tag{2}$$

where j is an index denoting the outer and inner layers; and $\sigma_{xx}^{(j)}, \sigma_{\varphi\varphi}^{(j)}, \sigma_{x\varphi}^{(j)}, \sigma_{xz}^{(j)}, \sigma_{\varphi z}^{(j)}, \varepsilon_{xx}^{(j)}, \varepsilon_{\varphi\varphi}^{(j)}, \varepsilon_{x\varphi}^{(j)}, \varepsilon_{xz}^{(j)}, \varepsilon_{\varphi z}^{(j)}$ are the elements of the stress and strain tensors for the bottom and top layers.

The structure's geometric nonlinearity is not accounted for in the model, and therefore the displacement-strain relations are considered to be linear. A treatment of geometrically nonlinear behavior in beam structures can be found in [14–16].

4 Results and Discussion

4.1 Global Buckling Under Longitudinal Compression

The global buckling of sandwich panels with the eight types of honeycomb imperfections shown in Fig. 3 is analyzed. The case of a "perfect" panel without imperfections in the honeycomb is studied too. The compression pressure acts on the simply supported top edge of the structure. The pressure is uniform. It is applied on the sandwich panels and on the honeycomb core. Table 1 lists the critical buckling pressures P_{buc} of the cylindrical sandwich panel for the perfect case and the eight imperfect cases, which are calculated numerically. The case Structure Number 0 is for the sandwich panel without imperfections.

The convergence of the buckling pressure solutions is analyzed by refining the finite element mesh. Table 1 contains the results of such analysis, where each calculation is identified by the parameter "element size" (ES). This parameter is used by ANSYS and is equal to the maximal finite element size. The buckling pressure solutions for different values of ES in Table 1 show that convergence is sufficiently achieved for ES = 0.5.

As follows from the numerical results, a single imperfection in the center of the honeycomb (Fig. 3, case 1) results in a decrease of the critical buckling load of 5.3%. Additional cell breakings along the horizontal direction, as in cases 2 and 3 in Fig. 3, result in an additional drop in the critical buckling load of about 2%, but increasing the number of imperfections from three to five has very little effect on the buckling

Table 1. Critical global buckling pressure of sandwich panels under longitudinal compression

Structure Number	ES	0	1	2	3	4	5	6	7	8
P_{buc}, MPa	2	159.82	151.41	148.24	147.65	151.27	151.12	150.44	150.60	146.34
	1	158.46	150.02	146.87	146.22	149.89	149.73	149.06	149.21	144.89
	0.5	158.44	150.01	146.81	146.21	149.88	149.70	149.04	149.19	144.88

pressure. If multiple imperfections on vertical and diagonal lines are considered, such as cases 4 and 5 in Fig. 3, then the critical buckling pressures are only very slightly decreased compared to the case of a single imperfection, which is remarkable. Similarly, honeycomb imperfections located on two horizontal and two vertical lines, cases 6 and 7 in Fig. 3, respectively, do not result in significant drops in the critical buckling pressure compared to the single cell breaking case. It is interesting to note that the smallest critical buckling pressure is obtained for the case of imperfections located on both horizontal and vertical lines, as case 8 in Fig. 3. Then the critical buckling pressure is 8.6% smaller than the one of the structure without imperfections.

The buckling mode shapes for all considered cases of sandwich panels with imperfections were found to be very similar. This makes sense because the buckling mode is global, such that the whole structure undergoes deformation. As an example, the buckling modes of the panels and honeycomb for case 5 of Fig. 3 are shown on Fig. 4.

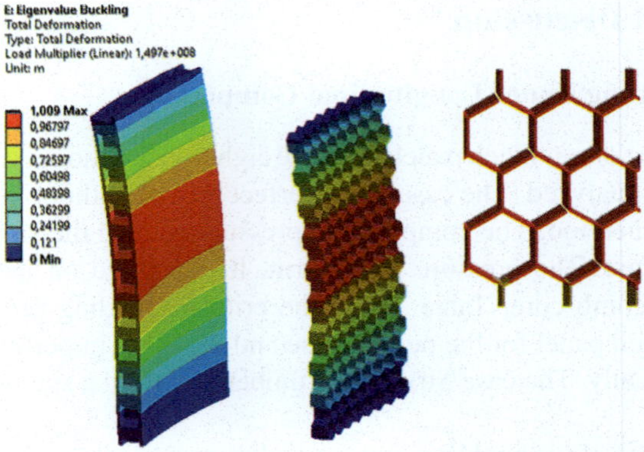

Fig. 4. Buckling mode under longitudinally compressed sandwich panel with imperfections

Thus in conclusion, while honeycomb cell imperfections can lead to a moderate decrease—no more than 8.6% in the cases considered here—of the critical buckling pressures of longitudinally compressed sandwich panels, they essentially do not affect the global buckling mode shapes. More research is required when the size of the imperfections is commensurate with the linear dimensions of the panel.

4.2 Local Buckling Under the Action of Radial Pressure

In this section, the local buckling of the sandwich structure is considered. The sandwich panel is compressed by a radial pressure which acts on the outer layer. Under the action of this pressure, some of the cells of the honeycomb core undergo plate-like buckling. The eight cases of the honeycomb core with cell breaking imperfections shown on Fig. 3 are analyzed, as well as the case of the perfect honeycomb. The critical buckling pressures P_{loc} are presented in Table 2. Again, the case Structure Number 0 is for the sandwich panel without imperfections.

The convergence of the buckling pressure is analyzed by refining the finite element mesh in all cases considered. It is found that the relative differences of the buckling pressures for ES = 2 and ES = 1 are smaller than 3.1%, and the relative differences for ES = 1 and ES = 0.5 are smaller than 1.3%. Thus, the convergence of the solutions is deemed satisfactory.

Table 2. Critical local buckling pressures of cylindrical sandwich panels under radial pressure

Structure Number	ES	0	1	2	3	4	5	6	7	8
P_{loc}, MPa	2	7.38	2.45	2.38	2.36	2.42	2.42	2.41	2.33	2.32
	1	7.31	2.4	2.32	2.3	2.36	2.34	2.35	2.27	2.25
	0.5	7.29	2.37	2.29	2.28	2.33	2.32	2.33	2.25	2.24

The first observation from Table 2 is that critical loads in the radial pressure case are about 21 times smaller than those for the longitudinal compression case in Sect. 3. Also, it is found that contrary to the longitudinal compression case, in the radial pressure case the critical buckling loads are highly sensitive to the presence of imperfections in the honeycomb core, such that even a single cell breaking (Fig. 3, case 1) results in a decrease of the critical buckling pressure by a factor greater than three. The buckling pressures for the various cases of imperfections are all close, showing that it is not the number and location of imperfections which affect the buckling pressures, but the mere presence of an imperfection in the honeycomb core. The smallest critical buckling load is observed in the panel with cell breakings located on horizontal and vertical lines (Fig. 3, case 8).

The buckling mode shape of the honeycomb core of the sandwich structure, which is compressed under the action of the radial loads, is investigated. It is found that the buckling mode is of a very local nature, contrary to the global buckling found in Sect. 3 for longitudinal compression. Finally, cases 1, 2, and 3 in Fig. 3 are considered, with one, three and five imperfect cells located on a horizontal line, respectively. It is shown in Table 2 that increasing the number of cell breakings from one to five results in a very small decrease of buckling pressure of 3.4%, while again the critical pressure of the structure with a single imperfect cell is 3.08 times smaller than that of the perfect structure. Figure 5a,b,c show the local buckling modes of the honeycomb core with imperfections in one, three and five cells, respectively. It is observed that the imperfect cells undergo plate-like buckling in all three cases.

a. *b.* *c.*

Fig. 5. Buckling of a honeycomb with imperfections: *a.* in one cell (Fig. 3, case 1); *b.* in three cells (Fig. 3, case 2); *c.* in five cells (Fig. 3, case 3)

4.3 Influence of Boundary Conditions on Buckling

Buckling of the sandwich panel structure with cell breakings under the action of radial pressure is discussed for different boundary conditions. As follows from the numerical results, the minimal local and global buckling pressures were observed in the case of cell breakings located on both horizontal and vertical lines (Fig. 3, case 8). Thus, the local buckling of this panel is considered for the following boundary conditions: the bottom and top edges of the shell are clamped and the lateral edges are free (see Fig. 2); the lateral edges are clamped and the bottom and top edges are free; all edges are clamped. Table 3 shows the results of critical buckling pressures calculations of the structure with different boundary conditions. As before, convergence of the solution is analyzed by refining the finite element mesh. The difference in buckling pressures for ES = 2 and ES = 1 is smaller than 3.9%, and for ES = 1 and ES = 0.5 it is smaller than 1.4%. Thus, sufficient convergence is achieved.

Table 3. Local honeycomb buckling pressures of sandwich panel under radial compression

Structures	ES	Clamped bottom and top edges	Clamped lateral edges	All edges clamped
P_{loc}, MPa	2	0.622	4.141	4.145
	1	0.599	3.992	3.994
	0.5	0.591	3.937	3.941

It is found that boundary conditions greatly affect the critical loads. In Sect. 4, the critical load for a structure with clamped bottom edge and simply supported top edge is 2.24 MPa, while Table 3 shows a critical load 3.8 times smaller when the top and bottom edges are clamped and 1.8 times larger when the lateral edges are clamped. In the latter cases, local buckling is observed in the cell breakings which are located close to a clamped edge. Therefore, the critical buckling pressures for the structure with clamped lateral edges and for the structure with four edges clamped are very close.

5 Conclusions

The buckling of cylindrical sandwich panels with perfect and imperfect honeycomb cores manufactured by additive technology from PLA material was investigated. Cases of global buckling under longitudinal compression and of local buckling under radial pressure were analyzed using the finite element method with refined meshes to ensure convergence. In each case, the effect of cell breakings on the buckling of the sandwich panel structure was analyzed, and eight configurations with different numbers and locations of imperfections were considered.

In the case of global longitudinal buckling of the compressed panel, the numerical analysis shows that a single cell breaking in the center of the honeycomb results in a drop of the critical buckling pressure of 5.3%. The presence of the multiple imperfections, located on vertical and horizontal lines, results in a further modest decrease of the buckling pressure, for a total drop of 8.6% compared to the perfect honeycomb.

In the case of local buckling of honeycomb cells under the action of radial pressure, finite element computations show that the presence of a single cell breaking in the honeycomb results in a decrease of more than three times in the critical pressure, indicating extreme sensitivity to imperfections. The presence of multiple breakings located on horizontal and vertical lines results in decreases in the critical buckling pressure that are only slightly greater than that for a single defect, indicating that the presence of any imperfection is sufficient to reduce critical loads drastically, and that the locations and numbers of such imperfections are not the dominant factors. Examination of the buckling modes further shows that the imperfect honeycomb cells undergo local buckling. Finally, it was observed that boundary conditions can greatly affect the critical buckling loads of sandwich panels with honeycomb imperfections.

Acknowledgements. This study was particularly funded by National Research Foundation of Ukraine (grant number 2023.04/0001) and NATO Science for Peace & Security Programme, Project G6176 – "Composite Metamaterials for Aerospace Structures – CoMetA".

References

1. Lan, X.-K., Huang, Q., Zhou, T., Feng, S.: Optimal design of a novel cylindrical sandwich panel with double arrow auxetic core under air blast loading. Def. Technol. **16**, 617–626 (2020). https://doi.org/10.1016/j.dt.2019.09.010
2. Ciccarelli, D., et al.: Buckling behavior of 3D printed composite isogrid structures. Procedia CIRP **99**, 375–380 (2021). https://doi.org/10.1016/j.procir.2021.03.053

3. Carlsson, J., Li, K., Deshpande, V., Fleck, N.: The in-plane, elastic-plastic response of a filled hexagonal honeycomb at finite strain. J. Mech. Phys. Solids **168**, 105047 (2022). https://doi.org/10.1016/j.jmps.2022.105047
4. Tankasala, H.C., Deshpande, V.S., Fleck, N.A.: The in-plane elastic-plastic response of an incompressible, filled honeycomb. J. Mech. Phys. Solids **155**, 04536 (2021). https://doi.org/10.1016/j.jmps.2021.104536
5. Yu, B., Aranguren van Egmond, H., Zurob, D.: The design of "Grain Boundary Engineered" architected cellular materials: the role of 5–7 defects in hexagonal honeycombs. Acta Materialia **243**(16), 118513 (2022). https://doi.org/10.1016/j.actamat.2022.118513
6. Chen, Y., Hu, H.: In-plane elasticity of regular hexagonal honeycombs with three different joints: a comparative study. Mech. Mater. **148**, 103496 (2020). https://doi.org/10.1016/j.mechmat.2020.103496
7. Franzoni, F., Gliszczynski, A., Baciu, T., Arbelo, M.A., Degenhardt, R.: Enhanced vibration correlation technique to predict the buckling load of unstiffened composite cylindrical shells. J. Sound Vib. **539**(2), 117280 (2022). https://doi.org/10.1016/j.jsv.2022.117280
8. Franzoni, F., Odermann, F., Wilckens, D., Skuk, E., Kalnin, K., Arbelo, M.A., Degenhardt R.: Assessing the axial buckling load of a pressurized orthotropic cylindrical shell through vibration correlation technique. Thin-Walled Struct. **137**, 353–366 (2019). https://doi.org/10.1016/j.tws.2019.01.009
9. Franzoni, F., Degenhardt, R., Albus, J., Arbelo, M.A.: Vibration correlation technique for predicting the buckling load of imperfection-sensitive isotropic cylindrical shells: an analytical and numerical verification. Thin-Walled Struct. **140**, 236–247 (2019). https://doi.org/10.1016/j.tws.2019.03.041
10. Salloomi, K.N., Sabri, L.A., Hamad, Y.M., Al-Sumaidae, S.: Nonlinear buckling analysis of steel cylindrical shell with elliptical cut-out subjected to longitudinal compressive load. Int. J. Automot. Mech. **16**(2), 6723–6737 (2019). https://doi.org/10.15282/ijame.16.2.2019.19.0506
11. Zhu, S., Du, W., Wang, X., Han, G., Ren, Z., Zhou, K.: Advanced additive remanufacturing technology. Chin. J. Mech. Eng.: Addit. Manuf. Front. **2**(1), 100066 (2023). https://doi.org/10.1016/j.cjmeam.2023.100066
12. Derevianko, I., Uspensky, B., Avramov, K., Salenko, A., Maksymenko-Sheiko, K.: Experimental and numerical analysis of mechanical characteristics of fused deposition processed honeycomb fabricated from PLA or ULTEM 9085. J. Sandwich Struct. Mater. **25**(2), 264–283 (2023). https://doi.org/10.1177/10996362221137292
13. Zhu, S., Du, W., Zhou, K.: Advanced Additive Remanufacturing Technology. Chin. J. Mech. Eng.: Addit. Manuf. Front. **2**(1), 100066 (2023). https://doi.org/10.1016/j.cjmeam.2023.100066
14. Avramov, K.V., Chernobryvko, M., Uspensky, B., Seitkazenova, K.K., Myrzaliyev, D.: Self-sustained vibrations of functionally graded carbon nanotubes-reinforced composite cylindrical shells in supersonic flow. Nonlinear Dyn. **98**(3), 1853–1876 (2019). https://doi.org/10.1007/s11071-019-05292-z
15. Avramov, K., Malyshev, S.: Bifurcations and chaotic forced vibrations of cantilever beams with breathing cracks. Engin. Fract. Mech. **214**, 289–303 (2019). https://doi.org/10.1016/j.engfracmech.2019.03.021
16. Avramov, K., Strel'nikova, E.A., Pierre, C.: Resonant many-mode periodic and chaotic self-sustained aeroelastic vibrations of cantilever plates with geometrical non-linearities in incompressible flow. Nonlin Dynam. **70**, 1335–1354 (2012). https://doi.org/10.1007/s11071-012-0537-5

An Improved Stress State Mathematical Model for the Adhesive Joint of Coaxial Tubes

Sergiy S. Kurennov, Kostiantyn P. Barakhov(✉), Olexandr G. Poliakov, and Daria V. Dvoretskaya

Kharkiv Aviation Institute, National Aerospace University, 17 Chkalova Str, Kharkiv 61070, Ukraine

kurennov.ss@gmail.com, k.barakhov@khai.edu

Abstract. An improved axisymmetric deflected mode analytical model of the adhesive joint of two cylindrical tubes is proposed. An important difference from classical mathematical models is that a more accurate deflected mode model of the adhesive layer is used. The paper is based on the hypothesis that the adhesive layer acts on shear and cleavage. It is also assumed that tangential stress values vary linearly over the thickness of the adhesive layer. To describe the normal stress values in the adhesive layer, the two-parameter Vlasov-Pasternak elastic base model is used. The proposed approach makes it possible to satisfy the boundary conditions at the load-free butt of the adhesive layer and to describe its deflected mode with high accuracy. The problem is reduced to a system of linear differential equations of the fourth order relative to displacements. The solution of the problem is obtained in an analytical form. The model problem is solved.

Keywords: Adhesive Joint · Analytical Model · Interlayer Stress

1 Introduction

Adhesive joints have significant advantages over mechanical ones. Gluing does not violate the integrity of the structure of fibrous composites and does not reduce their strength [1, 2].

Composite pipes and rods are widely used in modern technology [3, 4]. However, there is a need to organize joints both when pairing individual composite elements, and when using metal assemblies. The strength of these structures is determined, as a rule, by the strength of the joint nodes.

According to most deflected mode mathematical models (DM) of lap joints, tangential stress values in the adhesive layer reach their maximum directly at the edge of the gluing area [5]. However, since the outer edge of the adhesive joint is load-free, then, due to the pairing law for tangential stress values, these stresses at the edge of the joint must be zero. Numerical studies and experiments show that the DM of the adhesive in the neighborhood of the glue line edge largely depends on the presence or absence of chamfers at the edges of the elements to be joined, as well as on the shape and relative size of the squeezed-out adhesive excess [6, 7]. One of the possible ways to get over

H. Altenbach et al. (Eds.): CAMPE 2023, pp. 207–216, LNME 2025.
https://doi.org/10.1007/978-3-031-82979-6_21

the above contradiction and satisfy the boundary conditions at the butt of the adhesive layer is to use the Vlasov-Pasternak multi-parameter elastic base model to describe the DM of the adhesive layer. This approach was proposed in [8] and further developed, for example, in papers [9, 10].

Deflected mode analytical models of lapped adhesive joints of cylindrical tubes can be separated into two groups. One approach is to use the theory of cylindrical rods [11–13] or of shells to describe the DM of the tubes to be joined [14–17]. Another, more complex approach is two-dimensional mathematical models [18], which are reasonable to apply to describe DM of the thick-wall structure joints.

Classical stress state models of joints of cylindrical structures [11–17] assume a uniform distribution of tangential stress values throughout the adhesive layer thickness. However, since the outer side of the adhesive layer has a larger area than the inner one, the equilibrium condition of the adhesive layer is not ensured in these models. To eliminate this contradiction, it was proposed in [14] to assume the tangential stress values being concentrated only in the middle surface of the adhesive joint. However, this approach does not allow to describe the DM of the adhesive layer in the neighborhood of the joint edge with high accuracy, while the stresses in these zones reach the highest values.

The purpose of this paper is to create a deflected mode mathematical model of joints, which is devoid of contradictions and allows one to determine the deflected mode of the adhesive layer with high accuracy, including in the most loaded area - in the neighborhood of the adhesive layer edge. This will make it possible to more accurately solve the problems of joint designing and strength computation of glued structures.

2 Problem Formulation

Consider two tubes joined by an internal tubular coupling with a length $2L$. The joint is symmetrical, the center of coordinates is located on the axis of the tubes at the junction. The structure is loaded with an axial force F (Fig. 1).

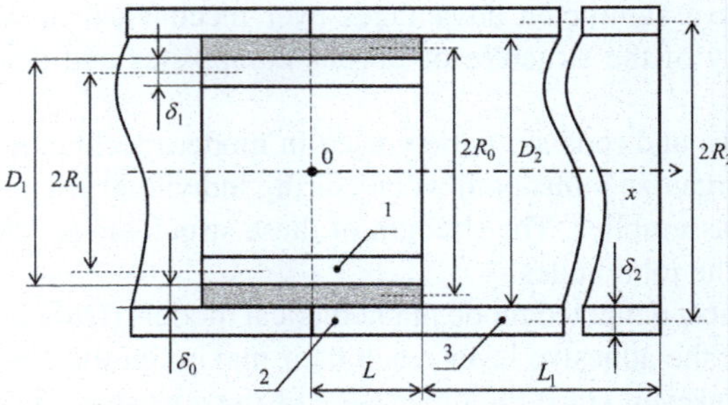

Fig. 1. Diagram of a structure

Bearing elements of the structure are considering as axisymmetric shells. Shell middle surface radii are R_1 and R_2. The shell thickness values are δ_1 and δ_2. An adhesive layer of the thickness δ_0 is located between shells, R_0 is a middle surface radius of the layer. Diameters of the inside and outside adhesive layer surface are D_1 and D_2, $D_1 = R_0 - \delta_0/2$, $D_2 = R_0 + \delta_0/2$.

Consider a gluing area $x \in (0;\ L)$. The inner and outer base layers in this section are denoted by numbers «1», «2», Fig. 1. The differential joint element is shown in Fig. 2.

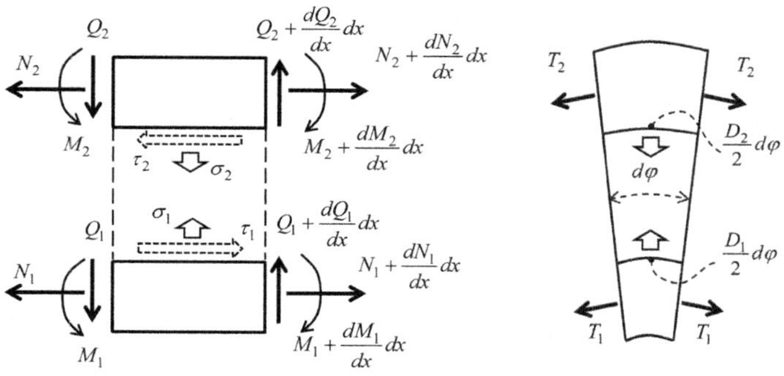

Fig. 2. Differential joint element

Due to the axial symmetry, the forces in the base layers do not depend on the angular coordinate, there are no tangential forces in the base layers.

Let us write the equilibrium equations for the elements of the base layers in the gluing area

$$R_1 dN_1/dx + D_1\tau_1/2 = 0; \quad R_2 dN_2/dx - D_2\tau_2/2 = 0, \tag{1}$$

$$R_1 dQ_1/dx + D_1\sigma_1/2 - T_1 = 0; \quad R_2 dQ_2/dx - D_2\sigma_2/2 - T_2 = 0, \tag{2}$$

$$R_1 dM_1/dx - R_1 Q_1 + D_1 s_1\tau_1/2 = 0; \quad R_2 dM_2/dx - R_2 Q_2 + D_2 s_2\tau_2/2 = 0, \tag{3}$$

where N_i, T_i and Q_i are axial, radial and shearing forces in base layers; M_i are bending moments in base layers; τ_i, σ_i are tangential and normal stresses, acting from the adhesive layer side on the corresponding base layer; s_i is a distance from a neutral surface of a corresponding shell to an outer shell surface which is faced to the adhesive joint, in the case of a symmetrical structure of the base layers $s_i = \delta_i/2$, $(i = 1,\ 2)$.

We write Hooke's law for base layers in the form

$$N_i = B_x^{(i)} du_i/dx + B_y^{(i)}\mu_{yx}^{(i)} w_i\Big/R_i, \quad T_i = B_x^{(i)}\mu_{xy}^{(i)} du_i/dx + B_y^{(i)} w_i\Big/R_i, \tag{4}$$

where u_i and w_i are displacements of the layer i in axial and radial directions correspondingly; $B_x^{(i)} = E_x^{(i)}\delta_i\Big/\left(1 - \mu_{xy}^{(i)}\mu_{yx}^{(i)}\right)$ and $B_y^{(i)} = E_y^{(i)}\delta_i\Big/\left(1 - \mu_{xy}^{(i)}\mu_{yx}^{(i)}\right)$ are membrane rigidity values of the base layer i, where $E_x^{(i)}$ and $E_y^{(i)}$ are elastic modulus values of the layer i in axial and circumferential directions correspondingly, $\mu_{xy}^{(i)}$, $\mu_{yx}^{(i)}$ are Poisson coefficients.

In formula (4), the deformations of the base layers are calculated as follows

$$\varepsilon_x^{(i)} = du_i / dx; \quad \varepsilon_y^{(i)} = w_i / R_i.$$

The equations for axisymmetric bending of cylindrical shells have the form

$$d^2w_1 / dx^2 = -M_1 / J_1; \quad d^2w_2 / dx^2 = -M_2 / J_2, \tag{5}$$

where J_1 and J_2 are bending rigidity values of the base layers. If the base layers have a homogenous structure in thickness, then $J_i = E_x^{(i)} \delta_i^3 / 12\left(1 - \mu_{xy}^{(i)} \mu_{yx}^{(i)}\right)$.

The equilibrium equation for the adhesive layer in the axial direction has the form

$$D_1\tau_1 - D_2\tau_2 = 0. \tag{6}$$

We consider that tangential stress values in the adhesive layer vary in thickness according to a linear law. In this case it is possible to introduce a stress value τ_0 in the middle surface, that allows to write

$$\tau_1 = 2R_0\tau_0 / D_1, \quad \tau_2 = 2R_0\tau_0 / D_2, \tag{7}$$

From (7) it follows, that $\tau_1 > \tau_2$. The equilibrium equation for the adhesive differential element in the radial direction has the form

$$d\tau_0 / dx = (D_1\sigma_1 - D_2\sigma_2) / 2\delta_0 R_0. \tag{8}$$

To simulate the DM of the adhesive layer, we use a two-parameter model of an elastic base in the membrane Filonenko-Borodich interpretation [8–10]. We consider the adhesive layer as a system of radial elastic elements and a membrane located in the middle surface of the adhesive layer, Fig. 3.

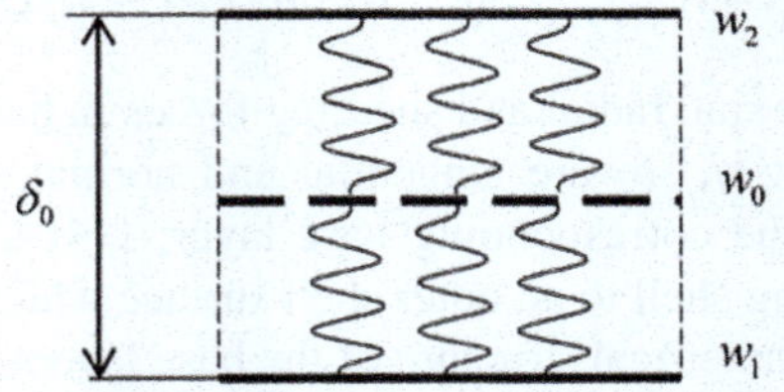

Fig. 3. Model of the adhesive layer

Normal stresses in the adhesive layer can be expressed in terms of radial displacements of the base layers (w_1, w_2) and a middle membrane (w_0)

$$\sigma_i = (-1)^i\left(k_1(w_i - w_0) - k_2\left(d^2w_i / dx^2 - d^2w_0 / dx^2\right)\right), \quad i = 1, 2. \tag{9}$$

where $k_1 = 2E_0 / \delta_0(1 - \mu_0^2)$ and $k_2 = E_0\delta_0 / 12(1 + \mu_0)$ are the first and second bed coefficient; E_0 and μ_0 are the elastic modulus and the Poisson coefficient of the adhesive.

Tangential stress values in the adhesive layer have the form [8]:

$$\tau_0 = P\left(u_2 - u_1 + s_2 dw_2 \big/ dx + s_1 dw_1 \big/ dx\right) + G_0 dw_0 \big/ dx, \tag{10}$$

where $P = G_0 / \delta_0$ is a shear rigidity of the adhesive layer.

The proposed DM model for the adhesive layer will make it possible to take into account the boundary conditions at the outer boundary of the adhesive layer. These conditions can be of two types: equality to zero of tangential stresses at the boundary in the absence of the glue excess that is squeezed out of the glue line, or achievement of a maximum by tangential stresses (i.e. $d\tau_0 / dx = 0$) in the presence of flash [10].

3 Constructing the Solution

3.1 Gluing Area

The system (1) – (10) is reduced to a system of differential equations relative to displacements.

$$\boldsymbol{A}_4 d^4\boldsymbol{V} \big/ dx^4 + \boldsymbol{A}_2 d^2\boldsymbol{V} \big/ dx^2 + \boldsymbol{A}_1 d\boldsymbol{V} \big/ dx + \boldsymbol{A}_0 \boldsymbol{V} = 0, \tag{11}$$

where $\vec{V} = \begin{pmatrix} u_1 \\ u_2 \\ w_1 \\ w_2 \\ w_0 \end{pmatrix}, \boldsymbol{A}_4 = \frac{1}{P_0}\begin{pmatrix} \cdot & \cdot & \cdot & \cdot & \cdot \\ \cdot & \cdot & \cdot & \cdot & \cdot \\ \cdot & \cdot & R_1 J_1 & \cdot & \cdot \\ \cdot & \cdot & \cdot & R_2 J_2 & \cdot \\ \cdot & \cdot & \cdot & \cdot & \cdot \end{pmatrix}, P_0 = R_0 P$

$$\boldsymbol{A}_2 = \frac{1}{P_0}\begin{pmatrix} R_1 B_x^{(1)} & \cdot & \cdot & \cdot & \cdot \\ \cdot & R_1 B_x^{(1)} & \cdot & \cdot & \cdot \\ \cdot & \cdot & -s_1^2 P_0 - D_1 k_2/2 & -s_1 s_2 P_0 & D_1 k_2/2 - s_1 \delta_0 P_0 \\ \cdot & \cdot & -s_1 s_2 P_0 & -s_1^2 P_0 - D_2 k_2/2 & D_2 k_2/2 - s_2 \delta_0 P_0 \\ \cdot & \cdot & D_1 k_2/2 - s_1 \delta_0 P_0 & D_2 k_2/2 - s_2 \delta_0 P_0 & -\delta_0^2 P_0 - (D_1 + D_2) k_2/2 \end{pmatrix},$$

$$\boldsymbol{A}_1 = \frac{1}{P_0}\begin{pmatrix} \cdot & \cdot & \mu_{yx}^{(1)} B_y^{(1)} + s_1 P_0 & s_2 P_0 & \delta_0 P_0 \\ \cdot & \cdot & -s_1 P_0 & \mu_{yx}^{(2)} B_y^{(2)} - s_2 P_0 & -\delta_0 P_0 \\ \mu_{yx}^{(1)} B_y^{(1)} + s_1 P_0 & -s_1 P_0 & \cdot & \cdot & \cdot \\ s_2 P_0 & \mu_{yx}^{(2)} B_y^{(2)} - s_2 P_0 & \cdot & \cdot & \cdot \\ \delta_0 P_0 & -\delta_0 P_0 & \cdot & \cdot & \cdot \end{pmatrix},$$

$$\boldsymbol{A}_0 = \frac{1}{P_0}\begin{pmatrix} -P_0 & P_0 & \cdot & \cdot & \cdot \\ P_0 & -P_0 & \cdot & \cdot & \cdot \\ \cdot & \cdot & D_1 k_1/2 + B_y^{(1)} R_1^{-1} & \cdot & -D_1 k_1/2 \\ \cdot & \cdot & \cdot & D_2 k_1/2 + B_y^{(2)} R_2^{-1} & -D_2 k_1/2 \\ \cdot & \cdot & -D_1 k_1/2 & -D_2 k_1/2 & (D_1 + D_2) k_1/2 \end{pmatrix}.$$

We search a particular solution of system (11) in the form $\boldsymbol{V} = \boldsymbol{h} e^{\lambda x}$, where $\boldsymbol{h}$ is some vector. Substituting this in (11) and simplifying, we get a system.

$$\boldsymbol{A}_s(\lambda)\boldsymbol{h} = 0, \tag{12}$$

where $\boldsymbol{A}_s(\lambda) = \boldsymbol{A}_4\lambda^4 + \boldsymbol{A}_2\lambda^2 + \boldsymbol{A}_1\lambda + \boldsymbol{A}_0$.

System (12) has a nontrivial solution provided that

$$\det(\boldsymbol{A}_s(\lambda)) = 0. \tag{13}$$

The characteristic Eq. (13) has a root $\lambda = 0$ of multiplicity two and twelve different nonzero roots λ_k, $k = 1, 2, ..., 12$. The general solution of system (11) can be written as

$$\boldsymbol{V} = \sum_{k=1}^{12} C_k \boldsymbol{h}_k e^{\lambda_k x} + \boldsymbol{H}_0 + \boldsymbol{H}_1 x, \tag{14}$$

where $\boldsymbol{H}_0$ and $\boldsymbol{H}_1$ are eigenvectors, corresponding to $\lambda = 0$; $\boldsymbol{h}_k$ are eigenvectors, corresponding to $\lambda_k \neq 0$; C_k are an arbitrary constants. Vectors $\boldsymbol{h}_k$ we obtain from the system (12) with accuracy up to an arbitrary factor, which in the solution (14) is denoted as C_k. In turn

$$\boldsymbol{H}_1 = \left(C_{14}\ C_{14}\ 0\ 0\ 0\right)^T; \quad \boldsymbol{H}_0 = \left(C_{13}\ C_{13}\ C_{14}\alpha_1\ C_{14}\alpha_2\ C_{14}\alpha_3\right)^T,$$

$$\alpha_i = -\left(k_1 R_2 R_1 \left(\mu_{yx}^{(1)} \big/ B_y^{(2)} + \mu_{yx}^{(2)} \big/ B_y^{(1)}\right) + 2\beta R_i \mu_{yx}^{(i)}\right) \Big/ \left(k_1 \left(R_2 \big/ B_y^{(2)} + R_1 \big/ B_y^{(1)}\right) + 2\beta\right)$$

$$i = 1, 2, \ \beta = 1/D_1 + 1/D_2, \ \alpha_3 = (D_1\alpha_1 + D_2\alpha_2)/(D_1 + D_2).$$

3.2 The Structure Out of the Gluing Area

Due to the symmetry of the structure, it is sufficient to consider only one area outside the gluing, $x \in (L;\ L + L_1)$, which is denoted in Fig. 1 with index «3». The DM of the structure at this area is described with the cylindrical shell deformation equations

$$R_3 dQ_3 / dx - T_3 = 0, \quad dM_3 / dx - Q_3 = 0, \quad d^2 w_3 \big/ dx^2 = -M_3 / J_3 \tag{15}$$

In the problem under considering,$R_3 = R_2$ and $J_3 = J_2$, because the outer tube outside the gluing area is the same as inside the gluing area. However, for convenience, we will keep the index «3».

Hooke's law

$$N_3 = B_x^{(3)} du_3 / dx + B_y^{(3)} \mu_{yx}^{(3)} w_3 \big/ R_3; \quad T_3 = B_x^{(3)} \mu_{xy}^{(3)} du_3 / dx + B_y^{(3)} w_3 \big/ R_3, \tag{16}$$

where $B_x^{(3)} = B_x^{(2)}$, $B_y^{(3)} = B_y^{(2)}$, $\mu_{yx}^{(3)} = \mu_{yx}^{(2)}$ and $\mu_{xy}^{(3)} = \mu_{xy}^{(2)}$.

Since the protruding part is loaded with the axial force F, we have

$$N_3 = F / 2\pi R_3 = const. \tag{17}$$

From (15) and (16) we obtain.

$$R_3^2 J_3 \big/ B_y^{(3)} \cdot d^4 w_3 \big/ dx^4 + \left(1 - \mu_{xy}^{(3)} \mu_{yx}^{(3)}\right) w_3 = -R_3 \mu_{xy}^{(3)} \big/ B_y^{(3)} \cdot F / 2\pi R_3$$

The general solution of this equation has form

$$w_3 = \sum_{j=1}^{4} S_j e^{\theta_j x} - \mu_{xy}^{(3)} \Big/ \delta_3 E_y^{(3)} F \Big/ 2\pi R_3, \tag{18}$$

where S_j are arbitrary constants; θ_j are roots of the equation $\theta^4 = -E_y^{(3)} \delta_3 \Big/ R_3^2 J_3$.

3.3 Boundary Conditions and Conjugation Conditions

Due to the symmetry of the structure, we assume that at the boundary of the coupling, the inner insert has a sliding seal, and the outer tube is load-free:

$$u_1(0) = 0, \quad dw_1 / dx\big|_{x=0} = 0, \quad Q_1(0) = 0, \quad M_2(0) = Q_2(0) = N_2(0) = 0$$

On the right border of the gluing area, the inner insert is load-free, and the outer tube from area «2» has a smooth continuation to area «3»:

$$M_1(L) = Q_1(L) = N_1(L) = 0, \quad w_2(L) = w_3(L), \quad u_2(L) = u_3(L),$$

$$dw_2 / dx\big|_{x=L} = dw_3 / dx\big|_{x=L}, \; M_2(L) = M_3(L), \quad Q_2(L) = Q_3(L), \quad N_2(L) = N_3(L).$$

The right edge of the outer tube will be considered loaded with a given longitudinal force (20), and having a sliding hinge seal:

$$M_2(L + L_1) = Q_2(L + L_1) = 0.$$

If the length of area «3» is large enough ($L_1 >> L$), then the type of seal does not affect the DM of the adhesive layer.

The outer border of the adhesive layer is load-free:

$$\tau_0(0) = 0, \quad \tau_0(L) = 0.$$

4 Model Problem

Consider the construction: $R_1 = 50$ mm, $\delta_1 = 2$ mm, $R_2 = R_3 = 52.1$ mm, $\delta_2 = \delta_3 = 2$ mm (main tube), $\delta_0 = 0.1$ mm, $D_1 = 2R_1 + \delta_1 = 102$ mm, $D_2 = 2R_2 - \delta_2 = 102.2$ mm, $L = 40$ mm, $L_1 = 10L$. The main tube is made of composite material: $E_x^{(2)} = E_x^{(3)} = 36.8$ GPa, $E_y^{(2)} = E_y^{(3)} = 8.27$ GPa, $\mu_{xy}^{(2)} = \mu_{xy}^{(3)} = 0.26$. Joint insert made of aluminum alloy: $E_x^{(1)} = E_y^{(1)} = 70$ GPa, $\mu_{xy}^{(1)} = 0.27$. Elastic characteristics of the adhesive $E_0 = 0.7$ GPa, $G_0 = 0.2414$ GPa, (i.e. $\mu_0 = 0.45$).

In Fig. 4 it is shown a graph of the tangential stress values τ_0 distribution (10) along the joint. Stresses are shown in dimensionless form, as a ratio of effective stresses to some hypothetical stresses $\tau^* = F \big/ 2\pi R_0 L$, which would act on the middle surface of the adhesive layer with a uniform load distribution over the area of this surface.

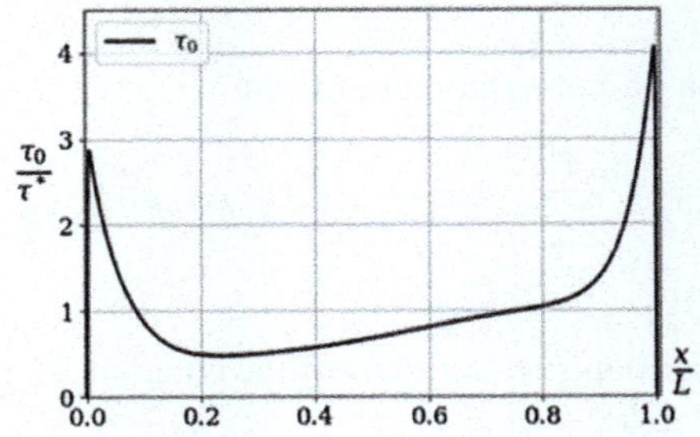

Fig. 4. Tangential stress values in the adhesive layer

Tangential stress values at the edges of the gluing area are equal to zero and reach a maximum at a distance of the order of the adhesive layer thickness from the edge of the gluing line. Computation shows, that normal stress values σ_1 and σ_2 are close to zero over the most of the joint length, but differ significantly from zero only near the ends of the gluing area. Therefore in Fig. 5 it is shown three graphs of σ_1, σ_2 and τ_0 in the neighborhood of the gluing area edges. Graphs are shown in dimensionless form as well.

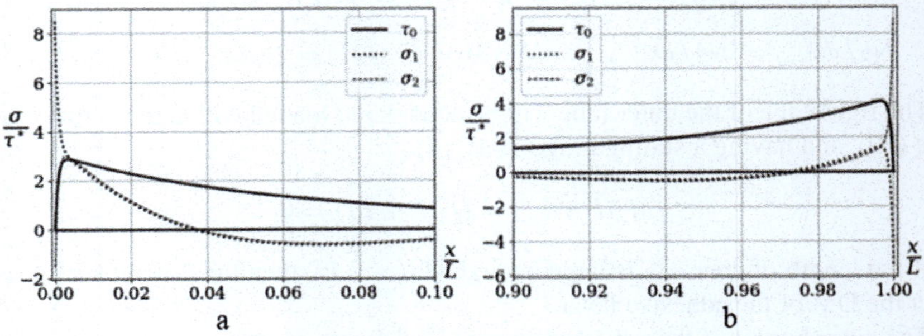

Fig. 5. Stress values in the adhesive at the left (a) and at the right (b) edge of the joint

The graphs in Fig. 5 show that over most of the joint length the stress values σ_1 and σ_2 are practically the same. However, near the edge of the adhesive layer, the stress values σ_1 and σ_2 differ significantly, and may even have a different sign. Simpler models [14] do not detect this effect and show an average value of σ_1 and σ_2, which can be more than two times less than the maximum values of normal stresses σ_1, σ_2 at the edge of the joint. Information about the DM of the adhesive in the most dangerous area at the edges of the joint is of paramount importance for the strength design of joints [10].

Under the action of an axial load, the structure sustains an axisymmetric bending. If the main tubes and the joint insert are uniform in thickness, then the normal stress values in the axial direction are expressed by the formula.

$$\sigma_m(x, z) = N_m(x) \Big/ \delta_m + 12 M_m(x) z \Big/ \delta_m^3$$

,

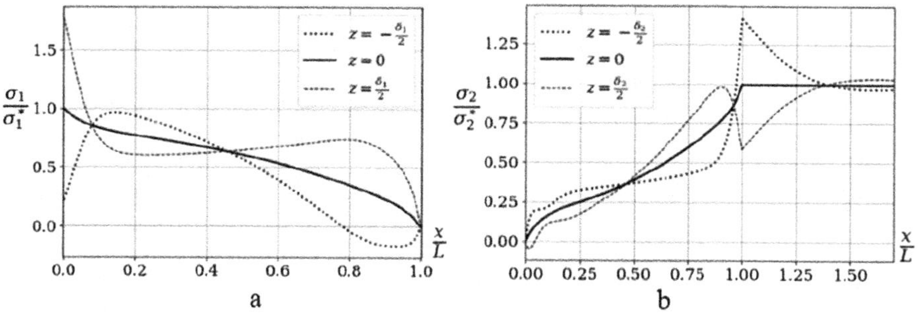

Fig. 6. Axial stress values in the insert (a) and in the main tube (b)

where $z \in \left[-\delta_m/2;\ \delta_m/2\right]$; index m takes the values 1, 2, 3.

In Fig. 6 it is shown the graphs of membrane stress values and stress values on the outer and inner surfaces of the insert and the main tube, i.e. $\sigma_x^{(m)}(x, z)$ with $z = -\delta_m/2$, $z = 0$ and $z = \delta_m/2$ for $m = 1, 2, 3$. The stress values on the graphs are shown in dimensionless form, as the ratio of effective stress values to stresses $\sigma_m^* = F/2\pi R_m \delta_m$.

5 Conclusions

An analytical solution to the axisymmetric DM problem for a structure consisting of two identical cylindrical tubes joined by a tubular insert is obtained. The difference from classical mathematical models is that a more accurate DM model of the adhesive layer is used, it makes it possible to take into account the boundary conditions at the ends of the joint and to describe the adhesive layer DM with high accuracy, including the irregular zone near the edge of the gluing area. This makes it possible to calculate the structure for strength using various criteria for the strength of the adhesive layer.

An internal pressure in the structure can be added to the equilibrium Eqs. (2), which will allow to use the proposed approach, for example, to analyze the strength of pipelines reinforced with a composite bandage [20]. The proposed axisymmetric model can include circumferential displacements and tangential stress values in the base layers [16]. This will allow us to find stress values in a structure that is loaded with torque. The proposed approach can also be used to find the DM of a telescopic joint of two tubes, a joint of two tubes with an external coupling, and also in the presence of a non-glued area in the adhesive layer [17]. Moreover, the proposed model of a three-layer structure with a pliable filler can be developed for the problems of dynamics [21, 22].

References

1. Dveirin, O.Z., Andreev, O.V., Kondrat'ev, A.V., & Haidachuk, V.Y.: Stressed State in the Vicinity of a Hole in Mechanical Joint of Composite Parts. Int. Appl. Mech. **57**(2), 234–247 (2021)
2. Kruse, A., Grunwald, C., Kaufmann, M., et al.: Einfluss konstruktiver Details auf Klebungen pultrudierter GFK-Lamellen. Adhaes Kleb Dicht **65**, 33–37 (2021)

3. Voß, M., Haupt, J.: Accelerated curing of adhesively bonded G-FRP tube connections – Part I: Experiments. Compos. Struct. **268**, 113999 (2021)
4. Thumitha Mandula Higgoda: Mohamed Elchalakani, Mehrdad Kimiaei, Adam Wittek, Bo Yang: Flexural behaviour of pultruded circular tubular GFRP composite truss bridges with novel non-corrosive connections. Structures **45**, 830–853 (2022)
5. da Silva, L.F.M., das Neves, P.J.C., Adams, R.D., Spelt, J.K.: Analytical models of adhesively bonded joints. Part I: Literature survey. Int. J. Adhes. & Adhes. **29**, 319–330 (2009)
6. Ferreira, L.R.F., Campilho, R.D.S.G., Barbosa, D.R., Rocha, R.J.B., Silva, F.J.G.: Static strength improvement of tubular aluminium adhesive joints by the outer chamfering technique. Procedia Manuf. **38**, 629–636 (2019)
7. Frostig, Y., Thomsen, O.T., Mortensen, F.: Analysis of adhesive-bonded joints, square-end, and spew-fillet—high-order theory approach. J. Eng. Mech. **125**(11), 1298–1307 (1999)
8. Wang, J., Zhang, C.: Three-parameter, elastic foundation model for analysis of adhesively bonded joints. Int. J. Adhes. Adhes. **29**(5), 495–502 (2009)
9. Veisytaba, M., Reza, A., Shekari, Y.: Stress analysis of adhesively-bonded single stepped-lap joints with functionally graded adherends based on the four-parameter fractional viscoelastic model. Eur. J. Mech. – A/Solids **98**, 104907 (2023)
10. Kurennov, S.S.: Refined mathematical model of the stress state of Adhesive lap joint: experimental determination of the Adhesive layer strength criterion. Strength Mater. **52**, 779–789 (2020)
11. Jäger, M., Wartzack, S.: Strength analysis of structurally optimized aluminium-composite tubular lap joints. Proceedings of the Design Society **2**, 2473–2482 (2022)
12. Parashar, A., Mertiny, P.: Adhesively bonded composite tubular joints: review. Int. J. Adhes. Adhes. **38**, 58–68 (2012)
13. Kumar, S., Scanlan, J.P.: On axisymmetric adhesive joints with graded interface stiffness. Int. J. Adhes. Adhes. **41**, 57–72 (2013)
14. Kurennov, S.S., Barakhov, K.P., Poliakov, A.G.: Stressed state of the axisymmetric Adhesive joint of two cylindrical Shells under Axial tension. Mater. Sci. Forum **968**, 519–527 (2019)
15. Goglio, L., Paolino, D.S.: Adhesive stresses in axially-loaded tubular bonded joints–Part II: Development of an explicit closed-form solution for the Lubkin and Reissner model. Int. J. Adhes. Adhes. **48**, 35–42 (2014)
16. Han, J., Yuan, H.: Full debonding process of adhesively bonded composite and metallic Pipe joints under torsion. Mathematical Problems in Engineering 1237493 (2016)
17. Shishesaz, M., Tehrani, S.: Interfacial shear stress distribution in the adhesively bonded tubular joints under tension with a circumferential void or debond. J. Adhes. Sci. Technol. **34**(11), 1172–1205 (2020)
18. Nemes, O., Lachaud, F., Mojtabi, A.: Contribution to the study of cylindrical adhesive joining. Int. J. Adhes. Adhes. **26**, 474–480 (2006)
19. Kutiová, K., Měsíček, J., Krzikalla, D.: Experimental examination of adhesive bonded carbon fibre tube to aluminum alloy AW 7075. Mater. Sci. Forum **994**, 125–132 (2020)
20. Larin, O., Barkanov, E., Vodka, O.: Prediction of reliability of the corroded pipeline considering the randomness of corrosion damage and its stochastic growth. Eng. Fail. Anal. **66**, 60–71 (2016)
21. Sukhanova, O., Larin, O., Naumenko, K., Altenbach, H.: Dynamics of curved laminated glass composite panels under impact loading. nonlinear mechanics of complex structures. Adv. Struct. Mater. **157**, 91–101 (2021). Springer, Cham
22. Ugrimov, S., Smetankina, N., Kravchenko, O., Yareshchenko, V., Kruszka, L.: A Study of the Dynamic Response of Materials and Multilayer Structures to Shock Loads. In: Altenbach, H., et al. Advances in Mechanical and Power Engineering. CAMPE 2021. Lecture Notes in Mechanical Engineering. Springer, Cham. pp. 304–313 (2023)

Analysis of Shape Distortion of the Composite Reflector Antennas During Assembly

Oleksandr Gaidachuk[1], Igor Taranenko[2], Tetyana Nabokina[2], and Andrii Kondratiev[3](✉)

[1] National University "Zaporizhzhia Polytechnic", 64 Zhukovsky St, Zaporizhzhia 69063, Ukraine

[2] National Aerospace University "Kharkiv Aviation Institute", 17 Chkalova St, Kharkiv 61070, Ukraine

[3] O.M Beketov National University of Urban Economy in Kharkiv, 17 Marshal Bazhanov St, Kharkiv 61002, Ukraine

andrii.kondratiev@kname.edu.ua

Abstract. Development of the modern communication systems necessitates the creation of mirror reflectors for antennas made of polymer composite materials. It allows to significantly reduce the weight of reflectors making them more precise and dimensionally stable. However, the actual process of manufacturing of the composite reflectors is often associated with distortions of their reflecting surface occurring because of process-induced deformations. The analytical dependences have been obtained that allowed predicting the regulated tolerance band for assembly by gluing or connection with screws of the composite products with frames such as parabolic antenna reflectors. These dependences take into account the product design and assembly process parameters, as well as strength characteristics of the adhesive, the rational combination of which allows achieving a high quality of the reflector. As shown by the example, recording of spring action depending on the unbalance of internal forces of the assembled product does not make a substantial correction to the tolerance for the theoretical contour of the mirror antenna reflector at the actual design parameters of its elements. Tolerance bands for gluing and connecting with screws are compared. It is proved that the frame to reflector connection with screws can be more effective at their rational number, but when this number is fairly large, the realizable tolerance band is almost the same. Our findings allow solving the new process tasks in order to determine the maximum permissible deviations of composite shells from their theoretical contour at the assembly stage.

Keywords: Paraboloid of revolution · Mirror · Tolerance band · Frame · Adhesive

H. Altenbach et al. (Eds.): CAMPE 2023, pp. 217–226, LNME 2025.
https://doi.org/10.1007/978-3-031-82979-6_22

1 Introduction

Reflector antennas are widely used in the various telecommunication systems [1] owing to main advantages such as high gain factor and wide bandwidth, along with simple and reliable design and relatively low cost [2]. Paraboloids of revolution of various designs are most often used as a mirror reflector in these antennas [3]. It is commonly known [4] that high operational requirements are imposed on this class of structures. Their characteristic feature is rather tight tolerance for deviation of the inlet section contour from the theoretical one [5, 6]. Sandwich structures with the honeycomb core and face sheets made of polymer composite materials (PCM) are often used in the reflector manufacturing [7]. The use of structures as above allows to reduce the weight of reflectors significantly, making them more precise and dimensionally stable [8, 9]. However, the actual process of manufacturing of PCM products, including reflectors, is often associated with distortions quite often occurring in the shape of the reflecting surfaces [10]. These distortions appearing after removal of the reflector from the mandrel are due to occurrence of process-induced deformations [11, 12]. In this case, it is difficult to predict the form of distortion of the reflecting surface and the realizable tolerance band. The need to solve this problem initiates numerous studies.

The purpose of this paper is to obtain the analytical dependences that allow to predict the regulated (normalized) tolerance band for assembly by gluing or connection with screws (bolts) of PCM products such as parabolic antenna reflectors with frames.

2 Literature Review

Prediction of the reflector mechanical properties and, above all, deviation of the shape of its reflecting surface from the specified one, is the primary objective of antenna designing. Distortion of shape of the reflecting surface in the reflector is determined by a number of factors. For example, the paper [13] proposes a model based on the improved theory of shells to predict the process-induced deformations of the antenna reflecting surface. The method of segmentation and assembly to improve the accuracy of the reflector antenna surface is developed. The method for optimization of the antenna reflector design, taking into account the deformations' compatibility conditions is proposed in [14]. Parameters of the structure with higher stability of the surface shape compared to the traditional design methods are obtained. The paper [15] develops the antenna reflector of ultra-high accuracy. Interrelations of process parameters with structural parameters are obtained. The use of sandwich structures made of PCM with honeycomb core for antenna reflectors significantly complicates the possibility of setting and implementing the optimal designing problem due to increase in the number of variables [16, 17]. For example, analysis of technological problems associated with the manufacturing of parabolic radio antennas of PCM is given in [18]. The peculiar features of the usage of these materials in the manufacturing of antenna reflectors are considered, and various methods for provision of the shape accuracy and stiffness of reflectors are analyzed. Specialized algorithm for constructing a finite element model of the antenna composite reflector is developed and applied in [19]. Using this algorithm, the impact of various design and process parameters on the maximum deflection and accuracy of the reflecting surface in the reflector is analyzed. The paper [20] examines the impact of process parameters of antenna manufacturing of PCM on the residual deformations of reflectors and the accuracy of their geometric parameters. The paper [21] proposes the simultaneous joint moulding of the shell and structural frame to eliminate the defects that occur during the sandwich structure assembly by gluing. Estimation of error of the shape of processed honeycomb surfaces to control the assembly process is given in [22]. An effective method for increasing the accuracy of assembly of honeycomb structures at accuracy of ± 7 mcm is proposed. The paper [23] describes the wide range of experimental studies showing a significant effect of process parameters of deformation of composite assemblies. The new method for assembling sandwich shells with the honeycomb core is developed in [24]. The paper [25] covers the experimental and numerical study of the process-induced deformation in the flat sandwich panels. As shown by the analysis of the technological capabilities to provide the accuracy of manufacturing PCM reflector antennas in the modern production conditions, most of papers deal with minimizing the deviations in their theoretical contour in the process of manufacturing of shape-generating moulding tools, development of the optimal moulding processes, and choice of materials and rational design and technology concepts. However, an interesting point is the analysis of the maximum permissible deviation of the inlet section of reflector antennas made of PCM from their theoretical contour at the stage of assembly with frames.

3 Research Methodology

It is assumed that the frame has a statistical average inner radius mating with the outer surface of the shell equal to R, while the outer (mating) statistical average shell radius is R_0 (Fig. 1). The frame width is B, and the height is h. The parabolic shell has the modulus of elasticity E_0 in the annular direction and cross-section area F_0. The shell is a sandwich structure with face sheets of δ_0 thick and honeycomb core of h_0 thick. Let us accept that by the time of assembly the inlet frame is produced of the material with the modulus of elasticity E in the annular direction and cross-section area F. It is assumed that the shell in the connection zone is not compressed under the action of tightening forces of screws (provided, for example, by inserting special bushings in the honeycomb core). The honeycomb core does not perceive the loads.

The section along the inner contour of the frame of the mirror antenna reflector to be assembled is an ellipse with semi-axes a, b, perimeter of which is equal to the perimeter of the equivalent circle of radius R

$$L = \pi\left[1.5\left(a + b + \sqrt{ab}\right)\right] = 2\pi R. \tag{1}$$

The dimensions of semi-axes a and b are determined as a result of N measurements with the accuracy providing the sampling variation coefficient of max. 10% and accuracy factor of max. 5% [26].

The section along the outer contour is also an ellipse with semi-axes a_0 and b_0, perimeter of which is equal to the perimeter of the equivalent circle of radius R_0

$$L_0 = \pi\left[1.5\left(a_0 + b_0 + \sqrt{a_0 b_0}\right)\right] = 2\pi R_0. \tag{2}$$

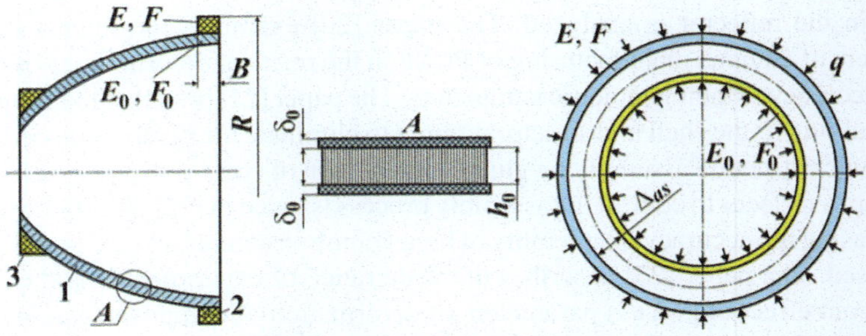

Fig. 1. Structural design of the PCM mirror antenna and schematic diagram of its loading in the assembly: 1 – parabolic shell with bearing layers of PCM and honeycomb core; 2 – inlet frame of PCM; 3 – pole frame of PCM.

Determination of semi-axes a_0 and b_0 is identical to that described above for semi-axes a and b of the frame section.

Statistical average gap Δ_{as} between the equivalent radii R and R_0, respectively, of the frame and shell to be assembled is:

$$\Delta_{as} = R - R_0. \tag{3}$$

The assembly by gluing suggests the uniform tightening of parts with the linear load q (Fig. 1) until they fit to contact surfaces completely. Assuming the sign of radial displacement of the mating parts positive, if it increases the initial radius, and negative in case of decrease in the initial radius, we can finally obtain from the solution of statically indeterminate problem by the force method, for the radial displacement of the frame Δ and shell Δ_0:

$$\Delta = -q(R + 0.5h)^2 (EF)^{-1}, \tag{4}$$

$$\Delta_0 = -q(R_0 - \delta_0 - 0.5h_0)^2 (E_0 F_0)^{-1}, \tag{5}$$

where E, E_0 – moduli of elasticity of the frame and shell material in the circumferential direction, $F = Bh$, $F_0 = 2B\delta_0 + Bh_0$ – frame and shell cross-section areas.

Since the total displacement during assembly by gluing Δ_{as} is

$$\Delta_{as} = \Delta_0 - \Delta, \tag{6}$$

then we determine the linear load q, required to ensure complete tightening, from (6) taking (4) and (5) into account

$$q = \frac{(R - R_0)E_0 F_0}{(R_0 - \delta_0 - 0.5h_0)^2 \left[1 + (R + 0.5h_0)^2 (R_0 - \delta_0 - 0.5h_0)^{-2} E_0 F_0 (EF)^{-1}\right]}. \tag{7}$$

Statistical average realizable tolerance for deviation of the theoretical contour of the shell inner radius is:

$$\Delta_{thc} = R_0 - R_{mx} + \Delta_0 - 2\delta_0 - h_0. \tag{8}$$

Substituting in (8) the values Δ_0 from (5), taking (3) and (7) into account, we get:

$$\Delta'_{thc} = R_0 - R_{thc} - 2\delta_0 - h_0 + \frac{R - R_0}{1 + (R + 0.5h)^2 (R_0 - \delta_0 - 0.5h_0)^{-2} E_0 F_0 (EF)^{-1}}. \quad (9)$$

The band of this tolerance is defined by expressions:

$$\Delta_{thc\ \max} = a_0 - R_{thc} - 2\delta_0 - h_0 + \frac{a - a_0}{1 + (a + 0.5h)^2 (a_0 - \delta_0 - 0.5h_0)^{-2} E_0 F_0 (EF)^{-1}}, \quad (10)$$

$$\Delta_{thc\ \min} = b_0 - R_{mk} - 2\delta_0 - h_0 + \frac{b - b_0}{1 + (b + 0.5h)^2 (b_0 - \delta_0 - 0.5h_0)^{-2} E_0 F_0 (EF)^{-1}}. \quad (11)$$

The allowable value of the total displacement during assembly Δ_{as} is determined by the level of linear load q, which depends on the tensile strength of the adhesive $\sigma_{ad}^{\lim}$ and allowable gluing stresses $[\sigma_{ad}] << \sigma_{ad}^{\lim}$ [27, 28].

The inequality below should be fulfilled for the technological linear pressing force q acting on the parts $q/L \le [\sigma_{ad}]$, where L – width of the frame and shell surface being bonded. Taking into consideration this and (7), we obtain a limitation on the gap of mating parts Δ_{as}

$$[\Delta_{as}] \le [\sigma_{ad}] L (R_0 - \delta_0 - 0.5h_0)^2 \left[1 + R^2 R_0^{-2} E_0 F_0 (EF)^{-1}\right] (E_0 F_0)^{-1}. \quad (12)$$

When the parts are assembled by connection with screws (bolts), the gap Δ_{as} between the equivalent radii of the frame (R) and shell (R_0) to be assembled is eliminated by displacements of the shell Δ_0 and frame Δ, determined from the statically indeterminate problem solution:

$$\Delta_0 = P[R_0 - \delta_0 - 0.5h_0]^3 (E_0 I_0)^{-1} \left(c - \alpha^{-1}\right) + P[R_0 - \delta_0 - 0.5h_0](E_0 F_0)^{-1} c; \quad (13)$$

$$\Delta = P(R + 0.5h)^3 (EI)^{-1} \left(c - \alpha^{-1}\right) + P(R + 0.5h)(EF)^{-1} c, \quad (14)$$

where; P – tightening force at the point; $\alpha = 2\pi/n$ – angle of sector between the tightening forces; n – even number of screws (bolts), $\mathrm{I} = \mathrm{Bh}^3/2$, $\mathrm{I}_0 = 2B\delta_0(0.5\delta_0 + 0.5h_0)^2$ – moments of inertia and cross-section areas of the frame and shell; $c = (\alpha + \sin\alpha)\left(8\sin^2\alpha/2\right)^{-1}$.

Tightening force P is determined from the equality (6). Expanding the equality (6), taking (1) – (3) into account, we get:

$$P = \frac{R - R_0}{\left\{\frac{[R_0 - \delta_0 - 0.5h_0]^3}{E_0 I_0} + \frac{(R + 0.5h)^3}{EI}\right\}\left(c - \alpha^{-1}\right) + \left\{\frac{R_0 - \delta_0 - 0.5h_0}{E_0 F_0} + \frac{R + 0.5h}{EF}\right\} c}. \quad (15)$$

After tightening of the shell and frame, there will be maximum gaps $\Delta_{\Sigma g}$ in the middle of gaps between screws (Fig. 2).

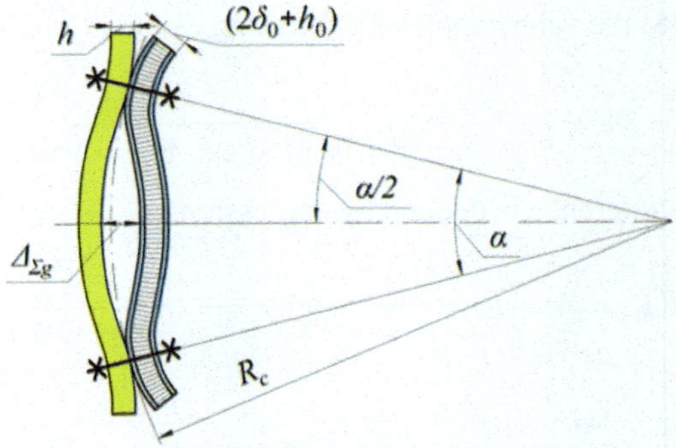

Fig. 2. Diagram for determination of the gap $\Delta_{\sum g}$

To determine the gap $\Delta_{\Sigma g}$, equal to

$$\Delta_{\sum g} = \Delta_{g\,0} + \Delta_g, \tag{16}$$

where Δ_{g0}, Δ_g – displacement of the contour of shell and frame at $\alpha/2$ relative to the radius of mating R_c of the shell and frame after tightening (Fig. 2), it is necessary to solve statically indeterminate problems for the frame and shell by the force method, applying the unit forces at the point of the expected displacement.

From the solution of the above statically indeterminate problems, we finally obtain

$$\begin{aligned}\Delta_{\sum g} &= 2(R - R_0)\left\{\frac{(R+0.5h)^3}{EI} + \frac{[R_0 - \delta_0 - 0.5h_0]^3}{E_0 I_0}\right\}\left(\frac{2\pi}{n}\cos\frac{\pi}{n} + \sin\frac{\pi}{n}\right) \\ &\times \left[\left\{\frac{[R_0 - \delta_0 - 0.5h_0]^3}{E_0 I_0} + \frac{(R+0.5h)^3}{EI}\right\}\right. \\ &\qquad \left(\frac{4\pi^2}{n} + \frac{2\pi}{n}\sin^2\frac{\pi}{n} - 1\right)\left(16\frac{\pi}{n}\sin^2\frac{\pi}{n}\right)^{-1} \\ &+ \left.\left\{\frac{[R_0 - \delta_0 - 0.5h_0]}{E_0 F_0} + \frac{(R+0.5h)}{EF}\right\}\left(\frac{2\pi}{n} + \sin\frac{2\pi}{n}\right)\left(8\sin^2\frac{\pi}{n}\right)^{-1}\right]^{-1}.\end{aligned} \tag{17}$$

The maximum statistical average tolerance for the deviation from the theoretical contour of the shell is:

$$\Delta^{c}_{thc\ \max} = (R_0 - R_{\text{thc}}) - (2\delta_0 + h_0) + \Delta_0, \tag{18}$$

while the minimum statistical average tolerance is determined by the dependence

$$\Delta^{c}_{thc\ \min} = \Delta^{c}_{thc\ \max} - \Delta_{g\,0}. \tag{19}$$

Maximum realizable tolerance $\Delta^{r}_{thc\ \max}$ is determined by replacement of R_0 in (14) and (18) by a_0, and minimum realizable tolerance $\Delta^{r}_{\min}$ is found by replacement of R_0

in (14) and (18) by b_0. In this case, a_0, a, b_0, b depend on the grade of accuracy, which regulates the accuracy of manufacturing of the shell and frame [29].

For the fairly large number of tightening points ($n \approx 100$), $\sin\alpha \to \alpha$. Then $c = 1/\alpha = R/dS$ and expressions (13) – (14) take the form of (5) and (4) obtained earlier for the assembly by gluing. Accordingly, expression for the intensity of q (7) takes place instead of (15). Displacements Δ_{g0}, Δ_g and $\Delta_{\Sigma g}$ decrease to negligible values, actually going to zero.

After assembly of the shell with the frame and removal of clamps, spring action occurs in the assembly section because of unbalance of the internal forces N and N_0, already acting in the solid cross-section. The algebraic sum of these forces N_{sp} is:

$$N_{sp} = N_0 + N = q(R_0 - R - 0.5h - \delta_0 - 0.5h_0). \tag{20}$$

The forces N_{sp} will lead to stretching forces of spring action q_{sp} in the assembly:

$$q_{sp} = N_{sp}R_{na}^{-1} = qR_{na}^{-1}(R_0 - R - 0.5h - \delta_0 - 0.5h_0), \tag{21}$$

where R_{na} – radius of neutral axis of assembly section

$$R_{na} = R_{as} + Y_{cg} = R_0 + \Delta_0 + Y_{cg} = R_0 + q(R_0 - \delta_0 - 0.5h_0)^2(E_0F_0)^{-1} + Y_{cg}. \tag{22}$$

Here Y_{cg} – ordinate of the center of gravity of assembly section reduced to one material.

When we bring the shell material to the frame material, the result after transformations is:

$$Y_{cg} = \frac{0.5h^2 - 2\delta_0 h_0 E_0 E^{-1}}{h + 2\delta_0 E_0 E^{-1}} R_0. \tag{23}$$

The radial displacement Δ_{sp} from the forces of spring action q_{sp} is determined from the solution of statically indeterminate problem for the assembly section using the force method

$$\Delta_{sp} = q_{sp}R_{na}^2[b(Eh + 2E_0\delta_0)]^{-1}. \tag{24}$$

Then the realizable tolerance Δ_{thc}^* for the shell bonding to the frame by gluing taking spring action into account is equal to:

$$\Delta_{thc}^* = \Delta_{thc} + \Delta_{sp}. \tag{25}$$

4 Results

In conclusion, let us have a look at the example. Mirror antenna reflector is assembled by gluing or connection with screws of structural elements with the following values: radius of the reflector theoretical contour $R_{thc} = 500$ mm; frame height $h = 10$ mm; frame width $B = 20$ mm; thickness of reflector face sheets $\delta_0 = 1$ mm; honeycomb core height $h_0 = 8$ mm; modulus of elasticity of the frame material in the annular direction E

= 100 GPa; modulus of elasticity of the reflector material in the annular direction $E_0 =$ 50 GPa; dimensions of semi-axes of the frame $a_0 = 509$ mm, $a = 512$ mm and reflector $b_0 = 508$ mm, $b = 510$ mm determined as a result of a series of measurements; number of screws $n = 16$. For the specified parameters, we determine the equivalent radii of the frame ($R = 511$ mm) and the shell ($R_0 = 508.5$ mm) to be assembled; statistical average gap $\Delta_{as} = 2.5$ mm. Calculation by formula (9) determines the tolerance for the reflector theoretical contour in case of gluing without consideration of the spring action $\Delta_{thc} = 0.795$ mm; in this case, radius of the neutral axis of the assembly section R_{na} = 514.6 mm; stretching forces of spring action $q_{sp} = 0.43$ N/mm; radial displacement from spring action forces $\Delta_{sp} = 0.005$ mm. Thus, the realizable tolerance for the frame to reflector connection by gluing, taking spring action into account, will be $\Delta = 0.79$ mm. As shown by the considered example, recording of spring action does not make a substantial correction to the tolerance for the theoretical contour of the mirror antenna reflectors at the actual design parameters of its elements, which confirms the conclusions of the work [21, 23]. For the assembly of mirror antenna by connection with screws we find the tightening force at the point $P = 188$ N, reflector radial displacement $\Delta_0 = 2$ mm, displacement of its contour relative to the mating radius after tightening $\Delta_{g0} = 0.19$ mm, maximum and minimum statistical average tolerance, respectively, $\Delta = 0.5$ mm, $\Delta =$ 0.31 mm.

Therefore, connection of the frame and reflector with screws can be more effective with the rational number of screws, which should be installed additionally [25, 29]. As is clear from the obtained results, with fairly large number of screws ($n \approx 100$) the realizable tolerance band for the deviation of the reflecting surface from the theoretical contour in cases of gluing and connection with screws is almost the same.

5 Conclusions

The analytical dependences have been obtained that allowed predicting the regulated tolerance band for assembly by gluing or connection with screws of the PCM products such as parabolic antenna reflectors. These dependences take into account parameters of the product design and assembly process, as well as strength characteristics of the adhesive, the rational combination of which allows achieving a high quality of the product. Our findings allow solving a number of new process tasks in order to determine the maximum permissible deviations of composite shells from their theoretical contour at the assembly stage. In addition, results of the study can be used in the manufacturing of precision products of various applications. However, results of this investigation are tentative, and they require a more profound further analysis with the use of new mathematical models describing the real deformation character of such structures and experimental data.

References

1. Lian, P.Y., et al.: Future research trend for improving large reflector antenna service performance. Eng. **7**(8), 1047–1050 (2021). https://doi.org/10.1016/j.eng.2021.05.005
2. Handbook of Antenna Technologies. Ed. Ning, C. Zhi, Duixian, L., Hisamatsu, N., Xianming, Q., Thomas, Z. – Springer Singapore (2016). https://doi.org/10.1007/978-981-4560-44-3
3. Chandra, M., Kumar, S., Chattopadhyaya, S., Chatterjee, S., Kumar, P.: A review on developments of deployable membrane-based reflector antennas. Adv. Space Research **68**(9), 3749–3764 (2021). https://doi.org/10.1016/j.asr.2021.06.051
4. E Morozov, E. V., Lopatin, A. V.: Design, analysis, manufacture and testing of the spacecraft mirror antenna with the composite high precision and size-stable solid surface reflector. Compos. Struct. **301**, 116185 (2022). https://doi.org/10.1016/j.compstruct.2022.116185
5. Tanaka, H., Kogiso, N., Sakano, F., Katsumata, N., et al.: Experimental demonstration of deformable reflector antenna system with high accuracy deformation measurement. Acta Astron. **194**, 93–105 (2022). https://doi.org/10.1016/j.actaastro.2022.02.002
6. Kurennov, S., Smetankina, N., Pavlikov, V., Dvoretskaya, D., Radchenko, V.: Mathematical model of the stress state of the antenna radome joint with the load-bearing edging of the skin cutout. International Conference on Reliable Systems Engineering (ICoRSE) - 2021, **305**, 287–295 (2022). https://doi.org/10.1007/978-3-030-83368-8_28
7. Park, T.Y., et al.: Thermal Design and analysis of unfurlable CFRP Skin-based parabolic reflector for spaceborne SAR antenna. Int. J. Aeron. Space Sci. **22**(2), 433–444 (2021). https://doi.org/10.1007/s42405-020-00301-7
8. Castanie, B., Bouvet, C., Ginot, M.: Review of composite sandwich structure in aeronautic applications. Compos. C. **1**, 100004 (2020). https://doi.org/10.1016/j.jcomc.2020.100004
9. Kondratiev, A., Potapov, O., Tsaritsynskyi, A., Nabokina, T.: Optimal design of composite shelled sandwich structures with a honeycomb filler. In: Ivanov,V., Trojanowska, J., Pavlenko, I., Zajac, J., Perakovi´c, D. (eds.) DSMIE 2021. LNME, pp. 546–555. Springer, Cham (2021). https://doi.org/10.1007/978-3-030-77719-7_54
10. Baran, I., Cinar, K., Ersoy, N., Akkerman, R., Hattel, J.H.: A review on the mechanical modeling of composite manufacturing processes. Archiv. Comput. Meth. Eng. **24**(2), 365–395 (2017). https://doi.org/10.1007/s11831-016-9167-2
11. Slyvynskyi, V.I., Alyamovskyi, A.I., Kondratjev, A.V., Kharchenko, M.E.: Carbon honeycomb plastic as light-weight and durable structural material. In: Proceed. of the 63rd International Astronautical Congress 2012, IAC 2012, Naples, Italy, **8**, 6519–6529 (2012)
12. Kivirenko, O., Kostenko, S., Sorokin, V., Tsegelnyk, Y., Plankovskyy, S., Kombarov, V.: Fiberglass pipeline continuous filament winding automation. Int. J. Mechatron.S Appl. Mech. **14**, 241–247 (2023) https://doi.org/10.17683/ijomam/issue14.28
13. Wu, Z.M., Li, H., Li, Q.: Novel thermally stable segmented solid paraboloid antenna for high surface accuracy. Aiaa J. **59**(11), 4669–4685 (2021). https://doi.org/10.2514/1.j060169
14. Zhang, S.J., Zhang, S.X., Zhong, W., Sun, Z.H.: Thermal design optimization method of mesh reflector antennas considering deformation compatibility conditions. J. Aerosp. Eng. **35**(4), 04022048 (2022). https://doi.org/10.1061/(asce)as.1943-5525.0001442
15. Wei, J.F., Zhang, C., Qi, X.L., Ma, X.F., Li, Y.Y.: Ultra-high-precision reflectors - design concepts, structural optimization and zero-expansion composites. Int. J. Comp. Mater. Sci. Eng. **7**(1–2), 1850009 (2018). https://doi.org/10.1142/s2047684118500094
16. Artner, G., Gentner, P.K., Nicolics, J., Mecklenbrauker, C.F.: Carbon fiber reinforced polymer with shredded fibers: quasi-isotropic material properties and antenna performance. Int. J. Anten. Propag. **2017**, 6152651 (2017). https://doi.org/10.1155/2017/6152651
17. Garbin, D.F., Tonatto, M.L.P., Amico, S.C.: Compressive and flexural behavior of fiberglass/Polyurethane sandwich panels: Experimental and numerical study. Mechan. Compos. Mater. **57**(4), 459–468 (2021). https://doi.org/10.1007/s11029-021-09969-8

18. Taygin, V. B., Lopatin, A. V.: The review of designs of mirror spacecraft antennas with solid high precision size stable reflector. Spacecr. Technol. **5**(1), 14–26 (2021). https://doi.org/10.26732/j.st.2021.1.02
19. Morozov, E.V., Lopatin, A.V., Khakhlenkova, A.A.: Finite-element modelling, analysis and design of anisogrid composite lattice spoke of an umbrella-type deployable reflector of space antenna. Compos. Struct. **286**, 115323 (2022). https://doi.org/10.1016/j.compstruct.2022.115323
20. Shah, D.B., Patel, K.M., Patel, A.I., Pariyal, V., Joshi, S.J.: Experimental investigation on spring-back deformation during autoclave curing of parabolic antenna reflectors. Compos. A Appl. Sci. Manuf. **115**, 134–146 (2018). https://doi.org/10.1016/j.compositesa.2018.09.017
21. Zebrine, D., Wadhwani, E., Nutt, S.: Surface porosity development in tool-side facesheets of honeycomb core sandwich structures during co-cure. Adv. Manuf. Polym. Compos. Sci. **8**(2), 43–55 (2022). https://doi.org/10.1080/20550340.2022.2056313
22. Qin, Y., Kang, R.K., Wang, Y.D., Sun, J.S., Dong, Z.G.: A form error evaluation method of honeycomb core surface. Measur. **201**, 111667 (2022). https://doi.org/10.1016/j.measurement.2022.111667
23. Moretti, L., Olivier, P., Castanie, B., Bernhart, G.: Experimental study and in-situ FBG monitoring of process-induced strains during autoclave co-curing, co-bonding and secondary bonding of composite laminates. Compos. A Appl. Sci. Manuf. **142**, 106224 (2021). https://doi.org/10.1016/j.compositesa.2020.106224
24. Wei, X.Y., Li, D.F., Xiong, J.: Fabrication and mechanical behaviors of an all-composite sandwich structure with a hexagon honeycomb core based on the tailor-folding approach. Compos. Sci. Techn. **184**, 107878 (2019). https://doi.org/10.1016/j.compscitech.2019.107878
25. Zhang, X.Y., Xu, F., Zang, Y.Y., Feng, W.: Experimental and numerical investigation on damage behavior of honeycomb sandwich panel subjected to low-velocity impact. Compos. Struct. **236**, 111882 (2020). https://doi.org/10.1016/j.compstruct.2020.111882
26. Nsengiyumva, W., Zhong, S.C., Lin, J.W., Zhang, Q.K., Zhong, J.F., Huang, Y.X.: Advances, limitations and prospects of nondestructive testing and evaluation of thick composites and sandwich structures: A state-of-the-art review. Compos. Struct. **256**, 112951 (2021). https://doi.org/10.1016/j.compstruct.2020.112951
27. Panchenko, S., Gerlici, J., Vatulia, G., Lovska, A., Pavliuchenkov, M., Kravchenko, K.: The analysis of the loading and the strength of the FLAT RACK removable module with viscoelastic bonds in the fittings. Appl. Sci. **13**, 79 (2023). https://doi.org/10.3390/app13010079
28. Fomin, O., Logvinenko, O., Burlutsky, O., Rybin, A.: Scientific substantiation of thermal leveling for deformations in the car structure. Int. J. Eng. Techn. **7**(4.3), 125–129 (2018). https://doi.org/10.14419/ijet.v7i4.3.19721
29. Li, Y., Xiao, Y., Yu, L., Ji, K., Li, D.S.: A review on the tooling technologies for composites manufacturing of aerospace structures: materials, structures and processes. Compos. A Appl. Sci. Manuf. **154**, 106762 (2022). https://doi.org/10.1016/j.compositesa.2021.106762

Investigation on Induced Intra/Interlaminar Damage Propagation in CFRP Subjected to Cyclic Tensile Loading After Impact (TAI)

Francisco Maciel Monticeli[1,2](✉), Felipe Ruivo Fuga[2], Mariano Andrés Arbelo[2], and Maurício Vicente Donadon[2]

[1] Department of Aerospace Structure Materials, Faculty of Aerospace Engineering, Kluyverweg 1, 2629HS Delft, Netherlands
F.M.Monticeli@tudelft.nl

[2] Department of Aeronautical Engineering, Technological Institute of Aeronautics (ITA), São José Dos Campos 12228-615, Brazil

Abstract. Impact damage to composite structures results in multiple, complex failure modes, often requiring the replacement of entire components and thereby escalating aircraft maintenance costs. To address this issue, the present study investigates the damage propagation behaviour with particular emphasis on intra- and interlaminar failure modes. Carbon fibre/epoxy composites were subjected to tensile after impact (TAI) fatigue tests at different energy levels to induce different damage modes and extents within the specimens. A non-destructive testing technique (C-scan) was used to assess the interlaminar damage propagation, while the intralaminar fracture toughness of the post-impact specimens was characterised using a finite fracture mechanics model. The results show that the crack propagation behaviour is strongly influenced by the initial impact damage characteristics, in particular the impact energy level. Lower impact energies tend to promote interlaminar failure modes leading to fatigue crack propagation by delamination. Conversely, higher impact energy levels induce fibre fracture, resulting in a self-similar relationship between intra- and interlaminar propagation.

Keywords: Fatigue · Crack Propagation · Finite Fracture Mechanics · Polymer Composite

1 Introduction

The progressive adoption of carbon fibre-reinforced polymer (CFRP) composites in the aeronautic and aerospace fields has led to a substantial weight reduction while ensuring optimal correlation with mechanical properties [1, 2]. This remarkable increase in structural application underscores its significance as a transformative material in achieving lightweight design objectives without compromising structural integrity [3].

Structural components within aircraft are designed to withstand the load per established airworthiness regulations [4]. Nevertheless, unexpected structural damage can compromise the stability of these components [5]. Such damage leads to a decay in

H. Altenbach et al. (Eds.): CAMPE 2023, pp. 227–236, LNME 2025.
https://doi.org/10.1007/978-3-031-82979-6_23

mechanical properties, rendering the component incapable of supporting the projected load. Aeronautical structures may be subjected to different impact scenarios whilst in service, including bird strikes and hail (impact damages), which can significantly affect the composite mechanical behaviour [6, 7]. Allan [8] also conducted a comprehensive study revealing that the repair or replacement costs of primary and secondary structures subject to bird strike-induced damage can reach US$ 1.2 billion annually. Moreover, impact damage initiates intra- and interlaminar fractures, further amplifying the complexity of analysing and controlling the damage progression.

The interlaminar fracture refers to the separation or delamination between adjacent layers of composite material [8]. Intralaminar fracture refers to fractures that begin and grow within each layer, and translaminar fracture of fibres should also be considered [9]. Measuring and distinguishing damage propagation induced by fatigue loading, particularly interlaminar and translaminar failure modes, imposes challenges for designing and inspecting composite aerostructures. Traditional non-destructive evaluation techniques may need help differentiating between these fracture types, as their resolution often falls short in detecting smaller-scale damage [10, 11]. Furthermore, the complex mechanisms involved in damage propagation within composite materials, including delamination and fractures within layers, demand advanced experimental techniques and physical-based models [12, 13].

Multi-scale damage further complicates the measurement of fatigue damage propagation [14]. Microcracks, matrix cracks, fibre breaks, and their interactions can occur simultaneously, demanding a comprehensive approach that combines several experimental techniques [15, 16]. Overcoming the challenges in distinguishing interlaminar and intralaminar damage provides a crucial advantage in effectively controlling the fatigue life of a structural component aiming at designing damage-tolerant structures.

In the search to comprehensively characterise the physical behaviour of crack propagation in laminated composites, a lack of current knowledge has been identified regarding the simultaneous consideration of intra/interlaminar fractures in crack propagation rate to assess the energy release during cyclic loading. Therefore, this work aims to investigate the behaviour of intra/interlaminar damage propagation in structural composites by subjecting them to post-impact cyclic tensile stress (TAI – tensile after impact). Additionally, the assessment of damage propagation will be conducted using a non-destructive testing technique (ultrasound microscopy) combined with the finite fracture mechanic model.

2 Experimental Procedure

2.1 Materials and Manufacturing

The composite laminate was manufactured by resin transfer moulding (RTM). The reinforcement used was carbon fibre plain weave fabrics, Hexcel® HexForce™ AGP193-P. The matrix was epoxy resin (Araldite® LY 5052) with a 100:32 weight ratio with hardener (Aradur® 5052). The stacking sequence was a quasi-isotropic lay-up configuration with eight layers $[0/45/-45/90]_S$. The laminate thickness was 2 mm, with eight layers of reinforcement (50% of fibre/matrix). The manufacturing procedure involved injecting resin at 5 bar injection pressure at 25 °C for 120 min. The 20 mbar vacuum was applied during injection. The cure was carried out at 25 °C for 24 h and then a post-cure at 100 °C for 4 h.

2.2 Experimental Test Configurations

The low-velocity impact (LVI) tests were performed according to the ASTM D7136M-15 standard, where a specimen geometry of 200 × 110 × 2 mm^3 was chosen, and a 1.46 kg semi-hemispherical impactor with 16 mm diameter was used. The impactor was dropped from a specific height and guided by lateral rails. The drop heights were set to achieve impact energy levels of 11 and 18 J. Detailed test information can be found in Ref. [16].

Tensile testing was carried out using an MTS servo-hydraulic machine with a 250 kN load cell. The static tests were conducted at a rate of 2 mm/min. The specimen dimensions were 200 × 50 × 2 mm^3, with the impact damage at the central region of the specimen. The fatigue tests were performed under load control with a sinusoidal loading spectrum with a maximum stress amplitude of 85% of the laminate tensile strength (σ_{max}) and stress ratio R = ($\sigma_{min}/\sigma_{max}$) = 0.1 and frequency of 5 Hz. A constant amplitude was maintained during the fatigue tests. The elastic properties at ply level are $E_1 = E_2 =$ 51.83 GPa, $G_{12} = 2.82$ GPa, $\nu_{12} = 0.064$, measured following ASTM D3039 and ASTM D3518 standards. C-scan images were obtained through acoustic ultrasound inspection using the Sonotron ISONIC 2006 (NDT, 2006) equipment with a 5 MHz probe.

Figure 1 illustrates the flowchart of each test step and the data analysis during each stage. Initially, impact tests were conducted at two distinct impact energy levels. Subsequently, a static tensile test was performed to evaluate the rupture stress corresponding to each impact energy level. For additional specimens, tensile tests were applied through cyclic loading, with intervals between cycles to facilitate C-scan analysis. Following 100,000 cycles, a static stress-to-fracture test was performed to determine the new post-fatigue rupture residual stress. The remote stress is the failure stress measured during the static test. Remote stress was measured for the pristine material (laminate without damage), after impact, and after impact and tensile cyclic loading. All collected data were analysed utilising the finite fracture mechanics (FFM) model, as proposed by Camanho et al. [14], which accounts for open-hole strength and follows Eqs. 1 and 2, which can be fully described in the supplementary material (Eqs. S1–S12) and also in Ref. [14]. The application of this methodology primarily involves assessing the crack propagation rate (*dL/dN*) relative to the released energy (*ΔK*) per cycle. In parallel, interlaminar growth analyses were conducted using C-scan techniques, recognising that the FFM model neglected the propagation of interlaminar damage.

$$\frac{1}{L}\int_R^{R+L} \sigma_{yy}(x,0)dx = X_t(N);\ \text{ for } X_t = X_t(N) = C_1 N^{-m_1} \tag{1}$$

$$\frac{1}{L}\int_R^{R+L} K_I^2(a)da = K_I^2(N);\ \text{ for } K_I = K_I(N) = C_2 N^{-m_2} \tag{2}$$

where, L is the crack size at the failure, R is the hole radius ($2R = D$), $\sigma_{yy}(x,0)$ is stress distribution along the *x-axis*, X_t is the strength of the laminate without damage, K_{IC} is the mode I intralaminar fracture toughness of the laminate, N is the cycle number, C_i and m_i are the constants coefficients. The remote and pristine stresses represent the failure stress of the material. The pristine stress is the failure stress of the material without impact damage, and the remote stress means the failure stress after impact and fatigue damage added to the material.

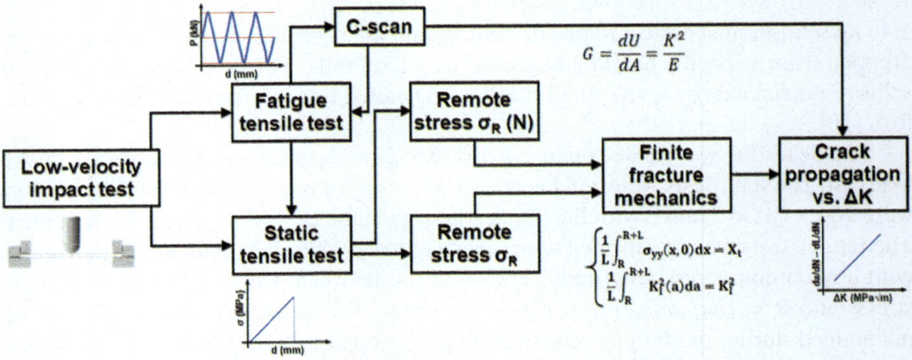

Fig. 1. Work flowchart following test procedure and data analysis.

3 Results and Discussion

Table 1 presents the summarised results of the experimental drop test, divided into low-energy (LE) and high-energy (HE) groups. Considering the impact mass and the standard procedure, the heights used were 76.58 and 125.32 cm to reach 11 and 18 J, respectively. The velocity at the impact (v_i) provides insight into the initial kinetic energy the impacting object possesses. On the other hand, v_{out} represents the impactor velocity after the collision, which was determined by analysing the rebound or post-impact motion of the impactor. The energy associated with the incident impact (E_i) was calculated using the mass and velocity of the impactor. It represents the total kinetic energy transferred to the specimen during the collision. The dissipated energy (E_d) quantifies the amount of energy absorbed and dissipated by the specimen due to the impact.

The difference in incidence velocities is attributed to variations in the positioning height of the impactor and friction effects. The output velocity corresponds to the dissipated energy within the material. Consequently, the low-energy (LE) level exhibits relatively lower energy dissipation due to induced internal delamination fractures and a limited occurrence of intralaminar fractures (Fig. 2b - 11J). On the other hand, samples subjected to higher impact energy levels (HE) increase the energy dissipation, characterised by the difference between the incident and rebound velocities. This disparity arises from the greater extent of material fracture resulting from the higher impact energy level (Fig. 2b - 18J), consuming more energy in the process.

Post-impact cyclic stress loading was applied after subjecting the laminate to impact at two levels of impact energy. Figure 2a exhibits the C-scan images of non-impacted

Table 1. Experimental impact results

Impact energy	v_i (m . s^{-1})	v_{out} (m • s^{-1})	E_i (J)	E_d (J)
Low (LE)	3.88 ± 0.03	2.77 ± 0.14	11.00 ± 0.17	5.43 ± 0.42
High (HE)	4.96 ± 0.08	2.69 ± 0.15	17.96 ± 0.36	12.71 ± 0.26

laminate and the impacted curve with respective C-scan images from each impact damage energy. Figure 2b shows the images of the backside of the impacted laminate and the ultrasound results after cyclic loading. Similar behaviour can be observed in both impact tests, where the maximum load reaches approximately 3000 N in both cases. However, a significant difference is observed in Table 1, indicating longer contact with the impactor for higher impact energy levels (HE), resulting in more intralaminar damage. Conversely, the low-energy (LE) laminates primarily exhibited internal damage, particularly associated with delamination. The results presented are consistent with published studies [16, 17] showing delamination as the primary fracture mechanism at low impact energy and increased fibre fracture at higher impact energy.

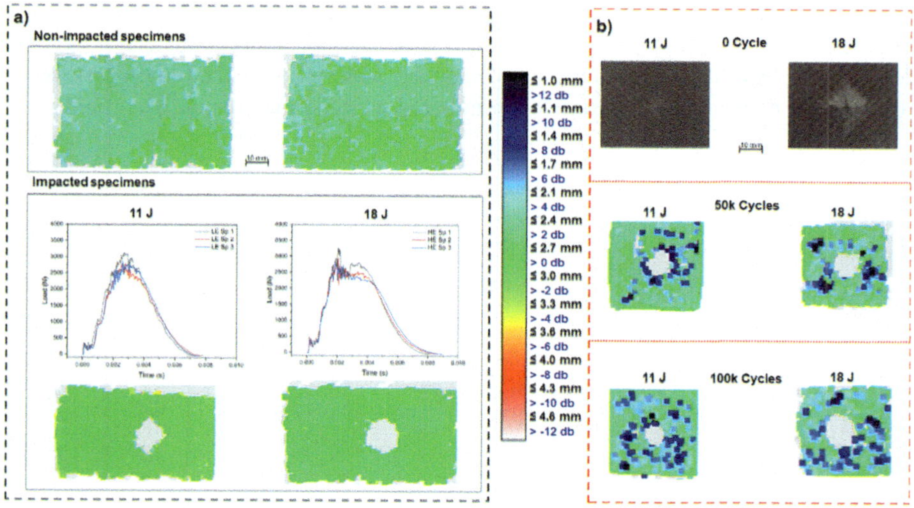

Fig. 2. Mechanical test analysis: a) impact test curve and C-scan images, and b) propagation of interlaminar fatigue damage (C-scan images).

The regions surrounding the hole after cyclic loading displayed fluctuations in attenuation values. This phenomenon can be attributed to internal damage. Ultrasound waves are emitted through the thickness, travel through the material and return to the receiving probe. In case of internal defect, there is attenuation/loss of the return signal, generating a change in the colour scale to identify internal damages. As a result, the wave direction deviates, leading to variations in the signal read by the probe.

During the cyclic loading of impacted samples, the crack initiates in the matrix and propagates along the fibre/matrix interface, following the fibre direction and forming splitting [18]. The static test promotes intralaminar fractures perpendicular to the fibre direction, reaching the required energy for fibre rupture. On the other hand, splitting and microcracking damage generate a preference for interlaminar propagation occurring between the laminae since fatigue loading provides sufficient energy for these fracture patterns. Interlaminar fractures can grow and lead to significant stress decay, potentially preventing translaminar fractures in the material [19]. Therefore, C-scan ultrasound images were used to measure the propagation of interlaminar damage after the cyclic loading. The increase in fracture radius was evaluated by considering the equivalent fracture area generated by the highlighted fluctuations in attenuation values associated with the equivalent delamination region, as shown in Fig. 2b.

After the cyclic test, a static tensile test was performed in the same specimen to evaluate the residual stress. According to Eqs. 1 and 2, as there is a decay on the remote stress resulting from the fatigue damage, the size of the crack length (L) also changes. Table 2 presents the experimental results of the static and cyclic tests (as shown in flowchart Fig. 1). The C-scan damaged radius is the internal damaged area (interlaminar), measured in terms of the equivalent radius with the C-scan images through the thickness, following Fig. 2b. L is the translaminar crack length provided from Eqs. 1 and 2. The static test after fatigue provided the remote stress (σ_r), considered the fracture stress. K_I was calculated following $K_I = \sigma F_1 F_2 \sqrt{\pi a}$, in which F_i are the corrections factor (detailed described in Eqs. S8 and S9 in the supplementary material), σ is the remoted measured stress, and a is the total crack size (open-hole radius R + crack length L).

Table 2. Experimental results of damage propagation for low and high impact energies.

Low-energy (LE)					
C-scan damage radius (mm)	L (mm)	σ_r (MPa)	Cycles ($\times 10^3$)	K_I (MPa$\sqrt{m}$)	X_t (MPa)
0	3.97	443.28	0	58.70	497.99
3.56	4.50	441.18	50	86.69	
12.81	4.89	431.93	100	96.20	
High-energy (HE)					
C-scan damage radius (mm)	L (mm)	σ_r (MPa)	Cycles ($\times 10^3$)	K_I (MPa$\sqrt{m}$)	X_t (MPa)
0	2.71	430.80	0	56.56	497.99
4.54	4.87	414.37	50	89.84	
9.28	5.04	402.17	100	101.18	

The pristine stress (X_t) was measured from a non-impacted in the same lay-up configuration. The HE impact provides a diameter of 20 mm, and the LE impact provides a damage diameter of 10 mm. After cyclic loading, the LE exhibits larger internal damage growth than the HE laminates, considering the delamination measured by the C-scan. HE presents lower damage propagation through delamination due to translaminar damage from L-crack growth. The low coefficient of variation presented ($\approx 3.74\%$) indicates that the specimen exhibited homogeneity behaviour. Therefore, the average values will be used to model finite fracture mechanics. The relationship between the results in Table 2 is observed in detail in Fig. 3.

The residual stress exhibits a decay as a function of cycles (Fig. 3a). The HE specimens exhibited a more abrupt stress decay than the LE specimens, indicating a dependency on the initial damage generated by the impact test. Along the cycles, the stress decay difference is 6.6% for the high-energy laminate, while for the low-energy laminate, it is 2.5%. This difference is directly associated with the initial decay generated by the impact fracture. In other words, HE specimens have a larger fracture region (i.e., 20 mm) and tend to show higher stress decay during cyclic loading than LE specimens, which present 10 mm of the fractured area.

The propagation of simultaneous intra/interlaminar damage is directly associated with the impact energy level. Even with the delamination present for both energy levels, significant intralaminar fracture only occurs for higher energy. Delamination is crucial in fatigue fracture analysis [20]. The interfacial region between plies is highly susceptible to delamination mechanisms due to its weaker mechanical and chemical interaction than fibre rupture. Consequently, microcracking between layers during cyclic loading can reduce stress [21]. Therefore, this work focused on measuring interlaminar damage progression through damage growth (C-scan) and intralaminar propagation (FFM model).

The crack growth (da) was considered analogous to the crack length (L) in the FFM model and the radius size (R) of the internal damage region measured by the *C-scan*. Consequently, the crack growth curve was determined by the derivative of the curve in Fig. 3b for each fracture mechanism concerning the number of cycles following the Paris curve model [22]. Figure 3b illustrates the growth rate of interlaminar and intralaminar damages as a function of cycles for each post-impact energy level. A significant increase in interlaminar crack growth is observed for the LE specimen, while the intralaminar (translaminar) crack growth presents a lower propagation. This difference arises because low-impact energy results in more damage between layers and less fibre breakdown [16]. The interlaminar damage propagation during the application of cyclic loading is more prominent for the LE specimen, leading to a non-significant stress decay. Consequently, there is a lower variation in the crack length (L) since it is directly proportional to the stress variation.

The HE specimen exhibits a higher initial rate of intralaminar damage growth. This is attributed to the higher energy impact-induced fracture, resulting in a higher fraction of fibre rupture and a larger fractured region than LE. Thus, the growth of interlaminar and intralaminar damage occurs in parallel behaviour, as the impact damage establishes a similar relationship between both types of damage, promoting parallel growth for both failure modes. These results suggest that the propagation and failure modes are influenced by the initial configuration of the impact damage [23].

Based on the experimental data presented in Fig. 3c, a second-order polynomial curve was fitted to each experimental data. Subsequent, those curves were used to develop the

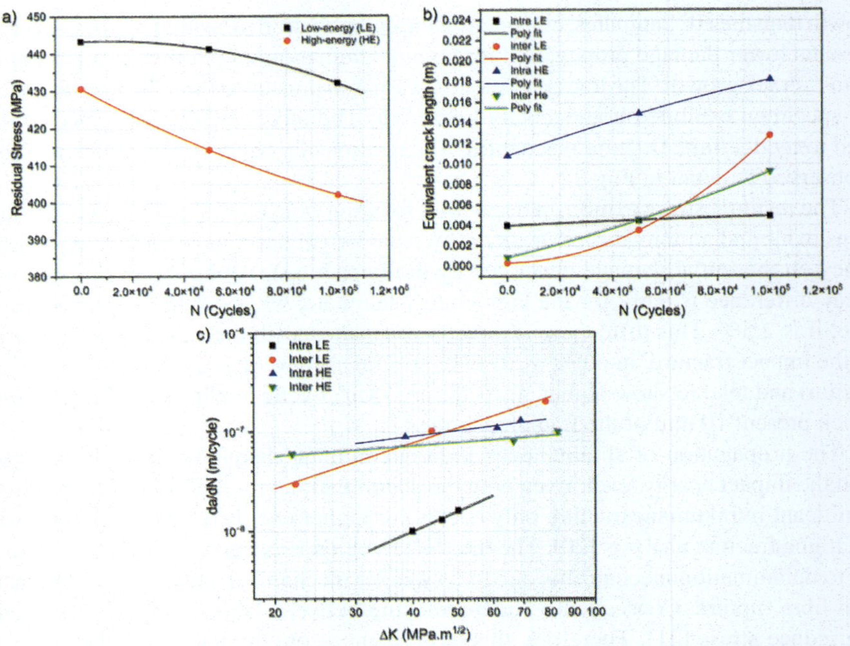

Fig. 3. a) SN curve, b) Compliance curve: equivalent crack length vs. cycles, and c) Paris curve: crack propagation rate versus stress intensity factor.

crack growth rate curve (*da/dN*), which was considered analogous to the crack growth rate of size *L* (*dL/dN*), as well as the growth of the internal damage radius *R* (*dR/dN*). The stress intensity factor (*K*) was calculated using Eq. 2.

Figure 3c shows the curves based on the power law curve, in which the crack propagation rate is related to the stress intensity factor (SIF – ΔK) as a similitude energy parameter. The following equation describes the crack propagation behaviour for each case: LE$_{intra}$ $(da/dN) = 6 \times 10^{-12}(\Delta K)^{2.01}$; LE$_{inter}$ $(da/dN) = 9 \times 10^{-12}(\Delta K)^{2.35}$; HE$_{intra}$ $(da/dN) = 2 \times 10^{-8}(\Delta K)^{0.47}$, HE$_{inter}$ $(da/dN) = 2 \times 10^{-8}(\Delta K)^{0.34}$. After interpolating the experimental data to determine the average crack propagation behaviour, the power function fitting coefficients C_i and m_i for each damage mode.

A comparative analysis of failure modes shows that the LE specimen exhibits lower intralaminar crack propagation rate levels, mainly due to the delamination-induced impact damage, favouring interlaminar propagation. In addition, the angular coefficient for interlaminar damage (2.35) is higher than that for intralaminar damage (2.01), suggesting that intralaminar damage suffers a more abrupt propagation than interlaminar propagation.

In contrast, the HE specimen demonstrates a similar behaviour between the two curves, indicating a close relationship between the propagation region for intra and interlaminar failure modes. The values of the constants for each curve support this observation. The damage caused by the initial impact directly affects the crack propagation under cyclic loading [24, 25]. In other words, lower impact energy levels result in a higher crack propagation rate of interlaminar fracture, making delamination more susceptible to growth during cyclic loading. Higher impact energy leads to more intralaminar initial damage features, creating an equal propagation behaviour for both modes.

4 Conclusion

The Tensile After Impact (TAI) fatigue test was conducted to investigate the damage propagation behaviour and the effects of different impact energy levels in carbon fibre/epoxy composites. Intra- and interlaminar crack propagation were measured and analysed using non-destructive ultrasonic techniques and a finite fracture mechanics model. A failure envelope based on the Paris model was derived by establishing a correlation between the different damage modes and the crack propagation rate.

The results show that the crack propagation behaviour is directly influenced by the amount and type of initial damage, which is determined by the impact energy level. In particular, damage zones caused by low energy impacts lead to a more rapid absolute increase in interlaminar cracking. As the severity of the impact damage increases, there is a shift in the initial failure modes, indicating a synergistic interaction between intra- and interlaminar failure mechanisms. Consequently, higher impact energy levels promote a parallel propagation rate for intra- and interlaminar fractures, highlighting the relationship between damage severity and crack propagation dynamics.

Acknowledgements. The authors acknowledge financial support from FAPESP (process number: 2021/05706-5), CNPq (process number: 301069/2019-0) and CAPES (financial code 001).

References

1. Shesan, Q., Stephan, A., Chioma, A., Neerish, R., Rotimi, S. Fiber-matrix relationship for composites preparation. Book: Renewable and sustainable composites, Elsevier, 1–30 (2019)
2. Al-Fatlawi, A., Jármai, K., Kovács, G.: Optimal Design of a Fiber-Reinforced Plastic Composite Sandwich Structure for the Base Plate of Aircraft Pallets In Order to Reduce Weight. Polymers **13**(5), 834 (2021)
3. Rajak, D.K., Wagh, P.H., Kumar, A., Sanjay, M.R., Siengchin, S., Khan, A.: Impact of fiber reinforced polymer composites on structural joints of tubular sections: A review ARTICLEINFO. Thin-Walled Structures **180**, 109967 (2022)
4. Dattoma, V., Nobile, R., Panella, F.W., Pirinu, A., Saponaro, A.: Damage investigation of aeronautical CFRP laminates under bearing tests. Procedia Structural Integrity **18**, 719–730 (2019)
5. Rozylo, P.: Experimental-numerical study into the stability and failure of compressed thin-walled composite profiles using progressive failure analysis and cohesive zone model. Compos. Struct. **257**, 113303 (2021)
6. Heimbs, S.: Computational methods for bird strike simulations: A review. Comput. Struct. **89**, 2093–2112 (2021)
7. Pierrat, E., Rupcic, L., Hauschild, M.Z., Laurent, A.: Global environmental mapping of the aeronautics manufacturing sector. J. Clean. Prod. **297**, 126603 (2021)
8. Allan, J.R.: The costs of bird strikes and bird strike prevention. Human Conflicts with Wildlife: Economic Considerations **1**, 147–153 (2000)

9. Monticeli, F.M., Cioffi, M.O.H., Voorwald, H.J.C.: Mode II delamination of carbon-glass fiber/epoxy hybrid composite under fatigue loading. Int. J. Fatigue **154**, 106574 (2022)
10. Monticeli, F., Fuga, F., Donadon, M.: A systematic review on translaminar fracture damage propagation in fiber-reinforced polymer composites. Thin-Walled Structures **187**, 110742 (2023)
11. Wang, F., Liu, J., Liu, L., Xu, L., Wang, Y., Chen, M.: Quantitative non-destructive evaluation of CFRP delamination defect using laser induced chirp-pulsed radar photothermal tomography. Opt. Lasers Eng. **149**, 106830 (2021)
12. Biagini, D., Pascoe, J., Alderliesten, R.: Investigating apparent plateau phases in fatigue after impact damage growth in CFRP with ultrasound scan and acoustic emissions. Int. J. Fatigue **177**, 107957 (2023)
13. Cornetti, P., Pugno, N., Carpinteri, A., Taylor, D.: Finite fracture mechanics: A coupled stress and energy failure criterion. Eng. Fract. Mech. **73**, 2021–2033 (2006)
14. Camanho, P.P., Erçin, G.H., Catalanotti, G., Mahdi, S., Linde, P.: A finite fracture mechanics model for the prediction of the open-hole strength of composite laminates. Compos. Part A Appl. Sci. Manuf. **43**, 1219–1225 (2012)
15. Roy, T., Chakraborty, D.: Delamination in hybrid FRP laminates under low velocity impact. J. Reinf. Plast. Compos. **25**, 1939–1956 (2006). https://doi.org/10.1177/0731684406069922
16. Fuga, F.R., Donadon, M.V.: Low velocity impact on pre-loaded composite plates: A novel standard-based experimental apparatus. Compos. Struct. **261**, 113315 (2021)
17. Davies, G.A.O., Zhang, X.: Impact damage prediction in carbon composite structures. Int. J. Impact Engno **16**, 149–170 (1995)
18. Yang, B., Chen, Y., Lee, J., Fu, K., Li, Y.: In-plane compression response of woven CFRP composite after low-velocity impact: Modelling and experiment. Thin-Walled Structures **158**, 107186 (2021)
19. Ramji, A., Xu, Y., Yasaee, M., Grasso, M., Weeb, P.: Influence of veil interleave distribution on the delamination resistance of cross-ply CFRP laminates under low velocity impact. Int. J. Impact Eng **157**, 103997 (2021)
20. Gutkin, R., Pinho, S.T., Robinson, P., Curtis, P.T.: A finite fracture mechanics formulation to predict fibre kinking and splitting in CFRP under combined longitudinal compression and in-plane shear. Mech. Mater. **43**, 730–739 (2011)
21. Llobet, J., Maimí, P., Essa, Y., Martin de la Escalera F. A continuum damage model for composite laminates: Part III - Fatigue. Mechanics of Materials 153, 103659 (2021)
22. Khan, R.: Fiber bridging in composite laminates: A literature review. Compos. Struct. **229**, 111418 (2019)
23. Zhang, D., Liu, L., Lan, X., Li, F., Liu, Y., Leng, J.: Experimental study on nonlinearity of unidirectional carbon fibre-reinforced shape memory polymer composites **16**, 107372 (2023)
24. Kebir, T., Benguediab, M., Imad, A.: A model for fatigue crack growth in the Paris regime under the variability of cyclic hardening and elastic properties. Fatigue of Aircraft Structures **9**, 117–135 (2017)
25. Hu, P., Pulungan, D., Tao, R., Lubineau, G.: An experimental study on the influence of intralaminar damage on interlaminar delamination properties of laminated composites. Compos. A Appl. Sci. Manuf. **131**, 105783 (2020)

Special Constraints of Packing Problems in Materials Structure Research

Andrii Chuhai[1,3](✉), Yuriy Stoyan[1], Tetyana Romanova[1,2], Georgiy Yaskov[1], and Olha Starkova[3]

[1] Anatolii Pidhornyi Institute of Mechanical Engineering Problems of the National Academy of Sciences of Ukraine, Pozharsky 2/10, Kharkiv 61129, Ukraine
chugay.andrey80@gmail.com

[2] University of Leeds, Maurice Keyworth Building, Leeds LS2 9JT, UK

[3] Simon Kuznets Kharkiv National University of Economics, Nauky Ave 9A, Kharkiv 61166, Ukraine

Abstract. The paper considers a relaxation technique in a sphere packing problem with a specific focus on study of materials structure. The objective is to maximize the number of non-overlapping spheres in a cylindrical container regarding a given ratio of the different radii spheres appearance. Two scenarios are discussed to examine an effect of relaxing constraints on the packing spherical objects. In the first scenario, spheres packed within the container while adhering to strict containment conditions. The second scenario violates the containment condition by allowing spheres intersect the container boundary. This relaxation provides a more accurate estimation of the material's porosity by eliminating artificial effects caused by boundary constraints. A fast heuristic approach is developed that combines a modification of the block coordinate descent algorithm and a new cylindrical lattice decomposition strategy. Computational results are presented and illustrated with examples for two scenarios. The findings highlight the importance of the proposed techniques in materials structure studies and can be extended to packing particles of various shaped.

Keywords: materials structure · sphere packing · cylindrical container · containment relaxation · cylindrical lattice decomposition · nonlinear optimization

1 Introduction

The investigation of material structure is a crucial aspect of materials science and engineering, as it directly impacts the properties and performance of various materials. To gain a deeper understanding of material structures, researchers have turned to mathematical modeling techniques and packing problems [1]. In this article review, we delve into the significance of employing mathematical modeling and packing problems in the investigation of material structure, exploring their applications, benefits, and challenges [1].

H. Altenbach et al. (Eds.): CAMPE 2023, pp. 237–245, LNME 2025.
https://doi.org/10.1007/978-3-031-82979-6_24

Mathematical modeling provides a powerful framework for studying material structures at different scales, from atomic and molecular arrangements to macroscopic architectures [2]. By utilizing mathematical equations and computational simulations, researchers can gain insights into the behavior and properties of materials [3]. These models can encompass a range of techniques, including finite element analysis [4], molecular dynamics simulations [5], and lattice-based approaches [5]. Mathematical models facilitate the prediction of material behavior, such as mechanical properties, thermal conductivity, and diffusion characteristics.

Packing problems are important in investigating the arrangement of particles or entities within a material. These problems involve determining the most efficient or optimal packing configurations to achieve specific goals, such as maximizing density or minimizing void spaces [6]. By addressing packing problems, researchers can uncover fundamental principles underlying material structures, leading to advancements in fields like materials synthesis, nanotechnology, and biomaterials. Packing problems also provide insights into the self-assembly behavior of materials and can guide the design of new materials with desired properties.

Packing problems are useful in wide-range of applications in materials science and engineering. One notable area is the study of crystalline structures, where mathematical models help elucidate crystal growth, defects, and phase transitions [7].

This class of geomertic problems play an important role in understanding granular materials [8], colloidal suspensions [9], and the behavior of nanoparticles [10]. These approaches are also used in various industrial applications, such as packing of grains in silos, efficient filling of containers, and designing materials with enhanced mechanical properties.

The utilization of packing problems offers several benefits. Firstly, it enables researchers to explore material structures in a cost-effective and time-efficient manner, reducing the need for extensive experimental investigations. Secondly, it provides a platform for predicting the behavior of materials under various conditions and identifying optimal configurations. However, challenges remain, including the accuracy and reliability of mathematical models, the complexity of packing problems in real-world scenarios, and the computational resources required for large-scale simulations.

When examining a cut of material to calculate porosity, relaxing the constraints on the belonging of particles to the domain can help to achieve a more accurate porosity estimation.

Here's how it can be beneficial.

Inclusion of boundary effects. By relaxing the strict boundaries that confine particle positions within the domain, researchers can account for the influence of particles that may partially extend beyond the cut surface. In reality, particles near the cut surface may have their volumes intersecting with the region of interest. By allowing for such partial inclusion, the calculated porosity becomes more representative of the actual material structure.

Consideration of edge effects. In some cases, the cut surface may introduce edge effects that impact the arrangement of particles near the boundaries. Relaxing the constraints on particle belonging to the domain enables researchers to account for these edge effects and capture the structural characteristics of particles near the cut surface accurately. This leads to a more realistic estimation of porosity.

Enhanced statistical representation. Relaxing the constraints on particle belonging to the domain allows for a larger sample size and statistical representation. Instead of strictly confining particles within the cut, researchers can include neighboring particles that may have a significant influence on the porosity calculation. This broader representation helps to capture the variability in particle distribution, ensuring a more reliable and statistically significant estimation of porosity.

Improved connectivity analysis. Porosity is not only determined by the individual particle positions but also by the connectivity of void spaces within the material. By relaxing constraints on particle belonging to the domain, researchers can accurately capture the connectivity of pores across the cut surface. This allows for a more comprehensive analysis of the pore network and its impact on material properties.

Overall, relaxing the constraints on particle belonging to the domain when examining a cut of material enables a more accurate estimation of porosity by considering boundary effects, edge effects, statistical representation, irregular particle shapes, and pore connectivity. This approach provides a more realistic understanding of the material's structure and porosity distribution, leading to improved predictions of material properties and behavior.

2 Problem Statement

Let there be a collection of spheres $\mathbf{S} = \{S_1, S_2, ..., S_N\}$ with different radii r_i, $i \in I_N = \{1, 2, ..., N\}$ and a cylinder:

$$C = \{X = (x, y, z) \in \mathbf{R}^3 : x^2 + y^2 \leq R^2,\ 0 \leq z \leq H\}$$

.

The spheres are grouped based on specific ranges of radii $r_i \in [r_m^1, r_m^2]$, $i \in I_N$, $m \in Q = \{1, 2, ..., K\}$. The proportions of spheres within each group are given $p_m = P_m/N$ where P_m, is a number of spheres in the m-th group, $m \in Q$.

Proportional Sphere Arrangement (PSA). Non-overlapping spheres from the set **S** should be arranged inside the cylinder C considering proportions p_m of the spheres appearance, while minimizing the porosity (or maximizing the packing factor).

The volume and sizes of the cylinder remaining unchanged, the packing factor is determined by the total volume of spheres to be packed.

Two variants of the problem are considered: PSA – when spheres have to be fully contained inside the cylinder C, i.e. $x_i^2 + y_i^2 \leq (R - r_i)^2$ and $r_i \leq z_i \leq H - r_i$; RPSA (relaxed PSA) – when centers of spheres have to belong to the cylinder C, i.e. $x_i^2 + y_i^2 \leq R^2$ and $0 \leq z_i \leq H$. Here (x_i, y_i, z_i) is the center of S_i, $i \in I_N$.

We develop a fast heuristic algorithm to solve these large-scale sphere packing problems. This approach combines the block coordinate descent technique (see, e.g., [11–13]) with a new cylindrical lattice decomposition strategy.

3 Block Coordinate Descent Algorithm

We arrange spheres from the set **S** by sequential addition, selecting the radii of the spheres according to their occurrence probabilities p_m, $m \in Q$. This approach ensures a sufficiently accurate adherence to the specified proportions p_m while placing a large number of spheres.

Let us describe *k-th* step of the algorithm. Assume that spheres $S_1, S_2, ..., S_{k-1}$ have been arranged in the previous steps. Let $(\hat{x}_i, \hat{y}_i, \hat{z}_i)$, be centers of the spheres S_i, $i \in I_k = \{1, 2, ..., k-1\}$. If $k = 1$, then $I_k=\emptyset$, i.e. there are no arranged spheres.

We consider x_k, y_k, z_k as a block variables and solve the following problem:

$$\min_{x_k, y_k, z_k} z_k \tag{1}$$

subject to

$$x_k^2 + y_k^2 \leq (R - r_k)^2, \ i \in I_k, \tag{2}$$

$$r_k \leq z_k \leq H - r_k, \ i \in I_k, \tag{3}$$

$$(r_k + r_i)^2 \leq (x_k - \hat{x}_i)^2 + (y_k - \hat{y}_i)^2 + (y_k - \hat{y}_i)^2, \ i \in I_k, \tag{4}$$

for PSA.

The model (1)–(4) takes the form

$$\min_{x_k, y_k, z_k} z_k \tag{5}$$

subject to

$$x_k^2 + y_k^2 \leq R^2, \ i \in I_k, \tag{6}$$

$$0 \leq z_k \leq H, \ i \in I_k, \tag{7}$$

$$(r_k + r_i)^2 \leq (x_k - \hat{x}_i)^2 + (y_k - \hat{y}_i)^2 + (y_k - \hat{y}_i)^2, \ i \in I_k, \tag{8}$$

for RPSA.

The sphere S_k (the center of S_k for RPSA) is arranged within C without overlapping with the previously packed spheres S_i, $i \in I_{k-1}$.

Then we search for a fast feasible solution of the problem (1)–(3). A point $v_k = (x_k, y_k, H - r_k) \in C$ (a point $v_k = (x_k, y_k, H) \in C$ for RPSA) is chosen as a starting point. Then z_k is minimized on W_k as it is described in [12].

The process is repeated until we fail to find a starting point $v_k \in C$. In this case, the placement of spheres S_i, $i \in I_k$, obtained at the previous step, is taken as an approximate solution to our problem.

To solve the large-scale sphere packing problem the solution technique proposed in [14] can be applied.

In order to decrease the number of inequalities (2) to be taken into account a cylindrical decomposition technique is applied.

4 Cylindrical Decomposition Technique

Let us consider a cylinder C in the cylindrical coordinate system:

$$C = \{(\rho, \varphi, z) \in \mathbf{R}^3 : 0 \le \rho \le R,\ 0 \le \varphi \le 2\pi,\ 0 \le z \le H\}.$$

We generate a partition of a cylinder with subsets (parts of sectors of the cylinder) by dividing segments of cylindrical coordinates into equal parts subject to $\rho \in [0, R]$, $\varphi \in [0, 2\pi]$, $z \in [0, H]$. For each placed sphere, the parts of the sectors with which it has a nonempty intersection are defined. We referee the technique to a cylindrical decomposition.

Note, that unlike the linear lattice, the cylindrical decomposition does not contain extra elements that do not belong to the cylinder.

Let us describe the corresponding subsets of the cylinder using a tensor denoted by A_{ijl} for $i = 1, 2, ..., s_1, j = 1, 2, ..., s_2, l = 1, 2, ..., s_3$. Indices (i, j, l) of the tensor specify a certain subset.

Each subset is characterized by a list $L(i, j, l)$, $i = 1, 2, ..., s_1, j = 1, 2, ..., s_2, l = 1, 2, ..., s_3$, of spheres. Let the sphere S_k have to be placed. Determine subsets of the cylinder associated with S_k and add the number k to the subset lists. While placing the sphere S_k, only spheres whose numbers are in the lists of subsets with which the sphere S_k has common interior points are monitored to check the non-overlapping conditions.

Consider a simple example in two-dimensional space. The circles S_1, S_2, S_3, S_4 are colored in blue while S_5 is colored in red in Fig. 1. When placing a circle S_5 we have $L(3, 5) = \{1, 2\}$,$L(1, 1) = L(2, 1) = \{3\}$,$L(3, 1) = L(4, 1) = L(3, 2) = L(4, 2) = \{4\}$, the rest of the lists are empty. Circle S_5 has a nonempty intersection with subsets with indices (3, 2),(4, 2),(3, 3) and (4, 3). List of circles numbers for S_5 is defined as $L(3, 2) \cup L(4, 2) \cup L(3, 3) \cup L(4, 3) = \{4\}$. Only the ratio of non-overlapping circles S_5 and S_4 has to be monitored. After placing the circle S_5 four lists: $L(3, 2)$, $L(4, 2)$, $L(3, 3)$, $L(4, 3)$ are updated by adding the number 5. After updating $L(3, 5) = \{1, 2\}$,$L(1, 1) = L(2, 1) = \{3\}$, $L(3, 1) = L(4, 1) = \{4\}$, $L(3, 2) = L(4, 2) = \{4, 5\}$ and $L(3, 3) = L(4, 3) = \{5\}$.

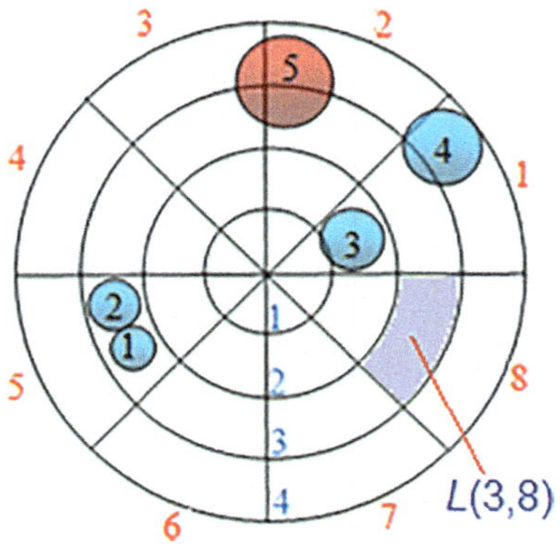

Fig. 1. Cylindrical decomposition.

5 Calculation of Porosity

For PSA, when all spheres are arranged fully inside the cylinder, the porosity is determined by the total volume of the spheres packed as follows:

$$porosity = 1 - \frac{\sum_{i \in I_k} Vol(S_i)}{Vol(C)}.$$

For RPSA, the centers of the spheres have to arranged within the cylinder, therefore some of the spheres can intersect the boundary of the container. The portion of spheres that extend beyond the cylinder's boundary should be excluded from the porosity calculation, i.e.,

$$porosity = 1 - \frac{\sum_{i \in I_k} Vol(C \cap S_i)}{Vol(C)}.$$

Assume that $S_i \not\subset C$, $v_i \in C$.

Let $0 \leq z_i \leq H$. Taking into account that the radius of C is considerably greater than the radii of the arranged spheres, the volume of $C^* \cap S_i$ is close to the volume of the spherical segment $P_i \cap S_i$, where P_i is a halfspace that touches the lateral surface of C and is perpendicular to the line passing through the points (0, 0, 0) and $(x_i, y_i, 0)$, i.e.

$Vol(P_i \cap S_i) = \frac{1}{3}\pi h_i^2(3R - h_i)$, where $h_i = (\sqrt{x_i^2 + y_i^2} - R) > 0$.

Let $z_i < r_i$ and $h_i < 0$, then $Vol(P_i \cap S_i) = \frac{1}{3}\pi \eta_i^2(3R - \eta_i)$, where $\eta_i = r_i - z_i$.

Let $z_i + r_i > H$ and $h_i < 0$, then $Vol(P_i \cap S_i) = \frac{1}{3}\pi \eta_i^2(3R - \eta_i)$, where $\eta_i = z_i + r_i - H$.

Let $z_i < r_i$ and $h_i < 0$, then $Vol(P_i \cap S_i) = \frac{1}{3}\pi \eta_i^2(3R - \eta_i)$, where $\eta_i = z_i + r_i - H$.

Let $z_i < r_i$ (or $z_i + r_i > H$) and $h_i > 0$, then $Vol(C \cap S_i) = \frac{1}{6}\pi r_i^3$.

6 Numerical Experiments

To carry out numerical experiments, experimental data on the structure of titanium powder for 3D printing is used. For a given powder, it is very important to choose a powder structure that will provide the required value of its porosity for 3D printing [14–17]. Data on the composition of the powder are taken from [14] and provided in the Table 1.

We consider the cylinder of dimensions R = 1500, H = 9000.

Two scenarios are considered to address the problem of sphere packing and determine the porosity of the examined powder. In the first scenario, the objective is to pack the maximum number of spheres within a specified cylindrical container while considering their arrangement. The result of this solution is shown in Fig. 2, where 1825 spheres are packed, resulting in a porosity of 0.4588.

In the second scenario, the containment constraints on the precise positioning of spheres within the cylindrical container are relaxed. The result of this solution is illustrated in Fig. 3, where 2397 spheres are packed, resulting in a porosity of 0.3987.

Table 1. A spectrum of the average powder particle diameters of VT20 titanium

Radii of spheres (particles) in the experiment	Number of spheres	Relative frequencies
120.65140	12	0.006814310
139.26935	88	0.049071607
155.67645	900	0.511073254
170.51205	625	0.354911982
184.15635	62	0.035207269
196.85720	50	0.028392959
208.78685	12	0.006814310
289.366	12	0.006814310

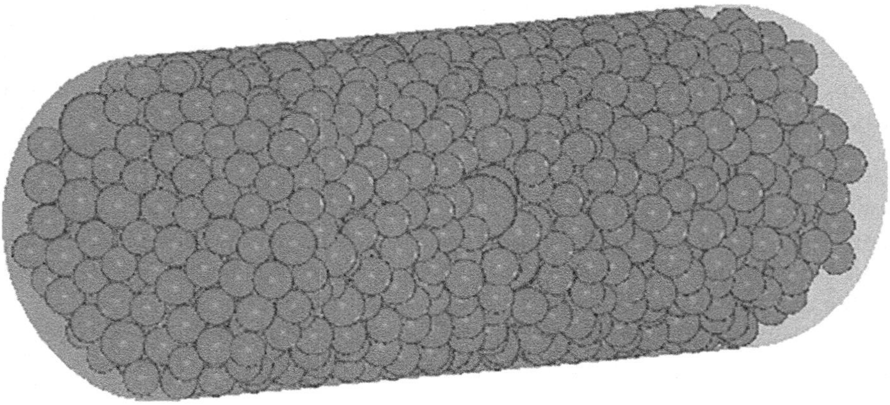

Fig. 2. Result of packing sphere for the first scenario.

The numerical modeling clearly demonstrates that by relaxing the restrictions on the strict placement of spheres inside the container, a more accurate estimation of the material's porosity is achieved. This relaxation effectively eliminated the inclusion of boundary effects, contributing to a more reliable measurement of porosity.

7 Conclusions

The research presented in this paper underscores the significant impact of relaxation techniques on solving materials structure problems, particularly in the context of packing spheres inside a cylindrical container. By exploring two scenarios, this study has shown that the application of relaxation methods leads to a more precise calculation of porosity, especially by reducing the artificial boundary effects that often skew results.

In the first scenario, where spheres are fully arranged within the container, the use of relaxation techniques resulted a porosity of 0.4588. The second scenario, which allowed

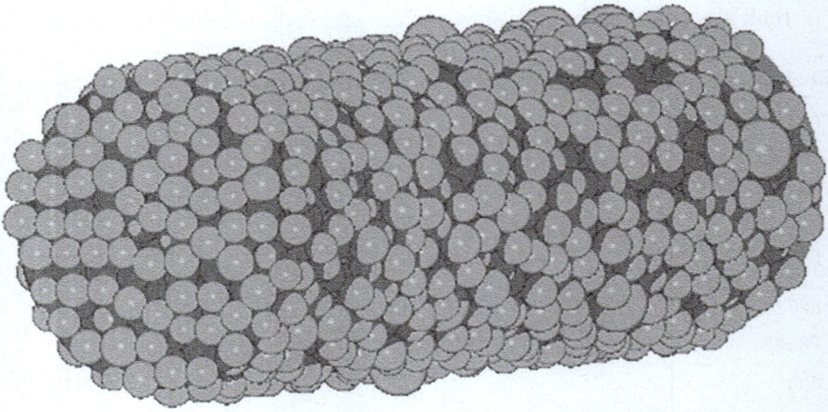

Fig. 3. Result of packing sphere for the second scenario.

for partial sphere arrangement, further demonstrated the robustness of the relaxation approach, yielding porosity of 0.3987. This represents an improvement in accuracy compared to traditional methods that do not account for boundary effects.

These findings highlight the necessity of integrating relaxation techniques into materials structure research. The enhanced accuracy in porosity calculations provided by these methods is crucial in scenarios where strict containment requirements may introduce significant boundary effects. Furthermore, the study opens new avenues for investigating the structural properties of materials with non-spherical particles, offering a more realistic and reliable approach to modeling complex material systems.

Future work could extend these techniques to a broader range of materials and geometries, potentially refining the models to achieve even greater precision in structural analysis.

Acknowledgements. The work is supported by National Research Foundation of Ukraine (#02.2020/167), Volkswagen Foundation (grant #97775), the British Academy (grant #100072).

References

1. Mizuno, H., Saitoh, K., Silbert, L.E.: Structural and mechanical characteristics of sphere packings near the jamming transition: From fully amorphous to quasiordered structures. Phys. Rev. Mater. **4**(11), 115602 (2020). https://doi.org/10.1103/PhysRevMaterials.4.115602
2. Zhang, D., Tian, Z., Chen, Z., et al.: Compaction effects on permeability of spherical packing. Eng. Comput. **37**(9), 3079–3096 (2020). https://doi.org/10.1108/EC-01-2020-0015
3. Liu, Y., Qiu, H., He, Z., et al.: A study of physical modeling and mathematical modeling on inclusion behavior in a planar flow casting process. Metals **12**(4), 606 (2022). https://doi.org/10.3390/met12040606
4. Sheikhi, M.R., Shamsadinlo, B., Ünver, Ö., et al.: Finite element analysis of different material models for polyurethane elastomer using estimation data sets. J. Braz. Soc. Mech. Sci. Eng. **43**, 554 (2021). https://doi.org/10.1007/s40430-021-03279-9

5. Liu, J., Zhang, L.: Molecular dynamics simulations of the impurity effect on packing structures, local stresses, and thermodynamics of Ih silver clusters. Phys. Chem. Chem. Phys. **24**(35), 21040–21051 (2022). https://doi.org/10.1039/D2CP02814J
6. Leao, A.A.S., Toledo, F.M.B., Oliveira, J.F., et al.: Irregular packing problems: a review of mathematical models. Eur. J. Oper. Res. **282**(3), 803–822 (2020)
7. Muhammad, S., Li, Y., Yan, Z., et al.: Phase-field crystal modeling of crystal growth patterns with competition of undercooling and atomic density. Phys. Chem. Chem. Phys. **22**(38), 21858–21871 (2020). https://doi.org/10.1039/D0CP03901B
8. Latham, J.P., Xiang, J., Farsi, A., et al.: A class of particulate problems suited to FDEM requiring accurate simulation of shape effects in packed granular structures. Comp. Part. Mech. **7**, 975–986 (2020). https://doi.org/10.1007/s40571-019-00294-5
9. Villada-Balbuena, A., Jung, G., Zuccolotto-Bernez, A.B., et al.: Layering and packing in confined colloidal suspensions. Soft Matter **18**(25), 4699–4714 (2022). https://doi.org/10.1039/D2SM00412G
10. Baig, N., Kammakakam, I., Falath, W.: Nanomaterials: a review of synthesis methods, properties, recent progress, and challenges. Mater. Adv. **2**(6), 1821–1871 (2021). https://doi.org/10.1039/D0MA00807A
11. Bertsekas, D.P.: Nonlinear programming, 2nd edn. Athena Scientific, Belmont, MA (1999)
12. Yaskov, G., Chugay, A.: Packing equal spheres by means of the block coordinate descent method. CEUR Workshop Proc. **2608**, 156–168 (2020). http://ceur-ws.org/Vol-2608
13. Lopes, R., Santos, S.A., Silva, P.J.S.: Accelerating block coordinate descent methods with identification strategies. Comput. Optim. Appl. **72**, 609–640 (2019). https://doi.org/10.1007/s10589-018-00056-8
14. Torres-Escobar, R., Marmolejo-Saucedo, J.A., Litvinchev, I.: Binary monkey algorithm for approximate packing non-congruent circles in a rectangular container. Wireless Netw. **26**, 4743–4752 (2020). https://doi.org/10.1007/s11276-018-1869-y
15. Duriagina, Z., Lemishka, I., Litvinchev, I., et al.: Optimized filling of a given cuboid with spherical powders for additive manufacturing. J. Oper. Res. Soc. China **9**(4), 853–868 (2021). https://doi.org/10.1007/s40305-020-00314-9
16. Duriagina, Z., Lemishka, I., Trostianchyn, A. et al.: The effect of morphology and particle-size distribution of VT20 titanium alloy powders on the mechanical properties of deposited coatings. Powder Metall. Met. Ceram. 697–702 (2019). https://doi.org/10.1007/s11106-019-00033-8
17. Kritskiy, D., Pohudina, O., Kovalevskyi, M., et al.: Powder mixtures analysis for laser cladding using OpenCV library. In: Nechyporuk, M., et al. (eds.) Integrated computer technologies in mechanical engineering—2021. LNNS, vol. 367, pp. 924–937. Springer, Cham (2022). https://doi.org/10.1007/978-3-030-94259-5_72

Optimized Designing of Parts for Additive Manufacturing

Georgiy Yaskov[1(✉)], Yuriy Stoyan[1], Tetyana Romanova[1,2], Andrii Chuhai[1,3], and Luis Gutierrez[4]

[1] Anatolii Pidhornyi Institute of Mechanical Engineering Problems of the National Academy of Sciences of Ukraine, Pozharsky 2/10, Kharkiv 61129, Ukraine
yaskov2004@gmail.com

[2] University of Leeds, Maurice Keyworth Building, Leeds LS2 9JT, UK

[3] Simon Kuznets Kharkiv National University of Economics, Nauky Ave 9A, Kharkiv 61166, Ukraine

[4] Nuevo Leon State University, Monterrey, Av. Universidad s/n, Col. Ciudad Universitaria, 66455 San Nicolas de los Garza, Nuevo Leon, Mexico

Abstract. The paper presents a novel technique for optimizing the design of parts intended for 3D printing, with a focus on enhancing structural integrity and material utilization. The study addresses the challenges of lightening of 3D parts bounded by cylindrical, spherical surfaces, and planes. A smart technique of generating spherical voids of varying diameters within 3D parts of complex geometry is proposed. The optimization procedure is aimed to maximizing the total volume of spherical voids while considering 3D printing standards. A mathematical model of the nonstandard packing problem is constructed, using normalized and adjusted phi-functions for describing distance constraints. A solution algorithm is developed reducing the problem to a sequence of subproblems with smooth functions. An illustrative example is provided for a 3D part with complex geometry. The findings emphasize the significant benefits of utilizing the proposed technique in designing 3D components, with a particular focus on achieving substantial weight reduction.

Keywords: topological optimization · mathematical modeling · spherical void structures · nonlinear programming · additive manufacturing

1 Introduction and Literature Review

Additive manufacturing (AM) [1, 2] enables the creation of parts with internal voids. This capability allows for the reduction of both part weight and material consumption (specifically, expensive powders).

This implies that when designing 3D parts for additive manufacturing (3D printing), several important considerations need to be taken into account. One of these considerations is the requirement to design the geometrical structure of the part to enable support-free 3D printing [3]. This geometrical structure is achieved through the solution of a

H. Altenbach et al. (Eds.): CAMPE 2023, pp. 246–254, LNME 2025.
https://doi.org/10.1007/978-3-031-82979-6_25

topological optimization problem [1]. In traditional additive manufacturing approaches, "supports" are typically added to the part geometry during the manufacturing process. However, this technology has the drawback of requiring the removal of these supports, which increases the duration and cost of production. Support-free technologies, on the other hand, eliminate this disadvantage.

Currently, there is a significant amount of research dedicated to the exploration of 3D printing techniques for producing parts with through holes [4]. For example, these parts frequently include polygonal perforations.

The paper [5] explores the potential of utilizing additive manufacturing techniques to achieve specific effects in product design. The authors present a range of case studies and examples to illustrate the practical application of this approach. The paper offers valuable insights into the integration of additive manufacturing into the product development process.

The paper [6] provides a comprehensive overview of recent trends and applications in additive manufacturing. It delves into the advancements and challenges within the field, highlighting the wide-ranging potential of additive manufacturing across various industries. Moreover, the paper offers valuable insights into future prospects and emerging technologies within the additive manufacturing domain.

In the review paper [7], a comparison is made between traditional processing methods and additive manufacturing technologies, focusing on performance aspects. Factors such as material properties, design flexibility, production efficiency, and cost-effectiveness are evaluated. By providing a comprehensive analysis of the strengths and limitations of both approaches, this paper assists in selecting suitable manufacturing methods for specific applications.

The research [8] focuses on cost estimation models for additive manufacturing processes. It presents a review classifying existing cost estimation models, discussing their methodologies and key factors taken into account. This paper offers valuable insights for estimating the costs associated with additive manufacturing, thereby facilitating better decision-making throughout the manufacturing process.

The review paper [9] investigates process phenomena and material properties involved in selective laser sintering of polymers. It offers an overview of the selective laser sintering (SLS) process, discussing key parameters, powder materials, and their effects on the final product's properties. The paper contributes to a better understanding of the SLS process and its implications for polymer-based additive manufacturing.

The work [10] explores the potential of additive manufacturing within the context of smart factories and Industry 4.0. It discusses the integration of additive manufacturing into smart factory systems, addressing the associated benefits, challenges, and future prospects. This paper provides valuable insights into leveraging additive manufacturing technologies in the era of advanced manufacturing.

In the paper [11], a review of mathematical models addressing irregular packing problems is presented. Various mathematical approaches used to solve packing problems in additive manufacturing are discussed, with an emphasis on the challenges posed by irregular-shaped objects. This paper offers a comprehensive overview of the mathematical foundations relevant to packing optimization in additive manufacturing processes.

The work [12] introduces a method for support-free hollowing in 3D printing utilizing Voronoi diagrams of ellipses. It presents an algorithm that generates lightweight hollow structures with improved structural integrity. The paper demonstrates the advantages of this approach in terms of material efficiency and printing speed, offering a practical solution for optimizing 3D printing processes.

The research [13] presents an analysis of irregular three-dimensional packing problems in additive manufacturing. It introduces a new taxonomy and dataset for evaluating the packing of irregular-shaped objects.

The paper [14] treats the radii of the spheres as variables. This approach allows for more flexibility in finding solutions to the sphere packing problem.

We propose a technique for designing 3D parts with spherical void structures assuming that these parts will not be subjected to any mechanical forces. Our objective is to minimize the volume of spherical voids within a (3D part) multiply connected set considering 3D printing standards.

2 Problem Formulation

Shapes of 3D parts for additive manufacturing (AM) can be approximated by 3D objects bounded by cylindrical, spherical surfaces and planes.

Let a part be given by a set of the form

$$M = \wedge \backslash \bigcup_{k=1}^{K} \operatorname{int} T_K, \ \wedge = \wedge^0 \backslash \bigcup_{j=1}^{m} \operatorname{int} \wedge_j \tag{1}$$

where Λ is a nonconvex polyhedron, Λ^0 is a convex polytope, Λ_j is a dihedral convex set, T_k is a solid 3D object shaped as an infinite right circular cylinder or an infinite right convex prism or a sphere. The set M is in general a multiply connected set.

We propose a technique to minimize the weight of 3D parts by generating spherical void structures through a sphere packing problem. The objective is to maximize the total volume of spherical voids placed inside M (model of 3D part) considering 3D printing standards.

Let us introduce a collection of spheres S_i, $i \in I = \{1, 2, \ldots, n\}$ with variable centers $v_i = (x_i, y_i, z_i)$, $i \in I$. Assume that $r_i, i \in I$, are also variables subject to $r_{\min} \leq r_i \leq r_{\max}$. The parameters $r_{\min}$ and $r_{\max}$ are given allowable limits of 3D printing standards in order to prevent the formation of excessively small or large dimensions of voids.

Denote a sphere under motion with variable radius by $S_i(v_i, r_i)$.

Thus, a tuple $(n, v_1, v_2, \ldots, v_n, r_1, r_2, \ldots, r_n)$ defines a variable geometric structure of the part

$$\Pi = M \cap \left(\bigcap_{i=1}^{n} S_i^*(v_i, r_i) \right), \tag{2}$$

where $S_i^*(v_i, r_i) = \mathbb{R}^3 \backslash \operatorname{int} S_i(v_i, r_i)$.

Further we refer to $S_i^*(v_i, r_i)$, $i \in I$, as a system of spherical voids.

In addition a minimal allowable distance δ between each pair of spheres as well as between each sphere and the boundary of M is given, i.e.

$$\rho(S_i(v_i, r_i), S_j(v_j, r_j)) \geq \delta \text{ for } i > j \in I,\ \rho(S_i(v_i, r_i), M^*) \geq \delta \text{ for } i \in I, \tag{3}$$

where ρ means the Euclidean distance between two objects, $M^* = \mathbf{R}^3 \backslash \text{int}\, M$.

Let us state the problem of generating spherical voids.

Find the number of spherical voids, their radii and centers arranged within the given domain M (1) meeting the distance conditions (3) to maximize the total volume of the voids.

We reformulate the problem in terms of packing spheres from the collection $S_i(v_i, r_i), i \in I$, inside the domain M taking into account distance conditions (3) to maximize the total volume of spheres.

3 Mathematical Modelling

To describe the distance conditions (3), we use the phi-function technique [15].

Denote a vector of variable radii of spherical voids by $r = \left(r_1, r_2, \ldots, r_{\mathbb{k}_{\max}}\right)$ and a vector of their variable centers by $v = \left(v_1, v_2, \ldots, v_{\mathbb{k}_{\max}}\right)$ where $\mathbb{k}_{\max} = \left[3 \cdot 0.74 \cdot V(\Pi)/(4 \cdot \pi \cdot (r_{\min} + \delta)^3)\right]$, $V(\Pi)$ is the volume of Π, $[A]$ is the integer of A. An upper bound for the problem dimension is $\varsigma = 4 \cdot \mathbb{k}_{\max}$.

A mathematical model of the packing problem can be formulated as a mixed integer programing problem

$$\max \frac{4}{3}\pi \sum_{i=1}^{\mathbb{k}_{\max}} r_i^3 sign\Phi_i(v_i, r_i) \text{ s.t. } X = (v, r) \in W \tag{4}$$

where

$$sign\, \Phi_i(v_i, r_i) = \begin{cases} 1 \text{ if } \Phi_i(v_i, r_i) - \delta \geq 0, \\ 0 \text{ otherwise,} \end{cases} \tag{5}$$

$$W = \left\{X \in \mathbf{R}^{4\mathbb{k}_{\max}} :\ \Phi_{ij}\left(v_i, r_i, v_j, r_j\right) \geq 0,\ i < j \in I_{\mathbb{k}_{\max}},\ r_{\min} \leq r_i \leq r_{\max},\ \ i \in I_{\mathbb{k}_{\max}}\right\}. \tag{6}$$

Here $I_{\mathbb{k}_{\max}} = \{1, 2, \ldots, \mathbb{k}_{\max}\}$, $\Phi_i(v_i, r_i)$ is the normalized phi-function for a sphere $S_i(v_i, r_i)$ and the object $M^* = \mathbb{R}^3 \backslash \text{int}\, M$, $\Phi_{ij}(v_i, v_j) = (x_i - x_j)^2 + (y_i - y_j)^2 + (z_l - z_j)^2 - (r_i + r_j + \delta)^2$ is an adjusted phi-function for spheres $S_i(v_i, r_i)$ and $S_j\left(v_j, r_j\right)$.

Since the objective function is not differentiable, gradient-based methods cannot be applied directly to tackle the problem (4)–(6). Therefore, the following approach is proposed.

The problem (4)–(6) is reduced to a sequence of at most $\mathbb{k}_{\max}$ optimization subproblems in the following formulation.

Subproblem. Pack spheres $S_i(r_i, v_i)$, $i \in I_{\mathbb{k}} = \{1, 2, \ldots, \mathbb{k}\}$, inside the domain M maximizing a total volume $V_{\mathbb{k}}^* = V\left(r^{\mathbb{k}*}\right)$ of the spheres subject to distance conditions (3).

The value $V^* = \max\{V_{\Bbbk}^*, 1 \leq \Bbbk \leq \Bbbk_{\max}\}$ corresponds to a local maximum of the problem (4)–(6).

A mathematical model of each subproblem (for $1 \leq \Bbbk \leq \Bbbk_{\max}$) can be presented as a nonlinear programming model

$$\max V\left(r^{\Bbbk}\right) = \max \frac{4}{3}\pi \sum_{i=1}^{\Bbbk} r_i^3 \text{ s.t. } X^{\Bbbk} = (v^{\Bbbk}, r^{\Bbbk}) \in \Omega_{\Bbbk} \subset \mathbf{R}^{4\Bbbk} \tag{7}$$

where

$$\begin{aligned} \Omega_{\Bbbk} = \{X^{\Bbbk} : \Phi_{ij}(v_i, r_i, v_j, r_j,) \geq 0, i < j \in I_{\Bbbk}, \\ \Phi_i(v_i, r_i) - \delta \geq 0, \ r_{\min} \leq r_i \leq r_{\max}, \ i \in I_{\Bbbk}\}. \end{aligned} \tag{8}$$

Note that the functions $\Phi_i(v_i, r_i)$, $i \in I_{\Bbbk}$, in (8) are piecewise smooth due to using operator *max*. Therefore, the feasible region (8) can be defined in the form

$$\Omega_{\Bbbk} = \bigcup_{\gamma=1}^{\mu} \Omega_{\Bbbk\gamma} \tag{9}$$

where $\Omega_{\Bbbk\gamma}$ is described by a system of inequalities

$$\varkappa_{\gamma\eta}(v, r) \geq 0, \ \eta = 1, 2, \ldots, \Xi \tag{10}$$

with smooth functions. Here $\Xi \geq 4\Bbbk + (m + K)\Bbbk$.

The problem (7), (8) is NP-hard [16]. Therefore the following heuristic approach is proposed.

4 Solution Scheme

Due to (9), (10) the problem (7), (8) is rewritten in the form

$$V\left(r^{\Bbbk *}\right) = \max\left\{V\left(r^{\Bbbk\gamma *}\right), \ \gamma = 1, 2, \ldots, \mu\right\} \tag{11}$$

where

$$V\left(r^{\Bbbk t *}\right) = \max V\left(r^{\Bbbk t}\right) \text{s.t. } \mathrm{X}^{\Bbbk} = \left(v^{\Bbbk}, r^{\Bbbk}\right) \in \Omega_{\Bbbk\gamma}. \tag{12}$$

In order to solve the subproblem (12), the following strategy is applied:

1) Construct a starting point $X^{\Bbbk 0}$ as follows: components $r_i^{\Bbbk} = \delta/2$, $i \in I_{\Bbbk}$ are randomly chosen such that $v_i^{\Bbbk 0} \in \Pi$, $i \in I_{\Bbbk}$, assuming $r_i^{\Bbbk} = \delta/2$, $i \in I_{\Bbbk}$, i.e. $X^{\Bbbk 0} = (v^{\Bbbk 0}, \delta/2) \in \Omega_{\Bbbk}$.
2) Single out a subregion $\Omega_{\Bbbk 0}$ containing the starting point $X^{\Bbbk 0}$.
3) Search for a local maximum point $X^{\Bbbk 1 *}$ solving the problem

$$V\left(r^{\Bbbk 0 *}\right) = \max V\left(r^{\Bbbk}\right) \text{ s.t. } X^{\Bbbk} \in \Omega_{\Bbbk 0}.$$

4) Single out a subregion $\Omega_{\Bbbk 1}$ containing the point $X^{\Bbbk 1*}$.
5) Search for a local maximum point $X^{\Bbbk 2*}$ solving the problem

$$V\left(r^{\Bbbk 2*}\right) = \max V\left(r^{\Bbbk}\right) \text{ s.t. } X^{\Bbbk} \in \Omega_{\Bbbk 1}.$$

The iterative procedure continues until $V\left(r^{\Bbbk\eta*}\right) = V\left(r^{\Bbbk(\eta-1)*}\right)$, where η is the iteration number. The point $X^{\Bbbk*} = X^{\Bbbk\eta*}$ is taken as a local maximum point of the problem (7), (8) (see Fig. 1).

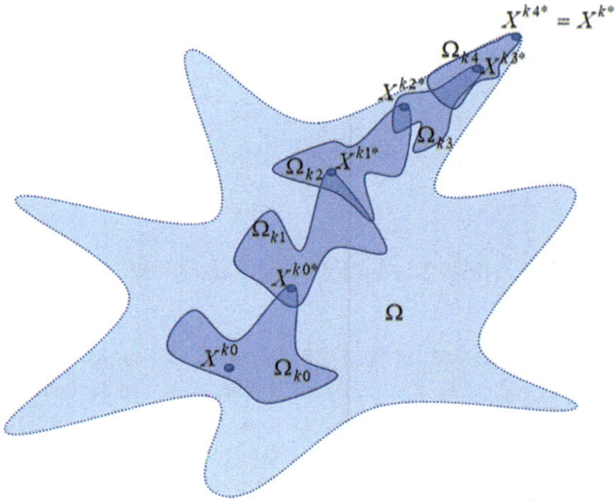

Fig.1. Local optimization scheme

A multistart approach is used for finding a good local maximum of the problem (7), (8).

The following solution scheme of the problem (4)-(6) is offered.

The number $\Bbbk^*$ is searched by bisection within $\Bbbk \in [1, \Bbbk_{\max}]$ considering $\Bbbk$ as an integer.

Step 1. Set $t = 0$. Take $\Bbbk_{low} = 1$ and $\Bbbk_{up} = \Bbbk_{\max} \geq 3$.
Step 2. Set $\Bbbk_t = [(\Bbbk_{low} + \Bbbk_{up})/2]$.
Step 3. If $\Bbbk_{low} = \Bbbk_{up}$ or $\Bbbk_{low} = \Bbbk_{up} - 1$, then set $\Bbbk^* = \Bbbk_{low}$ and stop the algorithm.
Step 4. Solve the subproblem (7), (8) for $\Bbbk_{t-1}$ and $\Bbbk_{t+1}$. Find points of local maximum $X^{\Bbbk_{low}*} = (v^{\Bbbk_{low}*}, r^{\Bbbk_{low}*})$ and $X^{\Bbbk_{up}*} = (v^{\Bbbk_{up}*}, r^{\Bbbk_{up}*})$ and the appropriate volumes $V_{low} = V(r^{\Bbbk_{low}*})$ and $V_{up} = V(r^{\Bbbk_{up}*})$.
Step 5. If $V_{low} < V_{up}$, then set $\Bbbk_{low} = \Bbbk_t$, otherwise set $\Bbbk_{up} = \Bbbk_t$.
Step 6. Increase t by 1 and go to Step 2.

As a result $V^* = \max\{V_{low}, V_{up}\}$ is taken as a feasible solution of the problem (4)–(6) found for t iterations.

Local optimal solutions for the subproblems (7), (8) can be found using the optimization techniques described in, e.g., [14, 15].

5 Numerical Examples

One of the benchmark instances is a thin 2D part investigated in, e.g., [4]. The part is rigidly fixed at one end and loaded in the lower part of the opposite side. We consider here 3D design part with spherical void structures, assuming that the part is not a subject to any mechanical loads.

Let us consider a model of the 3D part approximated by a multiply connected set $M = (\Lambda \cap D^*) \cap (S^* \cap C^* \cap P^*)$ shown in Fig. 2.

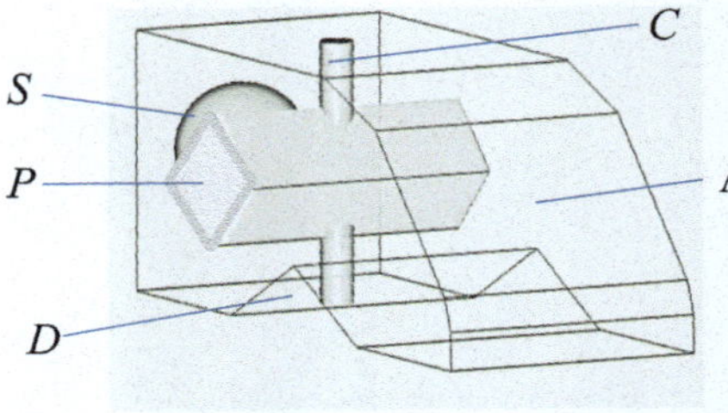

Fig. 2. Multiply connected set M

The polyhedron Λ is defined by its vertices $v_i = (x_i, y_i, z_i)$, $i = 1, 2, \ldots, n_v = 14$, (see Table 1) and faces $F_i, i = 1, 2, \ldots, n_f = 9$. Each face is associated with a list of vertices: $F_1(v_1, v_2, v_3, v_4, v_{5,}, v_6, v_7)$, $F_2(v_8, v_9, v_{10}, v_{11}, v_{12,}, v_{13}, v_{14})$, $F_3(v_1, v_2, v_9, v_8)$, $F_4(v_7, v_6, v_{13}, v_{14})$, $F_5(v_1, v_8, v_{14}, v_7)$, $F_6(v_6, v_5, v_{12}, v_{13})$, $F_7(v_5, v_4, v_{11}, v_{12})$, $F_8(v_4, v_3, v_{10}, v_{11})$ and $F_9(v_3, v_2, v_9, v_{10})$.

Table 1. Coordinates of vertices of Λ

i	1	2	3	4	5	6	7
x_i	40	40	40	40	40	40	40
	0	100	100	98	76	60	0
z_i	0	0	6	11	34	40	40
i	8	9	10	11	12	13	14
x_i	0	0	0	0	0	0	0
y_i	0	100	100	98	76	60	0
z_i	0	0	6	11	34	40	40

The dihedral set D is defined by: points $d_1(40, 30, 0)$ and $d_2(40, 70, 0)$ residing in the corresponding half-planes; the point $d_3(40, 50, 10)$ belonging to the edge of D; the vector $(1, 0, 0)$ characterized by the direction cosines along the edge of D.

The sphere S is centered at $(20, -5, 20)$ and has radius 10.

The cylinder C is defined by: the base radius 10; the point $(20, 25, 0)$ belonging to its axis; the direction cosines $(0, 0, 1)$ along the cylinder axis.

The prism P is described by: the points $(40, 25, 10), (40, 35, 20), (40, 25, 30), (40, 20, 25), (40, 15, 15)$ belonging to the appropriate edges of P and the direction cosines $(1, 0, 0)$ along the edges.

Our algorithm yields the following outcomes: $\Bbbk^* = 33$ and $V^* \approx 28592.2592$.

The packing is illustrated in Fig. 3.

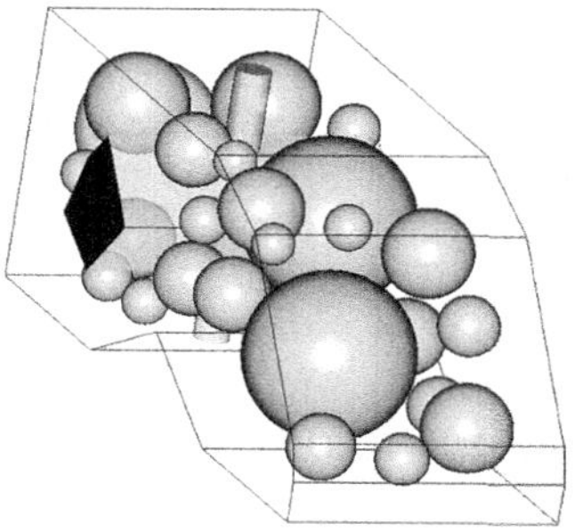

Fig. 3. Packing spheres corresponding to the local optimal solution

6 Conclusions

The paper considers a problem of generating spherical void structure inside a 3D irregular part bounded by cylindrical, spherical surfaces, and planes. Minimal allowable distance between spherical voids are be taken into account. Due to the phi-function technique, an optimization of the part structure is presented in form of nonsmooth nonlinear programming model. A solution algorithm employs the reduction of the problem to a sequence of subproblems with smooth functions. An illustrative example is provided to show results of applying the proposed technique.

The proposed approach allows designing a more complex shaped part structure including, e.g., conical, ellipsoidal, parabolic, hyperbolic surfaces.

This approach can be useful for support free technologies of 3D printing in producing components of complex geometry.

Acknowledgements. The work is supported by National Research Foundation of Ukraine (#02.2020/167), Volkswagen Foundation (grant #97775), British Academy (grant #100072).

References

1. Leary, M., Merly, L., Torty, F., Mazur, M., Brandt, M.: Optimal topology for additive manufacture: a method for enabling additive manufacture of support-free optimal stucture. Mater. Des. **63**, 678–690 (2014). https://doi.org/10.1016/j.matdes.2014.06.015
2. Araújo, L.J.P., Özcan, E., Atkin, J.A.D., Baumersumers, M.: Analysis of irregular three-dimentional packing problems in additive manufacturing: a new taxonomy and dataset. Int. Prod. Res. **57**(18), 5920–5934 (2019). https://doi.org/10.1080/00207543.2018.1534016

3. Jing, X., Lv, D., Xie, F., Zhang, C., Chen, S., Mou, B.: A robotic 3D printing system for supporting-free manufacturing of complex model based on FDM technology. Ind. Robot. **50**(2), 314–325 (2023). https://doi.org/10.1108/IR-05-2022-0136
4. Romanova, T., et al.: Optimal layout of ellipses and its application for additive manufacturing. Int. J. Prod. Res. **59**(2), 560–575 (2021). https://doi.org/10.1080/00207543.2019.1697836
5. Ehlers, T., Meyer, I., Oel, M., Bode, B., Gembarski, P.C., Lachmayer, R.: Effect-engineering by additive manufacturing. In: Lachmayer, R., Bode, B., Kaierle, S. (eds.) Innovative Product Development by Additive Manufacturing 2021. Springer, Cham (2023). https://doi.org/10.1007/978-3-031-05918-6_1
6. Srivastava, M., Rathee, S.: Additive manufacturing: recent trends, applications and future outlooks. Prog. Addit. Manuf. **7**, 261–287 (2022). https://doi.org/10.1007/s40964-021-00229-8
7. Kaikai, X., Yadong, G., Qiang, Z.: Comparison of traditional processing and additive manufacturing technologies in various performance aspects: a review. Archiv. Civ. Mech. Eng **23**, 188 (2023). https://doi.org/10.1007/s43452-023-00699-3
8. Kadir, A.Z.A., Yusof, Y., Wahab, M.S.: Additive manufacturing cost estimation models—a classification review. Int. J. Adv. Manuf. Technol. **107**, 4033–4053 (2020). https://doi.org/10.1007/s00170-020-05262-5
9. Lupone, F., Padovano, E., Casamento, F., Badini, C.: Process phenomena and material properties in selective laser sintering of polymers: a review. Materials **15**(1), 183 (2022). https://doi.org/10.3390/ma15010183
10. Mehrpouya, M., Dehghanghadikolaei, A., Fotovvati, B., Vosooghnia, A., et al.: The potential of additive manufacturing in the smart factory Industrial 4.0: a review. Appl. Sci. **9**(18), 3865 (2019). https://doi.org/10.3390/app9183865
11. Leao, A., Toledo, F., Oliveira, J., Carravilla, M., et al.: Irregular packing problems: a review of mathematical models. Eur. J. Oper. Res. **282**(3), 803–822 (2020). https://doi.org/10.1016/j.ejor.2019.04.045
12. Lee, M., Fang, Q., Cho, Y., Ryu, J., et al.: Support-free hollowing for 3D printing via Voronoi diagram of ellipses. Comput. Aided Des. **101**, 23–36 (2018). https://doi.org/10.1016/j.cad.2018.03.007
13. Araújo, L., Özcan, E., Atkin, J., Baumers, M.: Analysis of irregular three-dimensional packing problems in additive manufacturing: a new taxonomy and dataset. Int. J. Prod. Res. **57**(18), 5920–5934 (2019). https://doi.org/10.1080/00207543.2018.1534016
14. Romanova, T.E., Stetsyuk, P.I., Chugay, A.M., et al.: Parallel computing technologies for solving optimization problems of geometric design. Cybern. Syst. Anal. **55**(6), 894–904 (2019). https://doi.org/10.1007/s10559-019-00199-4
15. Stoyan, Y., Yaskov, G.: Optimized packing unequal spheres into a multiconnected domain: mixed-integer non-linear programming approach. Int. J. Comput. Math.: Comput. Syst. Theory **6**(1), 94–111 (2021). https://doi.org/10.1080/23799927.2020.1861105
16. Wäscher, G., Haußner, H., Schumann, H.: An improved typology of cutting and packing problems. Eur. J. Oper. Res. **183**(3), 1109–1130 (2007). https://doi.org/10.1016/j.ejor.2005.12.047

Thermal Stress Analysis of Multilayer Glazing Elements Subjected to Interlayer Heat Sources

Natalia Smetankina[1](✉), Serhii Misiura[1], Anatolii Vasiliev[1], Ievgeniia Misiura[2], and Kateryna Kren[3]

[1] Anatolii Pidgornyi Institute of Mechanical Engineering Problems of the National Academy of Sciences of Ukraine, 2/10 Pozharskogo Street, Kharkiv 61046, Ukraine
nsmet@ipmach.kharkov.ua

[2] Simon Kuznets Kharkiv National University of Economics, 9a Nauky Avenue, Kharkiv 61166, Ukraine

[3] Eötvös Loránd University, 1-3 Egyetem tér, Budapest 1053, Hungary

Abstract. The problem of studying the thermal stress state of structures arises in many areas of mechanical engineering. The paper deals with a method for studying multilayer glazing elements of aircrafts under temperature effects. We consider a glazing element as the multilayer shell with constant thickness isotropic layers and various physical properties. The number of layers and their layout is arbitrary. Convective heat exchange occurs on the shell surfaces. The multilayer shell with a non-canonical form in plan is heated with interlayer film heat sources. Temperature fields are obtained by solving the nonstationary heat conduction problem. The problem of thermoelasticity of shells is solved by the extension method. This method is based on extending a non-canonical-shaped shell to a simple domain. It enables to present the problem solution in analytical form. We describe a behavior of the shell by using of the of the first-order refined theory with account transverse shear strains in each layer. The thermal elastic equilibrium equations and the boundary conditions on the contour are obtained using Lagrange's variational principle. The method feasibility has tested on five- and seven-layer glazing. Results of calculation of stresses in layers are compared with the data obtained by the finite difference method. The method proposed can be used for defining thermal stresses in the multilayer structural elements and designing anti-icing systems.

Keywords: Multilayer shell · Heating element · Thermal stresses

1 Introduction

In modern engineering, the problem of valid identification of the thermal condition of structural components is a topical one [1]. This problem is often difficult for homogeneous [2] and especially inhomogeneous structures [3]. Models and methods of solving heat conduction and thermal elasticity problems are reviewed in [4]. Homogeneous structures under stationary thermal effects are the most investigated ones [5]. This is due to the difficulties in solving problems of thermal conductivity and thermoelasticity [6].

H. Altenbach et al. (Eds.): CAMPE 2023, pp. 255–263, LNME 2025.
https://doi.org/10.1007/978-3-031-82979-6_26

It has been multilayer structures are being extensively used in areas of aircraft, aerospace, transportation, and energy systems because of their incredible features, such as high stiffness and low weight, sound- and heat-insulating properties, protection against the influence of aggressive environment, and radiation [7]. Although multilayer structures offer advantages over other structures, the saltatory change of material properties across the interfaces can result in large interlaminar stresses, often leading to failure [8].

The issues of thermal conductivity and thermal elasticity of multilayer plates and shells having a canonical shape inplan (a rectangle, a circle, a cylindrical panel, a closed cylindrical shell) have been studied quite fully [9]. As a rule, there are no internal heat sources. Therefore, the development of methods for calculating temperature fields and stresses in multilayer plates and shells with a complex shape in plan, with convective heat exchange on the outer surfaces and the presence of distributed film interlayer heat sources is an urgent problem.

2 Literature Review

A number of studies have been performed to analyze a behavior of multilayer structures under thermomechanical loads [10]. From the literature it follows that the calculation of multilayer structures in the conditions of stationary heating under the influence of temperature distributions, which are set by the piecewise linear law along the thickness of the structure, is most often considered [11].

It should be noted that to solve the problems of deformation of multilayer structural elements under the influence of nonstationary temperature fields, methods based on grid discretization of complex domains or boundaries are most often used, for example, the finite difference method [12] or finite element method [13].

In the work [14], the authors studied the effect of temperature distribution through plate thickness on the normal deflection. The composite laminate plate is subjected to temperatures on surfaces with the significant temperature gradient through the plate thickness. The analysis has been done using finite element method. A temperature distribution is in the diesel engine design is investigated with account its durability and reliability.

Analytical solutions of the problem of thermoelasticity were obtained only for structures of a simple shape. In the paper [15], thermal stresses in a simply supported laminated composite plate with porosity under hygrothermal rising conditions are analyzed. The solution of the problem is obtained by the Navier method. The influence of various values of moisture, temperature, and the stacking sequence of layers on the distribution of stresses in multilayer plates has been studied. The work [16] proposes an analytical form Navier solution for composite plate structures with simple shape subjected to thermal loads.

Thus, the urgent task is to develop research methods of the thermal stress state of multilayer elements of structures of complex shape, taking into account their inhomogeneous structure, which allow to present the solution of the problem in analytical form.

When building models of multilayer shells, two main approaches are introduced [17]. The first approach is based on unified hypotheses for all layers of the structure. The second approach is based on different hypotheses for each layer of the shell. It is in this case that all peculiarities of deformation of a multilayer shell composed of essentially heterogeneous layers are taken into account. This is the approach used in this study.

A refined model of thermoelastic equilibrium is developed for multilayer glazing of aircrafts in terms of the different materials and the layer thickness variation as well as the influence of interface heat sources. Within the framework of the refined linear theory of multilayer structures, the differential equations are derived for quasi-static equilibrium of multilayer shells.

3 Research Methodology

Let's consider the glazing element as a constant thickness non-closed cylindrical multilayer shell composed of I isotropic layers of constant thickness h_i. The shell occupies an area Ω on the outer surface of the first layer (coordinate surface) bounded by the contour Γ. Temperature and mechanical loads $\boldsymbol{P} = \{p_j(x, y)\}(j = \overline{1,\ 3I + 3})$ act on the shell. The coordinate x changes along the generator, the coordinate y changes along the arc of the shell cross section. The positive direction of the Z-axis coincides with the direction of the outer normal to the coordinate surface. The temperature loads are obtained while solving the nonstationary heat conduction problem [18].

The behavior of the shell is described in terms of a refined first-order theory, which takes into account the strain of the transverse shear and compression of the normal element in each layer. It is assumed that the contact between the layers eliminates their stratification and mutual slippage. The displacements of the point of the i-th layer have the form

$$u_k^i = u_k + \sum_{j=1}^{i-1} h_j u_{3+I(k-1)+j} + (z - \delta_{i-1}) u_{3+I(k-1)+i},$$
$$k = 1,\ 2,\ 3,\quad i = \overline{1, I} \tag{1}$$

where $u_k = u_k(x, y)$, $k = 1,\ 2,\ 3$, are displacements of the point of the coordinate surface in the direction of the coordinate axes, $u_{3+I(k-1)+j} = u_{3+I(k-1)+j}(x, y)$, $k = 1,\ 2$, are the angles of rotation of the normal element relative to the coordinate axes in the i-th layer, $u_{3+2I+i} = u_{3+2I+i}(x,\ y)$ is compression of the normal element within the i-th layer, $\delta_i = \sum_{j=1}^{i} h_j$, $\delta_{i-1} \le z \le \delta_i$, $i = \overline{1,\ I}$.

The strains of the layers are determined according to Cauchy's formulas, and the stresses and strains in the layers are related by Hooke's law. From the variational Lagrange principle we obtain the equation of thermoelastic equilibrium of the multilayer shell

$$\Lambda \mathbf{U} = \mathbf{P}^T - \mathbf{P}, \quad (x, y) \in \Omega, \tag{2}$$

and boundary conditions on the contour Γ

$$[\mathbf{B}^\Gamma]\mathbf{U} = \mathbf{P}^\Gamma,\ (x, y) \in \Gamma. \tag{3}$$

The matrices $[\Lambda]$ and $[\mathbf{B}^{\Gamma}]$ in Eqs. (2) and (3) are square symmetric ones; $\boldsymbol{U}$ is the vector of displacements, $\boldsymbol{U} = \{u_i\}$ $\left(i = \overline{1,\ 3I+3}\right)$.

The solution of the boundary value problem (2), (3) is carried out by the method of extension of the given system [19]. The method takes into account the geometry of the area at the analytical level. The original shell is expanded to an auxiliary simply supported shell with a rectangular plan shape, for that it is possible to obtain a solution of the problem in the form of trigonometric series expansions. The auxiliary shell has the same composition of layers, and within the region Ω is loaded in the same way as a given shell.

To satisfy the specified boundary conditions on the contour Γ, and for the identity of the thermal stress state of the auxiliary shell in the region Ω and the state of the original shell, compensating loads $\mathbf{Q}^{\text{comp}} = \left\{q_j^{\text{comp}}(x, y)\right\}$ $\left(j = \overline{1,\ 3I+3}\right)$ are added. Loads are continuously distributed along the contour Γ and applied to the auxiliary shell. In the equation of thermoelastic equilibrium of the shell (2) compensating loads are included in the form of curvilinear distributions

$$p_j^{\text{comp}}(x, y) = \sum_{k=1}^{3I+3} \oint_{\Gamma} L_{jk}\, q_k^{\text{comp}}(s)\, \delta(x - x_{\Gamma}, y - y_{\Gamma})\, ds,$$

$$j, k = \overline{1,\ 3I+3}. \tag{4}$$

Here $\delta(x - x_{\Gamma}, y - y_{\Gamma})$ is two-dimensional Dirac delta function; s is the current arc length.

Boundary conditions (3) taking into account (4) lead to a system of integral equations, from which the unknown intensities of compensating loads are determined

$$[\mathbf{B}^{\Gamma}]\mathbf{U}[\mathbf{Q}^{\text{comp}}(x, y)] = \mathbf{P}^{\Gamma}, x, y \in \Gamma. \tag{5}$$

To convert integral equations into their algebraic analogue, the functions of displacements, of specified and compensating loads are expanded into trigonometric series by functions that satisfy the boundary conditions of the shell with a rectangular plan.

The boundary functions included in the initial boundary conditions (3) are also expended into a series along Γ. As a result, the system (5) is transformed into a system of linear algebraic equations for the coefficients of expansion of compensating loads in a series along the contour Γ.

4 Results

To confirm the reliability of numerical results, stress values for a five-layer non-closed cylindrical shell under the thermal field that varies along the thickness were compared with the data obtained by the finite difference method [12].

The contour of the shell is rigidly fixed, and its shape is given by the equations of the Lamé curves: $x(\varphi) = \alpha \cos^{2/c}(\varphi)$, $y(\varphi) = \beta \sin^{2/c}(\varphi)$, $2\alpha = 0.555$ m, $2\beta = 0.42$ m, $c = 10$. The shell has the following parameters: $h_1 = h_5 = 0.005$ m, $h_2 = 0.002$ m, $h_3 = 0.008$ m, $h_4 = 0.003$ m (thicknesses of the layers), $R = 0.7$ m (radius of the shell),

$E_i = 6.3 \cdot 10^4$ MPa ($i = 1, 3, 5$), $E_2 = 1.6 \cdot 10^3$ MPa, $E_4 = 10$ MPa (Young's modulus of the i-th layer material); $\nu_i = 0.22$ ($i = 1, 3, 5$), $\nu_i = 0.4$ ($i = 2, 4$) (Poisson's ratio), $\alpha_T^i = 8.8 \cdot 10^{-6}$ °C^{-1} ($i = 1, 3, 5$), $\alpha_T^2 = 7 \cdot 10^{-5}$ C^{-1}, $\alpha_T^4 = 2 \cdot 10^{-4}$ C^{-1} (linear thermal expansion coefficient of the i-th layer material).

Figure 1 shows the distribution of temperature T and tensile stress σ_y along the thickness of the shell. The given temperature T distribution within each layer is approximated by a quadratic parabola as shown in Fig. 1(a). Figure 1(b) shows the stress σ_y changing at the point with coordinates $x_0 = 0$ and $y_0 = 0$. Figure 1(c) shows the stress changing at the point with coordinates $x_0 = \alpha/2$ and $y_0 = 0$. The dashed lines show the stress distributions obtained by the finite difference method.

The present closed-form analysis solution proved to be in good agreement with the results of finite-difference analysis. Thus, this analytical solution can serve as a benchmark for approaches that require discretization of the complex domain under study and subsequent arrangement of the physical equations to solve the problem.

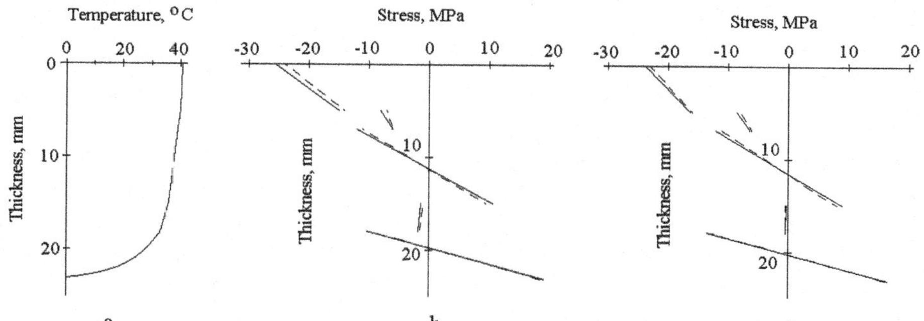

Fig. 1. The distribution of temperature and stress along the thickness of the five-layer shell.

The efficiency of the developed method is confirmed by the example of solving the problem of thermoelasticity of a seven-layer aircraft glazing. The calculation scheme of the glazing is shown in Fig. 2.

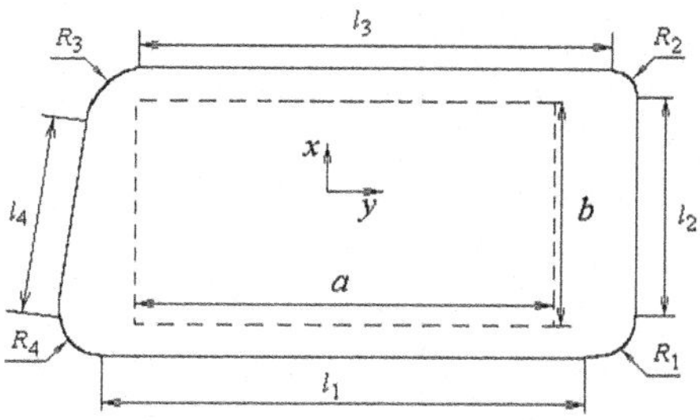

Fig. 2. Calculation scheme of the seven-layer glazing.

The shell is under the influence of a film heat source, which is located between the first and the second layers of the shell. The location of the heat source on the surface of the layers is shown by a dashed line. Its specific power is 0.84 W/cm^2. The field of temperature loads is obtained as a result of solving the nonstationary problem of thermal conductivity of the shell. Dimensions of the heat source are $a = 0.6$ m, $b = 0.49$ m.

The geometric and mechanical parameters of the shell are $R = 1.34$ m, $l_1 = 0.7$ m, $l_2 = 0.54$ m, $l_3 = 0.61$ m, $l_4 = 0.55$ m, $R_1 = 0.05$ m, $R_2 = 0.3$ m, $R_3 = 0.04$ m, $R_4 = 0.045$ m; $h_1 = 0.005$ m, $h_2 = 0.003$ m, $h_3 = h_5 = h_7 = 0.01$ m, $h_4 = h_6 = 0.002$ m; $E_i = 6.8 \cdot 10^4$ MPa, $i = 1, 3, 5, 7$, $E_i = 2.2 \cdot 10^2$ MPa, $i = 2, 4, 6$; $\nu_i = 0.22$, $i = 1, 3, 5, 7$, $\nu_i = 0.38$, $i = 2, 4, 6$; $\rho_i = 2520$ kg/m^3, $i = 1, 3, 5, 7$, $\rho_i = 1060$ kg/m^3, $i = 2, 4, 6$ (the i-th layer material density); $\alpha_T^i = 9 \cdot 10^{-6}$ °C^{-1}, $i = 1, 3, 5, 7$, $\alpha_T^i = 8.3 \cdot 10^{-5}$ °C^{-1}, $i = 2, 4, 6$; $k_i = 1.08$ W/(m·°C), $i = 1, 3, 5, 7$, $k_i = 0.22$ W/(m·°C), $i = 2, 4, 6$ (thermal conductivity of the i-th layer material); $c_i = 870$ J/(kg·°C), $i = 1, 3, 5, 7$, $c_i = 2344$ J/(kg·°C), $i = 2, 4, 6$ (the i-th layer thermal capacity).

The side surface of the shell was considered thermally insulated. Convective heat exchange occurred on the outer surfaces of the first and seventh layers. The convective heat exchange coefficients are $H_1 = 150$ W/(m^2·°C) and $H_I = 20$ W/(m^2·°C).

According to the results of the calculation, the distributions of the principal stresses at the time when they reach their highest values are obtained.

We considered two cases with the following flight aircraft modes: (a) the altitude is 4100 m, the velocity is 500 km/h, the outside air temperature is −10 °C; (b) the altitude is 7000 m, the velocity is 610 km/h, the outside air temperature is −30 °C.

Figure 3 shows the distribution of the principal stress σ_1 in the location of the heat source on the surface $z = \delta_1$, containing this source. Stress values are given in megapascals at time when they reach their highest values are obtained.

Figure 3(a) shows the stress distribution for the first mode, and Fig. 3(b) shows results for the second mode, respectively. In the corners and along the boundary of the heat source there is a concentration of thermal stresses. It was found that the stresses did not exceed their admissible values. The nature of stress distribution agrees well with the distributions given in the study [20].

Numerical modelling makes it possible to qualitatively and quantitatively assess the effects of heat sources on aircraft multilayer glazing [21, 22], as well as to select the conditions of optimal heating [23, 24]. This enables to avoid full-scale experiments and design correctly the units affected by such action, and the study of thermal stresses in various composite structures [25, 26] in the field of mechanical engineering and aeronautical technology [27].

5 Conclusions

A method for calculating the thermal stress state of multilayer glazing of aircraft as the shells by a complex shape in plan under the influence of film heat sources has been developed. It allows to solve the problem in the form of a trigonometric series and investigate shells with different boundary conditions and arbitrary arrangement of layers. Temperature fields are determined from the solution of a nonstationary heat conduction

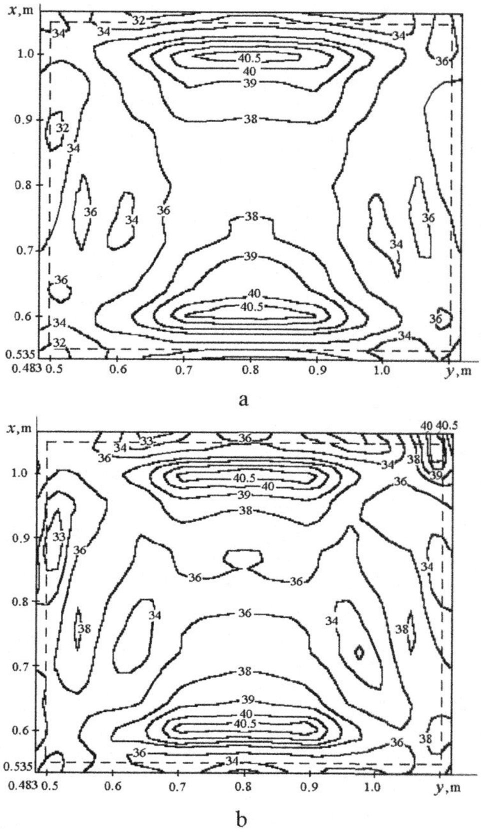

Fig. 3. Stress distributions on the surface $z = \delta_1$ of the glazing element for different flight mode.

problem. This approach makes it possible to study the thermal stress state of a glazing under the influence of various heat sources, including interlayer ones, without additional hypotheses about temperature distributions.

The possibilities of the method are illustrated by the example of calculating the deformations of five-layer and seven-layer simply supported shells. Good consistency of the results of calculation of temperature stresses by different methods confirmed the reliability of the results obtained using the proposed method. This indicates that this analytical solution can serve as a benchmark for difference and finite element solutions.

The effect of stress concentration in the corners and along the boundary of the heat source region is established. Since stress concentration can lead to the destruction of a glazing during operation, further research involves the selection of rational parameters of heat sources, namely, the shape of the heated area and the power of the source, as well as the selection of layer materials with the necessary properties.

The method can be used to analyze thermal stresses in various elements of structures, as well as to select optimal heating modes under the influence of heat sources in vehicle heating systems.

References

1. Zaitsev, B.P., Shvetsov, V.L., Hubskyi, O.M., Palkov, S.A., Protasova, T.V.: Investigation of the thermal strength of steam turbine diaphragms with reduction of axial dimension. J. Mech. Eng.—Problemy Mashynobuduvannia **24**(2), 37–49 (2021). https://doi.org/10.15407/pmach2021.02.037
2. Liu, L., Wu, J., Hu, Z., Jin, X., Lu, P., Zhang, T., Fan, X.: Thermal analysis of turbine blades with thermal barrier coatings using virtual wall thickness method. CMES-Comput. Mod. Eng. Sci. **134**(2), 1219–1236 (2023). https://doi.org/10.32604/cmes.2022.022221
3. Kondratiev, A., Píštěk, V., Vambol, O., Kučera, P.: Effect of heating conditions during moulding on residual stress-strain behaviour of a composite panel. Polymers **14**(9), 1660 (2022). https://doi.org/10.3390/polym14091660
4. El Abdi, R., Leite Pinto, R., Guérard, G., Capena, C.: Lifetime of optical fibers submitted to thermo-mechanical stresses. In: Lecture Notes in Networks and Systems, vol. 305, pp. 23–31, Springer, Cham (2022). https://doi.org/10.1007/978-3-030-83368-8_3
5. Plankovskyy, S., Shypul, O., Tsegelnyk, Y., Pankratov, A., Romanova, T.: Amplification of heat transfer by shock waves for thermal energy method. In: Nechyporuk, M., Pavlikov, V., Kritskiy, D. (eds.) ICTM 2020. Lecture Notes in Networks and Systems, vol. 188, pp. 577–587. Springer, Cham (2021). https://doi.org/10.1007/978-3-030-66717-7_49
6. Rusanov, A., Shubenko, A., Senetskyi, O., Babenko, O., Rusanov, R.: Heating modes and design optimization of cogeneration steam turbines of powerful units of combined heat and power plant. Energetika **65**(1), 39–50 (2019). https://doi.org/10.6001/energetika.v65i1.3974
7. Sierikova, O., Strelnikova, E., Gnitko, V., Degtyarev, K.: Boundary calculation models for elastic properties clarification of three-dimensional nanocomposites based on the combination of finite and boundary element methods. In: 2021 IEEE 2nd KhPI Week on Advanced Technology (KhPIWeek), pp. 351–356 (2021). https://doi.org/10.1109/KhPIWeek53812.2021.9570086
8. Shapovalova, M., Vodka, O.: Image microstructure estimation algorithm of heterogeneous materials for identification their chemical composition. In: 2019 IEEE 2nd Ukraine Conference on Electrical and Computer Engineering, UKRCON 2019—Proceedings (2019). https://doi.org/10.1109/UKRCON.2019.8879861
9. Qian, H., Qiu, Y., Yang, Y., Xie, F.: An exact analysis for heat conduction in laminated infinite cylindrical arches subjected to Dirichlet boundary conditions. Int. J. Appl. Electromagnet Mech **64**(1–4), 509–516 (2020). https://doi.org/10.3233/jae-209358
10. Kondratiev, A., Píštěk, V., Purhina, S., Shevtsova, M., Fomina, A., Kučera, P.: Self-heating mould for the composite manufacturing. Polymers **13**(18), 3074 (2021). https://doi.org/10.3390/polym13183074
11. Pendhari, S., Harwande, S., Kulkarni, S., Vora, T., Visariya, J.: Quick review and assessment of thermal stress analyses for exact and assumed temperature distribution for composite and sandwich laminates. Mech. Adv. Composite Structures **10**(2), 343–362 (2023). https://doi.org/10.22075/macs.2023.24592.1359
12. Apalak, M.K., Reddy, M.K.: Thermal stress formation in a functionally graded Al2O3-adhesive single lap joint subjected to a uniform temperature field. Math. Comput. Approach. Appl. Mech. **28**(4), 1–17 (2023). https://doi.org/10.3390/mca28040082
13. Pernica, M., Letal, T., Losak, P., Nad, M., Reppich, M., Jegla, Z.: Transient thermal stress calculation of a shell and tube condenser with fixed tubesheet. Chem. Ing. Tec. **93**(10), 1590–1597 (2021). https://doi.org/10.1002/cite.202100036
14. Jameel, A.N., Yousuf, L.S.: Temperature distribution on rectangular composite plate under thermal gradient loading environment: experiment and simulation. In: IOP Conf. Ser. Mater. Sci. Eng. 1094, 012016 (2021). https://doi.org/10.1088/1757-899X/1094/1/012016

15. Yuksel, Y.Z., Akbas, S.D.: Hygrothermal stress analysis of laminated composite porous plates. Struct. Eng. Mech. **80**(1), 1–13. (2021) https://doi.org/10.12989/SEM.2021.80.1.001
16. Azzara, R., Carrera, E., Filippi, M., Pagani, A.: Vibration analysis of thermally loaded isotropic and composite beam and plate structures. J. Therm. Stresses **46**(5), 369–386 (2023). https://doi.org/10.1080/01495739.2023.2188399
17. Zappino, E., Carrera, E.: Refined one-dimensional models for the multi-field analysis of layered smart structures. In: Altenbach, H., Carrera, E., Kulikov, G. (eds) Analysis and Modelling of Advanced Structures and Smart Systems. Advanced Structured Materials, vol. 81, pp. 343–366. Springer, Singapore. (2018). https://doi.org/10.1007/978-981-10-6895-9_15
18. Smetankina, N.V., Postnyi, O.V., Merkulova, A.I., Merkulov, D.O.: Modeling of nonstationary temperature fields in multilayer shells with film heat sources. In: 2020 IEEE KhPI Week on Advanced Technology (KhPIWeek 2020), pp. 242–246. IEEE (2020). https://doi.org/10.1109/KhPIWeek51551.2020.9250139
19. Merculov, V., Kostin, M., Martynenko, G., Smetankina, N., Martynenko, V.: Force simulation of bird strike issues of aircraft turbojet engine fan blades. In: Cioboată, D.D. (eds) International Conference on Reliable Systems Engineering (ICoRSE)-2021. ICoRSE 2021. Lecture Notes in Networks and Systems, vol. 305, pp. 129–141. Springer, Cham. (2022). https://doi.org/10.1007/978-3-030-83368-8_13
20. Wang, Y., Li, X., Bisby, L.A.: Comparative study of thermal breakage of annealed and tempered glazing with different thicknesses under uniform radiation conditions. Fire Saf. J. **140**, 103867 (2023). https://doi.org/10.1016/j.firesaf.2023.103867
21. Plankovskyy, S., Shypul, O., Tsegelnyk, Y., Zaklinskyy, S., Bezkorovaina, O.: Numerical simulations of mixture formation to ensuring the quality of thermal deburring. In: Advances in Mechanical and Power Engineering. CAMPE 2021. Lecture Notes in Mechanical Engineering, pp. 323–333. Springer, Cham (2023). https://doi.org/10.1007/978-3-031-18487-1_11
22. Kondratiev, A., et al.: Effects of the temperature–time regime of curing of composite patch on repair process efficiency. Polymers **13**(24), 4342 (2021). https://doi.org/10.3390/polym13244342
23. Kurennov, S., Barakhov, K., Taranenko, I., Stepanenko, V.: A genetic algorithm of optimal design of beam at restricted sagging. Radioelectronic and Computer Systems **2**, 83–91 (2022). https://doi.org/10.32620/reks.2022.1.06
24. Kondratiev A.V., Kovalenko V.O. Optimization of design parameters of the main composite fairing of the launch vehicle under simultaneous force and thermal loading. Space Sci. Technol. **25**(4(119)), 3–21 (2019). https://doi.org/10.15407/knit2019.04.003
25. Miroshnikov, V., Younis, B., Savin, O., Sobol, V.: A: linear elasticity theory to analyze the stress state of an infinite layer with a cylindrical cavity under periodic load. Computation **10**(9), 160 (2022). https://doi.org/10.3390/computation10090160
26. Kushnir, R.M, Zhydyk, U.V., Flyachok, V.M: Thermoelastic analysis of functionally graded cylindrical shells. J. Math. Sci. **254**(9), 1–13 (2021). https://doi.org/10.1007/s10958-021-05287-5
27. Balitskii, A.I., et al.: Hydrogen-containing "green" fuels influence on the thermal protection and formation of wear processes components in compression-ignition engines modern injection system. Energies **16**(8), 3374 (2023). https://doi.org/10.3390/en16083374

The Efficiency and Durability of a Nuclear Power Plant Turbine Without the Stage of Damaged Blades

Vitalii Peshko[1(✉)], Olexandr Usatyi[2], and Olga Chernousenko[1]

[1] Igor Sikorsky Kyiv Polytechnic Institute, 37 Beresteiskyi ave., Kyiv 03056, Ukraine
vapeshko@gmail.com
[2] Kharkiv Polytechnic Institute, 2 Kyrpychova str., Kharkiv 61002, Ukraine

Abstract. Events that lead to damage to the blades can occur during the operation of turbine equipment at nuclear power plants. In some cases, it may not be possible to quickly replace damaged parts. The purpose of this study is to investigate the possibility of operating the K-1000-60/3000 turbine with dismantled nozzle and working blades of the 5th stage. A mathematical model of the flowing part of a high-pressure cylinder based on the object-oriented approach is developed. The main thermodynamic and kinematic parameters of the flow are investigated. A significant change in the steam parameters in the extraction III, which feeds the main deaerator of the unit, has been established. The steam pressure in this extraction decreases by 37% and is insufficient to ensure the nominal deaeration of the feeding water. At the same time, the total power of the cylinder without the 5th stage is reduced by 3%. The thermal and stress-strain state of a high-pressure rotor in the project version and without a damaged stage was investigated. A decrease in the overall stress level and migration of the areas of maximum stresses were observed. The low-cycle fatigue of the rotor was investigated at typical operating conditions. It was found that the individual rotor life without the 5th stage increases by 5.6%.

Keywords: Flow Part · Turbine Stage · Efficiency · Thermal State · Stress Intensity · Damage · Fatigue · Residual Service Lifetime

1 Introduction

Nuclear power plants (NPPs) generate a large amount of electricity in the power systems of the United States, France, Japan, Belgium, and Ukraine [1]. Continuity and reliability of their operation is a strategically important task.

Steam turbine blades are often damaged during operation. Usually, the banding part of the blade is damaged due to radial contact between the rotor and stator. This event occurs due to uneven heating or cooling of the turbine flow part during start-up and shutdown. In particularly serious cases, the upper peripheral part of the blades may be damaged [2]. A less common, but more dangerous case is the detachment of the blade in the area of the tail attachment. Such events are associated with the accumulation of

H. Altenbach et al. (Eds.): CAMPE 2023, pp. 264–274, LNME 2025.
https://doi.org/10.1007/978-3-031-82979-6_27

fatigue damage in the base metal under the influence of mechanical and thermal stresses [3].

During the continuation of the operation of a turbine that has suffered such an event, its rotors and cylinders should be inspected [4]. Damaged blades must be dismantled and replaced with new ones [5]. In some cases, it may not be possible to replace the damaged blades and the turbine should continue to operate without blade stage. But such exploitation requires scientific justification.

2 Literature Review

The authors of [6] investigate various scenarios of operation of a cogeneration power plant where the blades of one of the stages were damaged. It is proposed to replace the damaged turbine stage with a special buffer board. This element provides hydraulic resistance similar to a regular turbine stage. Obviously, this part reduces the efficiency of power plant operation. However, the authors' calculations showed that such work is more cost-effective in the long term than shutting down the power unit for emergency maintenance with the production and installation of new blades.

It is possible to operate a turbine with dismantled blades without the use of special throttling buffer boards. The authors of [7] investigate the peculiarities of operating the T-250/300-23.5 turbine without the last stage of the low-pressure cylinder. This turbine is often operated in a heat recovery mode. In this mode, a small amount of steam enters the last blades of the turbine. Therefore, the generating company decided to dismantle the last turbine blades to reduce the ventilation capacity of the low-pressure cylinder. During several years of operation of this turbine after dismantling the blades, no accidents were detected. The turbine was operated both in the heating and condensing exploitation modes.

To scientifically justify the possibility of operation of the turbine without one of the stages, it is necessary to establish the conditions and parameters of steam flow in the flow part of the turbine without one of the stages, as well as the reliability and durability of the corresponding turbine rotor.

The characteristics of steam flow in the turbine flow part are often studied using CFD (computational fluid dynamics) methods. In particular, the authors of [8] considered a three-stage intermediate-pressure turbine cylinder. To solve these problems, the authors used their own CFD code developed at Tohoku University. The authors took into account secular-changed blades and the leakage flows in labyrinth seals. For the cylinder under study, it was possible to determine the reasons for the increase in the outlet steam static temperature after long-term operation of the turbine.

Quite often, flow parameters in a turbine are studied using object-oriented programming methods. Paper [9] is devoted to the study of the 4CK465 turbine performance parameters. Three formulations are considered: stationary, simplified dynamic and stage-by-stage dynamic. The developed models have been applied to various practical problems and have given a good agreement between the calculated results and the measured ones. A similar stage-by-stage model for studying the K-1000-60 turbine was developed in [10]. The model takes into account moisture separation at the stages and steam extraction. This model allows monitoring the performance indicators of a steam turbine.

Information on the steam parameters in a turbine is required both to assess the efficiency of its operation and to study its reliability and durability. The first step in assessing the durability of turbine components is to establish the heat transfer patterns. To do this, the boundary conditions for heat transfer are calculated and the thermal state of the parts at different moments of operation is determined [11].

The next step is to determine the stress distribution in turbine elements during operation. In particular, the authors of [12] investigated various types of stresses arising in the turbine rotor from thermo-mechanical loads during turbine startup. The authors were able to compare stress von Mises, principal stresses and stress tensors in a complex structure under a realistic loading condition.

The last and key stage in the study of turbine durability is the assessment of base metal damage during operation and the prediction of the permissible service life. The authors of [13] estimate the durability of a supercritical turbine rotor in different modes using the Neuber rule and Manson-Coffin equation. The obtained calculation data is necessary for training a neural network, which subsequently estimates the durability of the equipment online.

3 Research Methodology

3.1 Purpose of the Study

The aim of the study is to provide a scientific justification for the possibility of operating the K-1000-60/3000 steam turbine without damaged nozzle and working blades of the 5th stage. To achieve this goal, the following tasks are investigated:

- study of the impact of the absence of 5th stage on the steam parameters after turbine stages, internal capacity and efficiency of the high-pressure cylinder (HPC);
- research of the durability of high-pressure rotor without 5th stage blades (permissible number of startups, accumulated damage, individual service lifetime).

3.2 Object of Research

The high-pressure rotor of the K-1000-60/3000 turbine is investigated in the paper. This is a condensing type turbine consisting of 1 high-pressure and 4 low pressure cylinders. It has unregulated steam extractions, moisture separation and intermediate steam superheating. The turbine operates in a unit with a WWER-1000 reactor.

The high-pressure rotor is a two-stream rotor with 5 pressure stages in each stream (Fig. 1). All rotor parts, except for the impellers, are forged together with the rotor. The rotor is made of 38CrNi3Mo1V alloy steel.

A part of the entire rotor is considered in this paper. Namely, from the central plane of symmetry to the last segment of the end seals (*M* in Fig. 1). This simplification will significantly reduce the computational time when studying the thermal and stress state of the rotor by the finite element method. At the same time, it was decided not to model the working blades, but to take into account their masses using boundary conditions.

The computational model of the rotor is divided into a million finite elements. They decrease in size from the rotor rotation axis to the outer surface according to the law of geometric progression. The size of the smallest element is 2 mm.

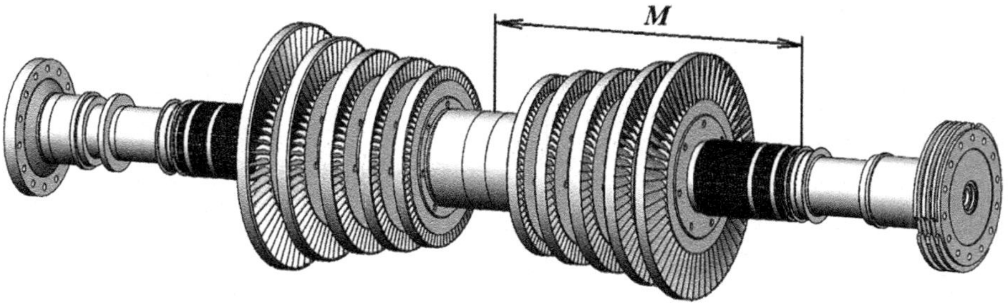

Fig. 1. High-pressure rotor of the K-1000-60/3000 turbine (M shows the part of the rotor that will be studied in this paper)

3.3 Mathematical Model of Steam Flow in the Turbine Flow Part

The steam flow in the turbine can be described by a system of equations containing the energy, continuity, process and state equations. The system also contains kinematic relations linking flow angles and velocities, as well as various closure relations. This system of equations is compiled for each of the turbine stages in the cross-section behind the nozzle apparatus, as well as in the cross-section behind the working blades [14].

Here is an example of a system of equations describing the flow of steam in one of the turbine stages. In a cross-section after the nozzle, the continuity equation, taking into account the energy, process, and state equations, is as follows:

$$G_1 = \rho_1 C_{1z} F_1 = \rho_1 \left(p_1 \left(\bar{h}_0 - \frac{C_1^2}{2\varphi^2}; S_0^* \left(\bar{p}_0; \bar{h}_0 \right) \right); h_0^* - \frac{C_1^2}{2} \right) C_1 \sin \alpha_1 F_1 \quad (1)$$

where G – steam flow rate, ρ – density, C – absolute velocity, F – area of the end part of the blades, p – pressure, h – enthalpy, φ – velocity coefficient of the nozzle apparatus, S – entropy, α – angle of steam flow outlet from the nozzle blades.

Similarly, in the cross-section behind the working blades:

$$G_2 = \rho_2 W_{2z} F_2 = \rho_2 \left(p_2 \left(H + \frac{u^2 - W_2^2}{2\psi^2}; S_1(p_1; h_1) \right); \right.$$
$$\left. H + \frac{u^2 - W_2^2}{2\psi^2} \right) W_2 \sin \beta_2 F_2 \quad (2)$$

where H – enthalpy of steam behind the nozzle blades, u – circumferential velocity, W – relative velocity, ψ – velocity coefficient of the working blades, β – angle of steam flow outlet from the working blades.

For the problem of direct one-dimensional calculation, the closing relation is:

$$p_2 \left(H + \frac{u^2 - W_2^2}{2\psi^2}, S_1^* \right) = p_{2set} \quad (3)$$

where p_{2set} – set value of the outlet pressure.

3.4 Mathematical Model for Calculating the Residual Rotor Service Lifetime

The temperature distribution in the rotor base metal is determined using the boundary value problem of unsteady heat conduction:

$$div[\lambda(T) \cdot grad(T)] = c(T) \cdot \gamma(T) \cdot \frac{\partial T}{\partial \tau} \tag{4}$$

where λ – coefficient of thermal conductivity of steel, c – specific heat capacity, which are functions of temperature T and coordinates. To solve this problem, heat transfer boundary conditions of the 2nd and 3rd kind are set on all rotor surfaces.

The stress-strain state is obtained by solving the equations of equilibrium, strain compatibility and the law of elasticity.

Permissible number of load cycles N_p under variable turbine operating conditions:

$$N_p = \left[1 - \left|\frac{1,25 \cdot \sigma_s}{\sigma_{l.s.}}\right|^q\right] \cdot \min\left\{\frac{N_1}{n_N}; N_2\right\},$$

$$N_{1,2} = \left[\frac{\frac{1}{4} \cdot \ln\frac{100}{100-\psi_{l.s.}}}{C \cdot \left(n_{1,2} \cdot \varepsilon_{a.r.} + \frac{1-2\cdot\nu}{3\cdot E} \cdot \sigma_i\right) - \frac{\overline{\sigma_N}}{E}}\right]^{\frac{1}{0,6}} \tag{5}$$

where σ_s – stress intensity in the state of steady creep, $\sigma_{l.s.}$ – long-term strength limit, n_N, n_1, n_2 – safety margins, $\psi_{l.s.}$ – long-term plasticity, $\varepsilon_{a.r.}$ – amplitude of strain intensity reduced to a symmetrical loading cycle, ν – Poisson's ratio, E – elastic modulus, σ_i – stress intensity, $\overline{\sigma_N}$ – fatigue limit under asymmetrical loading cycle.

Damage accumulated in the base metal [15]:

$$D' = D'_{static} + D'_{cyclic} = \sum\nolimits_{j=1}^{q'} \frac{t'_j}{t'_{pj}} + \sum\nolimits_{l=1}^{k'} \frac{n'_l}{N'_{pl}} \tag{6}$$

where D'_{static}, D'_{cyclic} – static and cyclic damage, t'_j – operating time at the j steady state mode, t'_{pj} – time until the limit state occurs, n'_l – number of cycles of the l type, N'_{pl} – permissible number of cycles of the l type loads, q' – number of different types of steady-state modes, k' – number of different types of cycles.

Residual service lifetime of the rotor:

$$R = \frac{1 - D'}{D''} \tag{7}$$

where D'' – the average hourly damage (resource exhaustion rate) predicted for the next period of operation, which will accumulate in the rotor when alternating q'' types of steady-state modes and k'' types of cycles with different cyclic loads [15].

4 Results and Discussion

4.1 Effect of the Modified Design of the HPC Flow Part on Steam Parameters

The synthesis of the HPC flow part was carried out at certain initial data, namely: steam pressure at the inlet and outlet of the cylinder, steam pressure in the extraction areas. Steam extraction for heat recovery is carried out in the turbine after each stage except the first (i.e., extraction I after the 2nd stage, extraction IV after the 5th stage).

Additionally, all geometrical parameters of the flow part were taken into account.

Based on these calculations, a variant of the K-1000-60/3000 turbine's HPC without the 5th stage was investigated. Due to the damage to the 5th stage blades, it was decided to dismantle these blades from both turbine flows. The nozzle blades are also dismantled along with the working blades. According to the results of calculations, the most important operating parameters are presented in Table 1.

Table 1. Technical characteristics of the HPC flow part of K-1000-60/3000 turbine

#	Technical characteristics	Project design of the HPC	HPC without 5th stage
1	Inlet steam pressure, kPa	5737	5737
2	Outlet steam pressure, kPa	573,7	573,7
3	Steam extraction *I* pressure, kPa	2261	2236
4	Steam extraction *I* inhibition temperature, °C	219,8	219,2
5	Steam extraction *II* pressure, kPa	1428	1367
6	Steam extraction *II* inhibition temperature, °C	197,2	195,2
7	Steam extraction *III* pressure, kPa	912,4	573,7
8	Steam extraction *III* inhibition temperature, °C	177,1	160,7
9	Steam extraction *IV* pressure, kPa	573,7	573,7
10	Steam extraction *IV* inhibition temperature, °C	157,1	160,7
11	Internal capacity of the HPC, MW	474,12	460,8
12	Internal relative efficiency, %	77,9	75,8

There are small changes in steam parameters at the 1st and 2nd stages for the HPC without 5th stage. The greatest changes are observed at the 4th stage. The steam pressure decreases by 37% (item 7 of Table 1), and the inhibition temperature decreases from 177.1 to 160.7 °C. There is extraction III after the turbine 4th stage for feeding main deaerator of the power unit. Reducing the pressure will completely disrupt the operating conditions of the deaerator and it will not be able to provide sufficient purification of water from corrosive gases. Therefore, it is necessary to find another steam source with sufficient parameters to feed the deaerator. This will require significant changes to the thermal scheme of the power unit.

The thermal energy that was previously used up at the 5th stage was redistributed to the previous 4 stages. As a result, the internal capacity of the HPC decreased slightly by

3% (item 11 of Table 1). However, the stages operate in a less cost-effective mode due to the increased stage energy, so the internal efficiency of the HPC decreases by 2.1% (item 12 of Table 1). At the same time, if we take into account future changes in the thermal scheme due to the lack of steam with the required parameters for the needs of deaerator, the internal efficiency of the HPC can be additionally reduced.

4.2 Effect of the Modified Design of the HPC on the Durability of Rotor

The thermal and stress-strain state of the high-pressure rotor was studied for the main operating modes: unit operation at 1000 MW, startup from a hot state of metal (up to 2 days of downtime) and startup from a cold state (more than 5 days of downtime).

The temperature distribution for both rotor designs is quite similar. The biggest differences are observed in the body of the 5th stage disc. Depending on the operating mode, the temperature in this area is 4–12 °C higher for the project design of the HPC.

The calculation of the rotor stress-strain state took into account the obtained temperature distributions, centrifugal forces, steam pressure and distributed masses of the blades on the stages. An example of the calculation results when the unit operates at rated power is shown in Figs. 2 and 3.

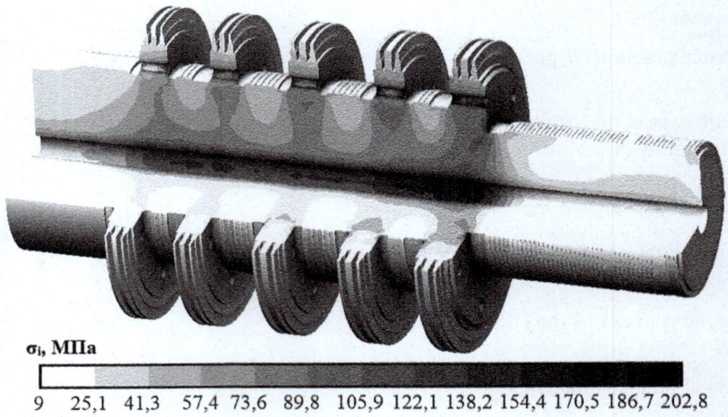

Fig. 2. Stress-strain state of the rotor of the project design at 1000 MW operation

The most stressed areas of the rotor of project design (Fig. 2) are the discharge holes of the 5th stage (202.8 MPa), the rotor axial hole below the 5th stage (195.2 MPa), and the 5th stage fillet on the side of the end seals (200.3 MPa).

For the rotor without the 5th stage, a shift in the maximum stress intensity to the area of the discharge holes of the 4th and 3rd stages, as well as the axial hole under the same stages, is observed (Fig. 3). The maximum stress value is 193.7 MPa.

Similar data were obtained for other operating modes of the turbine. For example, Fig. 4 shows the dynamics of changes in stress intensity in some of the studied areas of the rotor without the 5th stage during startup from a cold state.

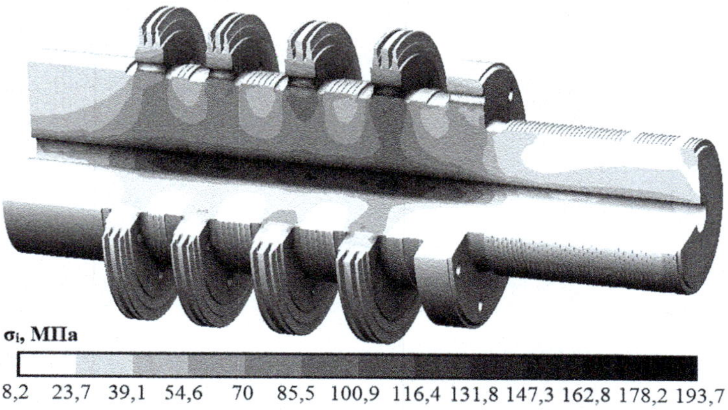

Fig. 3. Stress-strain state of the rotor without the 5th stage at 1000 MW operation

To justify the possibility of extending operation of the rotor with such design changes, it is necessary to calculate its service life. In this case, we consider a rotor that has been in operation for 25 years. It is assumed that the annual operating time is 6000 h with 10 startups (6 from cold state and 4 from hot). The results of the calculations are shown in Table 2.

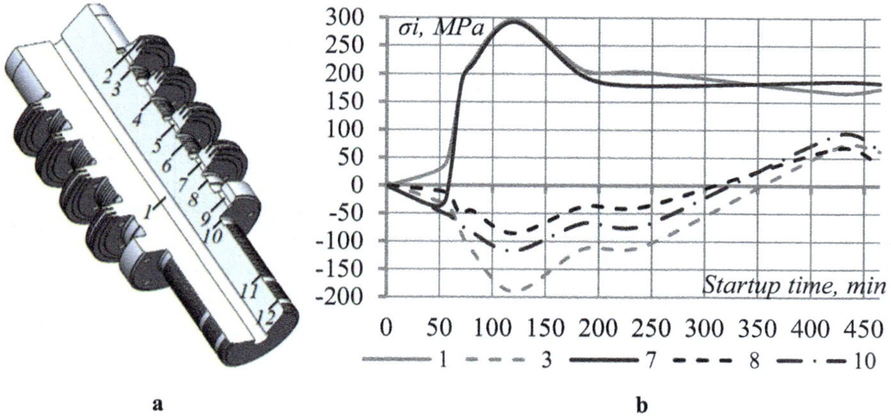

Fig. 4. Areas of study (a) and dynamics of stress intensity change (b) in the rotor without 5th stage blades during cold start

Depending on the rotor design, the area that determines its residual service life changes (items 4, 8 of Table 2). For the project design, the most loaded area is the discharge holes of the 5th stage (zone 9 in Fig. 4a). For a rotor without 5th stage blades, the most loaded areas are the discharge and axial holes of the 4th stage (zones 7 and 1 in Fig. 4a). The individual service lifetime of the rotor without the 5th stage increases by 5.6% compared to the project design rotor (item 14 of Table 2). This is due to an increase

Table 2. Resource indicators of the high-pressure rotor of different designs

#	Service lifetime indicators	Project design of the HPC	HPC without 5th stage
1	Operating time, h	150 000	150 000
2	Number of startups	250	250
3	Stress intensity at rated power, MPa	202,8	193,7
Startup from a cold state of metal			
4	Most loaded area in this mode (Fig. 4a)	9	1
5	Stress amplitude, MPa	188,6	148,9
6	The reduced strain amplitude, %	0,1188	0,0106
7	Permissible number of cycles	950	1050
Startup from a hot state of metal			
8	Most loaded area in this mode (Fig. 4a)	9	7
9	Stress amplitude, MPa	190	167,1
10	The reduced strain amplitude, %	0,1204	0,108
11	Permissible number of cycles	870	890
12	Accumulated damage in the metal, %	75,67	71,66
13	Residual service lifetime, h	48 200	59 300
14	Individual service lifetime, h	198 200	209 300

in the permissible number of cycles for different operating modes and a decrease in the overall stress level in the rotor.

Thus, the absence of the 5th stage does not have a negative effect on the resource indicators of the K-1000-60/3000 turbine. At the same time, it is advisable to investigate the vibration state of the rotor, since the absence of a blade device on both of its flows can seriously affect the critical frequencies of the shaft and will require taking this fact into account during further operation.

5 Conclusions

The high-pressure cylinder of the K-1000-60/3000 turbine was studied in two design versions: the project design and without nozzles and working blades of the 5th stage of both flows.

A mathematical model of the flow part of the HPC based on the object-oriented approach has been developed. It is established that in the absence of the 5th stage, the thermal energy used by this stage is distributed among the other four stages. This redistribution leads to a reduction in the internal power of the HPC by 3% and the relative internal efficiency by 2.1%. However, it also significantly reduces the steam pressure and temperature in the third extraction stage, which feeds the main deaerator of the unit. The new steam parameters will not be sufficient to ensure the quality of feed water.

The service lifetime indicators of a high-pressure rotor for different designs of the HPC is investigated. The migration of the maximum stress zones and a decrease in the overall stress level have been established. As a result, the individual service life of the rotor without the 5th stage increases by 5.6%.

Thus, the high-pressure rotor of the K-1000-60/3000 turbine without the 5th stage can be allowed for further operation. At the same time, it is necessary to ensure that the power unit deaerator is supplied with steam with sufficient parameters. It is also necessary to investigate new critical frequencies of the high-pressure rotor due to changes in its design and mass.

References

1. Wang, Q., Guo, J., Li, R., Jiang, X.: Exploring the role of nuclear energy in the energy transition: a comparative perspective of the effects of coal, oil, natural gas, renewable energy, and nuclear power on economic growth and carbon emissions. Environ. Res. **221**, 115290 (2023). https://doi.org/10.1016/j.envres.2023.115290
2. He, Q., Xue, S., He, H., Hu, F., Gao, H., Hu, W.: Fatigue fracture failure analysis of 12Cr12Mo steam turbine blade. Eng. Fail. Anal. **150**, 107356 (2023). https://doi.org/10.1016/j.engfailanal.2023.107356
3. Hua, H., Zhunbei, Z., Zhanjun, Y., Wang, X., Yuxing, W.: Failure analysis of steam turbine blade roots. Eng. Fail. Anal. **115**, 104629 (2020). https://doi.org/10.1016/j.engfailanal.2020.104629
4. Rivaz, A., Mousavi Anijdan, S., Moazami-Goudarzi, M., Nazari Ghohroudi, A., Jafarian, H.: Damage causes and failure analysis of a steam turbine blade made of martensitic stainless steel after 72,000 h of working. Eng. Fail. Anal. **131**, 105801 (2022). https://doi.org/10.1016/j.engfailanal.2021.105801
5. Krechkovska, H., Hredil, M., Student, O., Svirska, L., Krechkovska, S., Tsybailo, I., Solovei, P.: Peculiarities of fatigue fracture of high-alloyed heat-resistant steel after its operation in steam turbine rotor blades. Int. J. Fatigue **167**(B), 107341 (2023). https://doi.org/10.1016/j.ijfatigue.2022.107341
6. Lee, C.-H., Huang, S.-C., Chang, C.-A., Chen, B.-K.: Operation of steam turbines under blade failures during the summer peak load periods. Energies **7**(11), 7415–7433 (2014). https://doi.org/10.3390/en7117415
7. Slabchenko, O., Zaytsev, M., Kozlokov, A., Zolotukhin, A.: Assessment of performance of LPC of T-250/300-23,5 steam turbine of Kharkiv CHPP-5. Bull. NTU "KhPI". Series: Power Heat Eng. Process. Equip. **9**(1231), 34–40 (2009)
8. Yamamoto, S., Uemura, A., Miyazawa, H., Furusawa, T., Yonezawa, K., Umezawa, S., Ohmori, S., Suzuki, T.: A numerical and analytical coupling method for predicting the performance of intermediate-pressure steam turbines in operation. Energy **198**, 117380 (2020). https://doi.org/10.1016/j.energy.2020.117380
9. Kulkowski, K., Grochowski, M., Duzinkiewicz, K., Kobylarz, A.: Nuclear power plant steam turbine—modeling for model based control purposes. Appl. Math. Model. **48**, 491–515 (2017). https://doi.org/10.1016/j.apm.2017.04.008
10. Sun, X., Song, F., Yuan, J.: Online estimation approach of the steam specific enthalpy for wet steam turbines in nuclear power plants. Appl. Therm. Eng. **229**, 120531 (2023). https://doi.org/10.1016/j.applthermaleng.2023.120531

11. Mambro, A., Galloni, E., Congiu, F.: Assessment of the heat transfer coefficient for the prediction of ventilation power in steam turbine last stages operating at low load. Thermal Sci. Eng. Progress **18**, 100542 (2020). https://doi.org/10.1016/j.tsep.2020.100542
12. Liu, Y., Wang, W.: Evolution of principal stress of a turbine rotor under cyclic thermo-mechanical loading. Eng. Fail. Anal. **109**, 104242 (2020). https://doi.org/10.1016/j.engfailanal.2019.104242
13. Zhao, X., Ru, D., Wang, P., Gan, L., Wu, H., Zhong, Z.: Fatigue life prediction of a supercritical steam turbine rotor based on neural networks. Eng. Fail. Anal. **127**, 105435 (2021). https://doi.org/10.1016/j.engfailanal.2021.105435
14. Avdieieva, O., Usatyi, O., Mykhailova, I.: Optimization of the Flowing Part of the Turbine K-310-240 Based on the Object-Oriented Approach. In: Machado, J., Soares, F., Trojanowska, J., Ottaviano, E. (eds.) Innovations in Mechanical Engineering. icieng 2021. Lecture Notes in Mechanical Engineering. Springer, Cham (2022). https://doi.org/10.1007/978-3-030-79165-0_20
15. Chernousenko, O., Rindyuk, D., Peshko, V., Chernov, O., Goryazhenko, V.: Development of a system for estimating and forecasting the rational resource-saving operating modes of TPP. East-Eur. J. Enterp. Technol. **3**(8), 14–23 (2020). https://doi.org/10.15587/1729-4061.2020.204505

Analysis and Selection of SMAW Welding Process Parameters of Cr-Mo P11 Steel for Power Plant Applications

Mohamed Farid Benlamnouar[1](✉), Nabil Bensaid[1], Tahar Saadi[1], Yazid Laib Dit Laksir[2], and Riad Badji[1]

[1] Research Center in Industrial Technologies CRTI, Cheraga, 16014 Algiers, Algeria
csc.farid@yahoo.fr

[2] Département de Génie Climatique, Université constantine1, Constantine 25000, Algérie

Abstract. This work presents an analysis and multi-criteria selection by hybrid operation combined of analytic hierarchy process (AHP) and Taguchi design for the selection of optimal welding parameters to produce shielded metal arc welding (SMAW) of Cr-Mo P11 low alloy steel used in power plant. In our investigation, three welding parameters i.e. intensity (I (A)), voltage (V (v)) and welding speed (S (mm/min)) have been tested to evaluate four output mechanical properties i.e. penetration depth and maximum strain (ε (%)), tensile strength (σ (MPa)) and impact energy (IE (MPa)). Taguchi L9 orthogonal array is employed to extract mathematical models and welding parameters effects on output properties. It was deduced that welding current has a significant influence on tensile strength, strain and penetration depth properties followed by welding speed and voltage. This study delves into the investigation of sub-criteria of welding process through the use of AHP method coordinately with Taguchi results to identify the optimal filler rod (Ri). Taguchi's biggest effect parameter is included in the AHP matrix to choose the best welding rod by adopting different welding parameters with preheating temperature (T °C).

Keywords: Selection · SMAW welding · AHP Method · Taguchi design · ANOVA · Cr-Mo P11 steel

1 Introduction

In contemporary material assembly, welding processes play a pivotal role in ensuring the integrity and seamless integration of structures for power generation systems [1, 2]. Among these welding techniques, shielded metal arc welding (SMAW) emerges as the preferred choice due to its exceptional cost-effectiveness, making it the most widely used method in the industry. SMAW not only guarantees metallic continuity but also provides a dependable solution for achieving robust structural bonds without diminishing on the overall quality of welded assembly.

SMAW welding has proven to be a valuable method in joining Cr–Mo low alloy steels, such as P11 steel, which play a crucial role in vital power plant components

H. Altenbach et al. (Eds.): CAMPE 2023, pp. 275–289, LNME 2025.
https://doi.org/10.1007/978-3-031-82979-6_28

like silos and injection tanks. These steels are prized for their exceptional resistance to corrosion and outstanding mechanical properties, surpassing those of conventional steels [3, 4]. However, SMAW welding is a multifaceted technique influenced by various thermal, mechanical, and chemical factors. During the welding process, the transfer of electrons near the welding electrode generates a substantial amount of thermal input energy, resulting in localized melting between the wire metal and base material edges. Consequently, the welded joint exhibits significant variations between the mechanical properties of its seamless regions and those affected by welding heat inputs [5, 6].

The weld joint quality in SMAW welding is profoundly influenced by key parameters as intensity (I), voltage (V), and weld speed (S). These parameters hold critical significance in determining the overall welding process quality. To ensure precise and reliable assessments, researchers and industrial experts employ various tools, including mathematical models and numerical methods [6, 7]. The evaluation of welding processes has been extensively explored in numerous research studies and widely adopted by industrialists. This enables them to assess their current techniques in various industries and also to enhance and optimize SMAW parameters through experimental designs [7, 8]. Where, several researches conducted on experimental designs like Taguchi design to solve welding problematic [9, 10]. This technique has powerful tools that use a special design and standard matrixes to evaluate welding parameter with better calculation rate compared to simulation techniques, whilst ensuring high process performance in few computational operations. Taguchi design focuses on maximizing or minimizing process parameters based on experimental designs and matrices, which is operating on to optimize several industrial problems [11]. Taguchi design can contribute improving and optimizing welding process; where it can govern welding parameters and analyzing link between input and output parameters. It reported that welding current and speed have a great effect on depth penetration of hybrid welding [12]. Moreover, Ramkumar et al. discuss optimizing welding process parameters, such as current and welding speed, in order to achieve maximum penetration by TIG Welding of steel [13]. Yousefieh et al. employed Taguchi design to optimize pulsed welding parameters for enhancing corrosion resistance. They revealed that pulse current had the most significant impact on corrosion resistance, followed by background current, and pulse frequency [14].

Multi-criteria decision models hold paramount importance in effectively managing and categorizing determinants influencing various industrial process factors. One such prominent model is the Analytical Hierarchy Process (AHP), which has gained research effectiveness due to its ability to handle complex methods across different scales and facilitate both qualitative and quantitative decision-making for selecting optimal industrial processes parameters. AHP employs specific mechanisms, including hierarchical modeling and prioritization techniques, to streamline decision-making processes [15]. AHP method proves instrumental in making right choices [16]. It is particularly useful when dealing with a several number of options, each with multiple sub-criteria. This method becomes a valuable resource for addressing manufacturing challenges and guiding decision-making processes by combining numerical values with informed judgments. It aims to reach wise and decisive conclusions for industrial processes [16, 17].

Recently, AHP approach is used to achieve Hybrid welding process prioritization through welding parameters. The AHP methodology has been used in the hazard determination, control and risk assessment studies. It has been employed to obtain the weight of the criteria of welding process and the interrelation between qualitative and quantitative factors of welding process [17, 18]. Furthermore, AHP method is concerned with organizing and analyzing parameters that contain complex decisions. It suggests a judgment matrix based on the comparison of the different characteristics of the industrial process, two by two. From the construction of a square matrix, it can assess the relative importance of one characteristic compared to another by using an appropriate scale that will determine the quality of the solution obtained. AHP is also employed to identify the most influential parameters in SMAW arc processing, expanding the scope of SMAW welding research by establishing interdependencies among SMAW factors for the evaluation of optimal welding process parameters [18].

In our investigation, we use coupling of two methods constituting a selecting of welding parameters through Taguchi's design approach, and making the best process decision by AHP method.

It focuses on the optimization of welding parameters and integrates them into Making-decision process to select best welding rod the most compatible with criteria. Taguchi and AHP methods were achieved using mechanical characteristics of weld joints. Mechanical tests were achieved in order to analyze precisely penetration depth, deformation and tensile strength in weld joints. The experiments are conducted according to Taguchi orthogonal array L9 and AHP explanation.

2 Experiments and Methods

2.1 Material and Process

In this work, Chemical composition of P11 steel was presented (Table 1). For Metallographic observation, samples from as-received base material P11 steel are etched with reactive Nital (3% concentration), and observed in ZEISS optical microscopy. The microstructure of P11 steel is composed of P-pearlite phase in α-ferrite matrix as shown in Figure (P appears dark and α-ferrite appears light) (Fig. 1).

Table 1. Chemical composition of base and filler metal

Element %	C	Si	Mn	Cr	Ni	Mo	Cu	Nb	Ti
P11	0.08	0.70	0.42	1.19	0.046	0.47	0.01	0.06	0.09
E 7018	0.1	0.80	0.90	0.01	0.01	0.5	–	–	–

P11 steel used in this study was received in the form of tubes of 12" (30,48 cm) diameter and 8 mm in wall thickness. Three weld passes were employed using a shielded metal arc welding process (SMAW) by ER E7018 filler metal according to ASME standards [19].

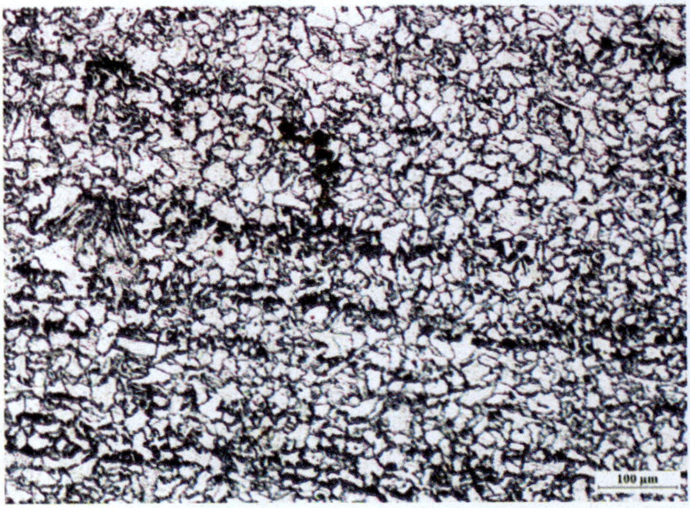

Fig. 1. Optical microstructure of P11 steel

Mechanical behavior of SMAW weld joints is performed by using tensile test and Charpy test. Tensile specimens were tested by a servo hydraulically controlled digital tensile testing machine. Each tensile specimen size was prepared in accordance with ASTM E08 standards [20]. The Charpy test was performed according to standard test methods for notch bar impact testing of metallic materials, while specimen length was 55 mm and its rectangular cross-section area is 10 mm × 7 mm. Charpy specimen has a V-shaped notch with a flank angle of 45° and depth of 2 mm (Fig. 3) in the fusion zone. Three tensile/Charpy specimens were tested from each welding parameters.

Penetration depth was measured by USN52L ultrasonic detector having 60 Hz rate and LCD screen resolution of 480 x 320 pixels.

2.2 Selection of Output and Input Parameters

Our experimental investigations are carried out to compare process parameter levels of different welding joints. The welding parameters used in this investigation are: (i) Average intensity I (A), (ii) Medium voltage V (V), (iii) Weld speed S (mm/min) (Table 2).

Table 2. Welding Process control parameters and their levels

Parameters	Level 1	Level 2	Level 3
I (A)	60.00	100.00	135.00
V (v)	25.00	30.50	35.00
S (mm/min)	60.50	85.20	101.60

2.3 Experimental Design

2.3.1 Taguchi Design

Taguchi design is used for determining the relationship between the process parameters with the desired responses, and exploring the effect of these parameters on responses. The second objective of Taguchi method is based on mathematical models of σ_m (MPa), εm, d (mm), and IE (KJ). The models were created using average intensity (A), V (V), S (mm/min) as industrial process parameters to reveal the relationship between the welding conditions and mechanical characteristics (Table 5).

The data required to develop Taguchi computation were collected by L9 designs and by varying each numeric factor over three levels. The experimental design consists of 9 runs as outlined in Table. The Design-Expert software is used for the analysis of responses and determining the empirical models with best fits. Then, Design-Expert facilities checking the adequacy of the models using analysis of variance (ANOVA) technique and signal-to-noise ratio (S/N). In Taguchi design, signal-to-noise (S/N) ratio was used to represent the best combination of welding parameters in order to provide best mechanical properties. The S/N ratio can be characterized into three categories i.e., nominal-the better, lower-the better, larger-the better [11–14]. In this research larger-the-better is chosen, which can be calculated with the equation:

$$S/N = -10\log\left[\frac{1}{n}\sum\nolimits_{k=1}^{n}\frac{1}{y_k^2}\right] \tag{1}$$

where S/N are the signal-to-noise ratio, n is the number of repetitions for an experimental combination, and a performance value of the kth experiment. Then, the mean S/N ratios at each level for various factors and the optimal level (the largest S/N ratio among all levels of the factors) could be calculated. Furthermore a statistical analysis of variance (ANOVA) is performed to check the statistical process parameters (Table 3).

Table 3. L9 orthogonal array

Exp	I (A)	V (v)	S (mm/min)	σ_m (MPa)	ε_m (Δl/l)	d (mm)	IE (KJ)
1	60	25	60.5	52,36	0,060	1,50	120.88
2	100	25	85.2	74,37	0,120	3,38	11.55
3	135	25	101.6	80,59	0,135	4,45	78.37
4	135	30.5	60.5	70,58	0,150	3,92	108.7
5	60	30.5	85.2	58,71	0,074	1,57	87.43
6	100	30.5	101.6	76,04	0,138	3,60	114.06
7	100	35	60.5	72,45	0,105	3,24	96.81
8	135	35	85.2	78,62	0,132	4,30	117.94
9	60	35	101.6	64,54	0,080	1,67	108.7

2.3.2 AHP Method

In Industrial systems, AHP method contains breaking down a complex decision problem into a hierarchy of criteria and alternatives [21, 22].

In order to ensure our judgment about the weights of the criteria, AHP process is used to estimate the impact of the alternative parameters on the global objective of the hierarchy. Thus, the decision depends on (Ci, Cj) criteria that are defined by (n x n) matrix as in the following (Eq. 2, Table 4):

$$A = \left(a_{jj}\right)_{nxn} = \begin{bmatrix} a_{11}\ a_{12}\ \ldots\ldots\ a_{1n} \\ a_{21}\ a_{22}\ \ldots\ldots\ a_{2n} \\ \cdot \\ \cdot \\ \cdot \\ a_{n1}\ a_{n2}\ \ldots\ldots\ a_{nn} \end{bmatrix} \tag{2}$$

Table 4. Pairwise Comparison Scale of AHP method

Intensity	Explanation
1	Two criterion contribute equally to the objective
3	Experience and judgment slightly favor one over another
5	Experience and judgment strongly favor one over another
7	Criterion is strongly favored and its dominance is demonstrated in practice
9	Importance of one over another affirmed on the highest possible order
2,4,6,8	Used to represent compromise between the priorities listed above

In the first step, each matrix must be normalized in order to find the relative weights. These later are defined by multiplying the right eigenvector (w) and corresponding largest eigen value ($\lambda_{\max}$):

$$A_w = (\lambda_{\max}) * w \tag{3}$$

2.3.3 AHP Eigenvector

The eigenvector characteristic guarantees that priority weights derived from pairwise comparisons should align with the dominant eigenvector of the comparison matrix. Thus, the priority weights must be consistent with decision making. The eigenvector property gives priority weights suitability with the dominant eigenvector of the comparison matrix, where the priority weights must be consistent with decision making.

2.3.4 AHP Consistency

In the Analytic Hierarchy Process (AHP), consistency refers to the degree of agreement or harmony in the pairwise comparisons made by decision making. Since the pairwise comparisons are subjective judgments, there is a potential for inconsistency if the decision maker's assessments contradict each other.

Ensuring consistency in the pairwise comparisons is important because inconsistent judgments can lead to unreliable results and undermine the validity of the decision-making process. To address this issue, the AHP method provides a consistency check using the principle of the eigenvector.

It noted that the quality of the output of the AHP is related to the consistency of the pairwise comparison judgments. The consistency is defined by the relation between the entries of A: aij ajk = aik, Consistency Index (CI) can be calculated, using the following formula [15–17]:

$$CI = \frac{\lambda_{\max} - n}{n - 1} \tag{4}$$

where, consistency index (CI) measures the level of inconsistency in the matrix. The CI is computed by comparing the average value of the matrix's principal diagonal with the expected average value derived from a randomly generated consistent matrix of the same size.

Consistency ratio (CR) provides a measure of the relative consistency by comparing the actual inconsistency with the expected inconsistency for a randomly generated consistent matrix. It can conclude whether the evaluations are sufficiently consistent. The CR is calculated as the ratio of the CI and the random index (RI), as indicated in Eq. 5.

RI is based on coefficient from 1 to 9, which take numerical values (Table 5).

$$CR = \frac{CI}{RI} \tag{5}$$

Table 5. Random Index

n	1	2	3	4	5	6	7	8	9	10
RI	0.00	0.00	0.58	0.90	1.12	1.24	1.32	1.41	1.45	1.49

3 Results

The main role of ANOVA analysis of Taguchi design is to determine the process parameters which have the significantly affect in weld performance. This analysis is measured by the sum of the squared deviations from the total mean of the multi response performance index, into contributions by each of the process parameters and the error. Depending on F value of ANOVA table, the analysis of variance was carried out for a 95% confidence level. Therefore, we make an assumption that the value of P less than 0.05 is found to be significant parameter.

The Analysis of the ANOVA variance of σ_m shows that welding intensity is the main factor affecting the tensile strength. The contribution of which is 70.45%, while welding

Table 6. ANOVA result of σ_m (as function of I (A), V (V), S (mm/min)

Source	DL	SomCarajust	CMajust	F value	Pvalue	Cont. %
Regression	3	628,35	209,45	11,88	0,010	–
I(A)	1	504,81	504,81	28,64	0,003	70.45
V(V)	1	10,32	10,32	0,59	0,479	1.44
S (mm/min)	1	113,22	113,22	6,42	0,052	15.80
Error	5	88,13	17,63			
Total	8	716,48				

Table 7. ANOVA result of ε_m (as function of I, V, S)

Source	DL	SomCarajust	CMajust	F Valus	P Valus	Cont. %
Regression	3	0,007230	0,002410	11,13	0,012	–
I(A)	1	0,007001	0,007001	32,34	0,002	96.84
V(V)	1	0,000004	0,000004	0,02	0,894	0.05
S(mm/min)	1	0,000224	0,000224	1,04	0,356	3.11
Error	5	0,001082	0,000216			
Total	8	0,008312				

speed comes in second order with an effect of 15.8%. The voltage has a weak effect on tensile strength, and his contribution is only 1.44% (Table 6).

ANOVA analysis of ε_m reveals that the main factor influencing strain is intensity, accounting for a substantial contribution of 96.84%. Welding speed emerges as the second most significant factor, contributing 3.11% to the overall variance. The voltage exhibits a very weak effect on strain, with a mere 0.05% contribution (Table 7).

Table 8. ANOVA result of d (as function of I, V, S)

Source	DL	SomCarajust	CM ajust	F Value	P Value	Cont.%
Regression	3	10,8343	3,6114	46,74	0,000	–
I (A)	1	10,6438	10,6438	137,75	0,000	94.82
V (V)	1	0,0029	0,0029	0,04	0,854	0.02
S(mm/min)	1	0,1876	0,1876	2,43	0,180	1.70
Error	5	0,3863	0,0773			
Total	8	11,2206				

The ANOVA analysis of depth penetration demonstrates that intensity is the predominant factor influencing d (mm), making a significant contribution of 94.82%, where welding speed emerges as the second most influential factor, contributing 1.7% to the overall variance. In contrast, the voltage exhibits a minimal effect on strain with a 0.02% contribution (Table 8).

Table 9. ANOVA result of IE (as function of I, V, S)

Source	DL	SomCarajust	CM ajust	F value	P value	Cont. %
Regression	3	945,17	304,89	15,85	0,011	–
I (A)	1	169.83	169,83	8,74	0,058	15.11
V (V)	1	11,47	11,47	0,73	0,61	1.03
S(mm/min)	1	942.52	942,52	15,63	0,007	83.87
Error	5	105.75	22.03			
Total	8	1123.57				

The ANOVA analysis of impact energy reveals that welding speed is the main factor influencing IE, making a great contribution of 83.87%. Intensity appears here as the second most influential parameters, contributing 15.11% to the overall variance. Conversely, the voltage exhibits a minimal effect on depth penetration, with a mere 1.03% contribution (Table 9).

From the Tables 6–8, we are able to deduce that welding current has a significant influence on σ_m, ε_m and d properties followed by speed and voltage. While the Table 9 shows that speed has the high contribution on impact energy property (IE).

Linear regression models were developed for tensile strength stain and depth using Minitab18 software. The responsible variables are current, speed and voltage. The equations of the fitted model for tensile strength stain and depth of P11 joints are given below, respectively:

$$\sigma_m = 20.6 + 24.44 * 10^{-2} I + 26.20 * 10^{-2} V + 21.00 * 10^{-2} S \tag{6}$$

$$\varepsilon_m = -0.85 + 0.09 * 10^{-2} I + 0.02 * 10^{-2} V + 0.03 * 10^{-2} S \tag{7}$$

$$d = -0.99 + 3.55 * 10^{-2} I - 0.44 * 10^{-2} V + 0.85 * 10^{-2} S \tag{8}$$

$$IE = 15.45 + 32.51 * 10^{-2} I + 39.42 * 10^{-2} V + 36.66 * 10^{-2} S \tag{9}$$

These results of ANOVA analysis underscore the significant influence of welding current on the mechanical properties of the welds in question. This finding contrasts with other welds involving similar base metals, where factors such as gas flow or welding speed tend to play a more dominant role, as demonstrated in references [4, 11, 23]. The observed differences can be attributed to several factors, including the relative variations in the chemical composition of the filler and base materials, the distinct welding techniques used, such as TIG applied in the previous references versus SMAW applied in this study. The differences in the microstructure of welds, with similar welds in this study and dissimilar welds in other studies [4, 23], also have a significant impact on the mechanical properties and the most influential welding parameters.

The preheating temperature refers to a specific temperature at which the material to be welded is kept for a few minutes before the welding process begins. This annealing ensures the proper melting of the wire metal, enhances the properties and quality of the welded joint, and minimizes potential defects that arise during welding [24, 25].

In SMAW process, although welding rod types (Ri) used in the welding of Cr-Mo P11 steel having roughly the same chemical composition on the market, they differ in terms of welding rod coating (protection powder), which affects the ability of wires to maintain their quality at preheating temperature, thus influencing weld joint quality.

In order to reach the order of these rods in terms of their response to the desired requirements, we will establish a matrix as follows (Fig. 2):

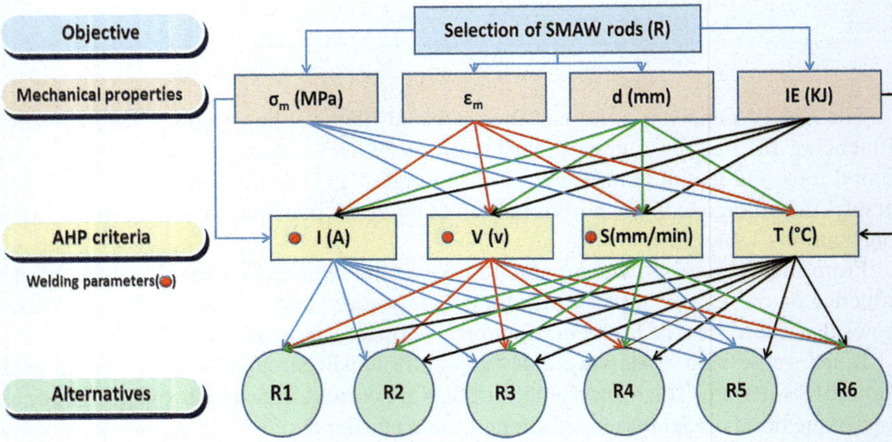

Fig. 2. Schema of AHP selection

In AHP analysis, the experimental values is represented in linguistic term, because it is more comprehensive, it provides us with the scope to interpret the degrees of uncertainty in processes and, consequently, enables us to arrange and determine a hierarchy of trade-off between experiments. According to the AHP making-decision process presented previously in Fig. 3, AHP loop is reiterating until the fulfillment of the conditions of RC less than 10%. According to Taguchi design, the index will be given to welding parameters in ascending order from high to low contribution in mechanical properties of weld joints.

The index 5 represents T parameter (Table 10), which can be defined as the indirect variable factor associated with SMAW rod efficiency. It can be, for example, post weld nuance, rod price or the amount of Rod energy consumption.

In our application, it was proposed T as preheating temperature for its high correlation with welding process parameters and mechanical properties of weld joints [25]. In this

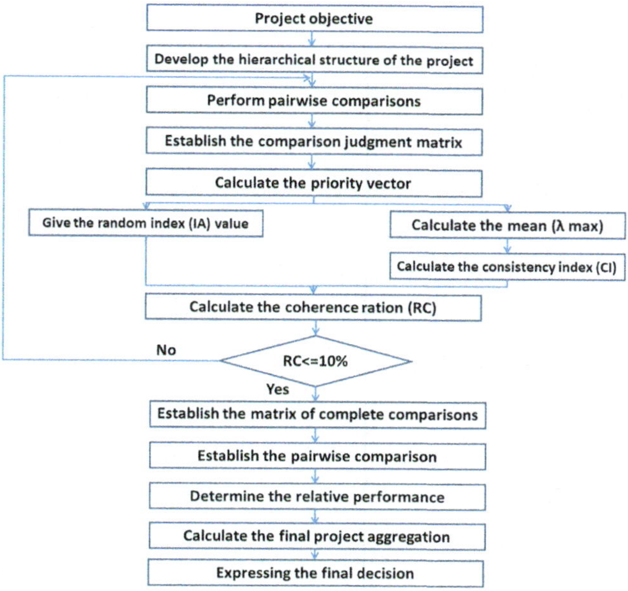

Fig. 3. AHP making-decision process

work, it is proposed five weld rods (R1, R2, R3, R4, R5) which associated with the parameters of I, S, T and V.

Table 10. Pairwise comparison matrix

	I	S	V	T
I	1	3	7	5
S	0,333	1	3	4
V	0,143	0,333	1	3
T	0,200	0,250	0,333	1

Based on Table 10 and the equation that computing largest eigenvalue (Eq. 3), it found that:

$$\lambda_{max} = \frac{4.421 + 4.338 + 4.129 + 4.059}{4} = 4.23$$

Pairwise comparison matrix prioritized intensity and speed followed by preheating temperature (Table 10), while voltage is ranks last in AHP priorities for its low contribution in the mechanical parameters (Tables 6–9).

The next crucial step entails normalizing the preference matrix, which involves transforming it into matrix B. This normalization process is pivotal as it allows us to bring the data into a standardized and more manageable form. The transformation is carried out as follows (Eq. 10, Tables 10–11):

$$B = [b_{ij}] = \frac{a_{ij}}{\sum_{i=i}^{n} a_{ij}} \quad (10)$$

Table 11. Calculation of priorities: row averages

	I	S	V	T
I	0,597	0,655	0,618	0,385
S	0,199	0,218	0,265	0,308
V	0,085	0,073	0,088	0,231
T	0,119	0,055	0,029	0,077

By utilizing Eq. 4, where 'n' represents the number of compared elements, which in this case equals 4, therefore:

$$CI = \frac{4.237 - 4}{3} = 0.079$$

Now we need to calculate the consistency index (CR) as follows (Eq. 5):

$$CR = \frac{CI}{RI} = \frac{0.079}{0.9} = 0.088$$

As CR is less than or equal to 0.1, the matrix is deemed sufficiently coherent. This implies that the judgments made during the pairwise comparisons are consistent and reliable. We can identify judgment matrix elements as follows (Table 12):

Table 12. Judgment matrix

	R1	R2	R3	R4	R5	R6
R1	1	0,333	0,2	0,111	0,125	0,143
R2	3	1	0,333	0,125	0,143	0,200
R3	5	3	1	0,143	0,167	0,250
R4	9	8	7	1	2,000	5,000
R5	8	7	6	0,500	1	2
R6	7	5	4	0,2	0,5	1

Table 13. Criteria judgment matrix

	I	S	T	V	Eigenvector
R1	0,015	0,006	0,003	0,002	0,026
R2	0,025	0,011	0,005	0,003	0,045
R3	0,044	0,019	0,009	0,005	0,078
R4	0,239	0,105	0,051	0,030	0,425
R5	0,148	0,065	0,031	0,018	0,264
R6	0,091	0,040	0,019	0,011	0,162

From the Table 12, it can calculate λ_{max} as:

$$\lambda_{\max} = \frac{6.131 + 6.063 + 6.253 + 6.833 + 6.803 + 6.777}{6} = 6.477$$

As a result, the criteria judgment matrix can be presented in Table 13:
Thus, depending on previous equations:

$$CI = \frac{6.477 - 6}{5} = 0.095$$

$$CR = \frac{CI}{RI} = \frac{0.095}{1.24} = 0.071 \leq 0.1$$

According to AHP criteria that ensure the adaptation to the established objectives of making-decision process to find the best welding rod, Table 13 concluded that SMAW rod ranking is as follow:

$$R4 \geq R5 \geq R6 \geq R3 \geq R2 \geq R1$$

It also shows that R4 is the most effective rod to meet the study requirements.

4 Conclusion

This study suggests an integrated approach to achieve high quality of weld joints and make the best decisions for selecting welding rods using the AHP method and Taguchi design. It examines the classification of welding parameters and aims to find the optimal decision-making process for the SMAW welding process. The optimization of welding parameters has been analyzed using the Taguchi design. The AHP method is then used as a tool to infer a consistent ranking of welding rod alternatives. From the Taguchi design, it was concluded that welding current has a significant influence on σ_m, ε_m, and depth penetration properties of P11 steel welded by the SMAW welding process, while welding speed has the highest contribution to the impact energy (IE) property. The most influential factors extracted by the Taguchi method were included to find the best criteria for SMAW welding rods using the AHP method. The application of both the Taguchi and AHP methods for the optimization and multi-criteria classification of SMAW welding has made it possible to optimize and specify the input parameters of the AHP through Taguchi models. Thus, precise criteria were established through the combination of Taguchi and AHP to make a best decision regarding the classification of SMAW welding criteria.

References

1. Olohunde Sadiq, T., Mohd Daud, L., Idris, J.: Investigation of microstructure and mechanical properties of A335 P11 main steam pipe in Stesen Janaelektrik Jambatan Connaught Power Plant, Malaysia.Trans. Indian Inst. Met.**71**, 2527–2540 (2018)
2. Sabouri, M., Faridi, H.R.: Investigation on corrosion failure of Cr–Mo P11 grade pipe in primary section of a superheated steam generation system. High Temp. **55**, 139–144 (2017)
3. Seob Lee, Y., Jung Lee, H., Hyeon Lee, J.: Effect of alloying elements and microstructure on the dynamic strain aging behavior of 1.25Cr-0.5Mo and 2.25Cr-1Mo steels. Korean J. Metal. Mat. **59**(11), 769–780 (2021)
4. Benlamnouar, M.F., Bensaid, N., Saadi, T., Badji, R.: Multi-objective optimization and evolution of dissimilar welding process between Cr-Mo steel and austenitic stainless steel for power plant application. Mater. Res. Express **11**(2), 026507 (2024)
5. Pushp Kumar Baghel: Effect of SMAW process parameters on similar and dissimilar metal welds. Heliyon **8**(12), 12161 (2022)
6. Tong, L.G., Gu, J.C., Yin, S.W., Wang, L., Bai, S.W.: Impacts of torch moving on phase change and fluid flow in weld pool of SMAW. Int. J. Heat Mass Transf. **100**, 949–957 (2016)
7. Benlamnouar, M.F., Saadi, T., Bensaid, N., Goussmine, M., Touggui, Y., Temmar, M.: Modelling and optimization of dissimilar welding between 304L and HSLA-X70 using response surface methodology. Lect. Note. Mech. Eng. 463–472 (2021)
8. Saadi, T., Benlamnouar, M.F., Bensaid, N., Boutaghane, A., Soualili, M.A.: Hachemi H 2021: Optimization of automatic TIG welding parameters of AISI 304L ASS welds using response surface methodology. Defe. Diff. Forum **406**, 319–333 (2021)
9. Effertz, P.S., Quintino, L., Infante, V.: The optimization of process parameters for friction spot welded 7050–T76 aluminium alloy using a Taguchi orthogonal array. Int. J. AdvManuf. Technol. **91**, 3683–3695 (2017)
10. Tarng, Y.S., Yang, W.H., Juang, S.C.: The Use of fuzzy logic in the Taguchi method for the optimisation of the submerged arc welding process. Int. J. Adv. Manuf. Technol. **16**, 688–694 (2000)
11. Benlamnouar, M.F., Bensaid, N., Azzoug, M.O., Saadi, T., Zidani, M., Badji, R.: Optimization and evaluation of mechanical and electrochemical properties of ferritic stainless steel welding using Taguchi design. Key Eng. Mater. **973**, 61–72 (2024)
12. Bang, H., Bang, H.S., Na, M.J., Jeon, G.H., Kim, G.S., Kim, B.R.: Application of Taguchi approach to optimize laser-arc hybrid welding parameters of galvanized steel. Strength Mater. **48**(1), 146–151 (2016)
13. Devendranath Ramkumar, K., Mishra, D., Ganesh Raj, B., Vignesh, M.K., Thiruvengatam, G., Sudharshan, S.P., Arivazhagan, N., Sivashanmugam, N., Maximus Rabel, A.: Effect of optimal weld parameters in the microstructure and mechanical properties of autogenous gas tungsten arc weldments of super-duplex stainless steel UNS S32750. Mat. Des. **66**, Part A, 356–365 (2015)
14. Yousefieh, M., Shamanian, M., Saatchi, A.: Optimization of the pulsed current gas tungsten arc welding (PCGTAW) parameters for corrosion resistance of super duplex stainless steel (UNS S32760) welds using the Taguchi method. J. Alloys & Compounds **509**(3), 782–788 (2011)

15. Liu, Y., Eckert, C.M., Earl, C.: A review of fuzzy AHP methods for decision-making with subjective judgements. Expert Syst. Appl. **161**, 113738 (2020)
16. Kokangül, A.: UlviyePolat, Cansu Dag suyu: A new approximation for risk assessment using the AHP and Fine Kinney methodologies. Saf. Sci. **91**, 24–32 (2017)
17. Sanchez-Lozano, J.M., Meseguer-Valdenebro, J.L., Portoles, A.: Assessment of arc welding process through the combination of TOPSIS–AHP methods with fuzzy logic. Trans. Indian Inst. Met. **70**, 935–946 (2017)
18. Vimal, K.E.K., Vinodh, S., Raja, A.: Modelling, assessment and deployment of strategies for ensuring sustainable shielded metal arc welding process – a case study. J. Clean. Prod. **93**, 364–377 (2015)
19. ASME: Qualification Standard for Welding and Brazing. An International Code of ASME Boiler and Pressure Vessel Committee on welding and Brazing, USA (2010)
20. E8/E8M: Standard Test Methods for Tension Testing of Metallic Materials. ASTM International, USA (2013)
21. Ruan, F., Chen, C.-H., Cheng, Y., Wang, J.-Y., Chen, L.-W.: Study on evaluation method for nuclear emergency rescue measures at containment vessel. Ann. Nucl. Energy **151**, 107942 (2021)
22. Ramezanzade, M., Saebi, J., Karimic, H., Mostafaeipour, A.: A new hybrid decision-making framework to rank power supply systems for government organizations: A real case study. Sustainable E. T. A **41**, 100779 (2020)
23. Benlamnouar, M.F., Hadji, M., Badji, R., Bensaid, N., Saadi, T., Laib dit Laksir, Y., Senouci, S.: Optimization of TIG welding process parameters for X70-304L dissimilar joint using Taguchi method. Solid State Phenom. **297**, 51–61 (2019)
24. Bunaziv, I., Langelandsvik, G., Ren, X., Westermann, I., Rorvik, G., Dorum, C.: Morten Hogseth Danielsen, Magnus Eriksson: Effect of preheating and preplaced filler wire on microstructure and toughness in laser-arc hybrid welding of thick steel. J. Manuf. Process. **82**, 829–847 (2022)
25. Mallick, R.K., Ghosh, M., Bahrami, A., Esmaeili, V.: Stress relaxation cracking failure in heat exchanger connection pipes in a petrochemical plant. Eng. Fail. Anal. **147**, 107156 (2023)

Boundary and Finite Element Methods in Crack Propagation Analysis

Kyryl Degtyariov[1(✉)], Vasyl Gnitko[1], Alexander Steinwolf[2], and Ivan Vierushkin[1]

[1] AST Consulting Ltd., Auckland, New Zealand
Kdegt89@gmail.com
[2] A. Pidhornyi Institute for Mechanical Engineering Problems NAS of Ukraine, Kharkiv, Ukraine

Abstract. The main goal of this paper is developing the advanced numerical methods for stress analysis of structures with crack-like defects. The novelty of proposed research consists in application of hypersingular integral equations for solutions of a benchmark test, and in stress analysis of the fuel tank under harmonic loads during transportation from the place of manufacture to the launch site. Boundary and finite element methods have been used for the benchmark test, whereas the finite element method is applied for the fatigue strength analysis of the tank. The fatigue analysis with finite amplitude cyclic loading was carried out. That has showed a high service life, which allows the intact structure to be successfully transported. When the model penny-shaped crack was placed in the zone of maximum stresses, an essential drop in the expected service life before destruction has been observed. From the point of view of practical applications, it was enough to transport such structures over short distances. However, a long-distance transportation can cause the mechanism of fatigue cracks propagation inside the shell with subsequent depressurization and failure.

Keywords: Elastic Bodies · Penny-Shaped Crack · Hypersingular Integral Equations

1 Introduction

Structural materials are always characterized by presence of various microdefects that arose either during manufacture or functioning. These defects can affect both safety and durability of structures. The fatigue cracks propagation is the main cause of catastrophic destruction on railway transport, of turbo and hydro generators, accidents of airplanes and helicopters, etc. In hydro-turbines, fatigue cracks tend to be located near the bearing surfaces. The newly formed crack begins to grow slowly, even if the loads do not exceed the nominal values. When the critical crack length is reached, the load-bearing element suddenly receives a huge amount of kinetic energy, and this leads to the catastrophic destruction. Currently, when a large part of the Ukraine energy equipment has practically exhausted its normative resource or has been damaged, the problems of assessing the

H. Altenbach et al. (Eds.): CAMPE 2023, pp. 290–300, LNME 2025.
https://doi.org/10.1007/978-3-031-82979-6_29

residual resource are particularly relevant. So, the issues of durability, performance, and residual life of the structure elements with cracks became very topical. This explains the increasing interest in the study of failure mechanics.

2 Literature Review

The study of dynamical processes in structure elements by experimental methods leads to significant material costs, unwanted dangerous consequences, and does not allow obtaining a detailed distribution of stress, deformations, displacements, temperatures, as well as describing the occurrence, cracks growth, and damage. Therefore, it becomes necessary to carry out computer simulations. A lot of numerical methods have been proposed to estimate stress intensity factors (SIF). Among them the Galerkin free element method [1], the finite element method for SIF simulations in cracked beams [2], the smoothed finite element method for 3D cracked bodies [3], the extended finite element method [4], P-Version of finite element method with contour integral methods [5], the boundary element method for 2D cracked bodies [6]. In [7] the boundary element metho is used for SIF calculation in functionally graded plates. Another important aspect in studying the stress-strain state of cracked structure elements is cracks propagation. The 2D formulation of the problem was considered in [8], experimental research of the crack propagation in beams was accomplished in [9]. Three-dimensional fatigue crack propagation analysis of welded steel beam based on global-local numerical mode was considered in [10]. As the literature analysis evidenced, the problem of the crack propagation is still far from being fully resolved. This requires developing advanced numerical methods that should show their effectiveness and reliability in studying the processes of crack propagation in real structures.

3 Research Methodology

3.1 Basic Equation for Stress Analysis of Cracked Bodies

To determine an elastic body stress-strain state, the resolving system of three elliptic type equations in partial derivatives of the second order (the Lame equations) is used

$$\mu \Delta U_j + (\lambda + \mu)\frac{\partial \vartheta}{\partial x_j} = 0,\ \Delta = \frac{\partial^2}{\partial x_1^2} + \frac{\partial^2}{\partial x_2^2} + \frac{\partial^2}{\partial x_3^2},\ \vartheta = \mathrm{div}\mathbf{U},\quad j = 1, 2, 3. \tag{1}$$

where $\mu = E/[2(1 + \nu)]$ is the shear modulus, $\lambda = E\nu/[(1 + \nu)(1–2\nu)]$ is Lame's coefficient, E is Young's modulus, ν is Poisson's ratio, and $\mathbf{U}$ is the displacement vector.

Introduce the surface differential stress operator of the classical elasticity theory as

$$\mathbf{T}^{\mathbf{n}(x)}\mathbf{U} = 2\mu\frac{\partial \mathbf{u}}{\partial n} + \lambda\mathbf{n}\mathrm{div}\mathbf{U} + \mu(\mathbf{n} \times \mathrm{rot}\mathbf{U}), \tag{2}$$

where $\mathbf{n}$ is a unit normal vector to the surface under consideration.

Let the kinematic conditions of the first boundary value problem be set at boundary Ω_1, whereas at boundary Ω_2, static loading conditions of the second boundary value problem are specified, and $\Omega = \Omega_1 \cup \Omega_2$. At crack surfaces, the normal and tangential stress components are usually considered to be known.

So, the following boundary value problem for the three-dimensional body, bounded by Ω and weakened by a system of cracks S_i ($i = 1,\ldots,n$), is formulated for defining **U**:

$$\mu\Delta U_j + (\lambda+\mu)\frac{\partial\vartheta}{\partial x_j}, j = 1, 2, 3, U_i(\mathbf{x}) = u_{i1}(\mathbf{x}), \mathbf{x} \in \Omega_1, .\mathbf{T}^{n(x)}\mathbf{U}(\mathbf{x}) = \mathbf{X}(\mathbf{x}), \mathbf{x} \in \Omega_2, \tag{3}$$

$$(\mathbf{T}^{n(x)}\mathbf{u}(x) \cdot \mathbf{n}(x)) = N_{3i}, (\mathbf{T}^{n(x)}\mathbf{u}(x) \cdot \tau_k(x)) = N_{ki}, \mathbf{x} \in, i = 1, 2.$$

Consider boundary value problem (3), and represent the solution of this problem as a sum of vector functions $\mathbf{U}(x) = \mathbf{u}_0(\mathbf{x}) + \mathbf{u}(\mathbf{x})$. Here $\mathbf{u}_0(\mathbf{x})$ are displacements that would occur in a solid body without cracks under the same surface loads. They are defined by

$$\mu\Delta u_{j0} + (\lambda+\mu)\frac{\partial\vartheta}{\partial x_j} = 0, u_{j0}(\mathbf{x}) = u_{i1}(\mathbf{x}), j = 1, 2, 3,$$
$$\mathbf{x} \in \Omega_1, \mathbf{T}^{n(x)}\mathbf{u_0}(\mathbf{x}) = \mathbf{X}(\mathbf{x}), \mathbf{x} \in \Omega_2 \tag{4}$$

The solution of boundary value problem (4) can be found using various numerical methods. Among them the finite (FEM) [11], and boundary element methods (BEM) [12], should be mentioned, both are based on using the method of weighted residuals. So, the unknowns are presented as series based on finite functions $\{\psi_n\}|_{n=1}^{N_1}$, and for test functions the same basis is involved.

Finite-Element Formulation. Firstly, the finite-element formulation is considered. So, approximation of displacements is applied in the form

$$\mathbf{u}_0 = \{N^u\}\{u_s\}, \tag{5}$$

where $\{N^u\}$, $\{u_s\}$ are shape functions for displacements and nodal values of the unknowns. After integrating the resolvong equations and reducing part of volume integrals to surface ones, the finite-element formulation of problem (4) is obtained as

$$[\mathbf{K}_s]\{u_s\} = \{\mathbf{f}_s\}. \tag{6}$$

Here $[\boldsymbol{K}_s]$ is the stiffness matrix, $\{\mathbf{f}_s\}$ is the known traction vector.

Boundary-Element Formulation. Next, the boundary-element formulation is involved. An essential BEM feature is in using singular test functions that satisfy differential Eq. (1) everywhere except for one singular point. So, fundamental and singular solutions of Eq. (1) are applied as test functions. The fundamental solution in elasticity is obtained considering Eq. (1) with the right part as a delta-function, and it has presented in the following matrix form:

$$\Gamma_{ij}(\mathbf{x}-\mathbf{y}) = \frac{\lambda+\mu}{8\pi\mu(\lambda+2\mu)}\left[\frac{\lambda+3\mu}{\lambda+\mu}\frac{\delta_{ij}}{|\mathbf{x}-\mathbf{y}|} + \frac{(x_i-y_i)(x_j-y_j)}{|\mathbf{x}-\mathbf{y}|^3}\right], \tag{7}$$

called the Kelvin–Somylyana matrix. This matrix is symmetric, and each of its columns and rows satisfy the equation of statics in displacements (1) at an arbitrary point $\mathbf{x} \in$

E^3, except for $\mathbf{x} = \mathbf{y}$. With fundamental solution (7) and the stress surface differential operator (2), the singular solution of Eq. (1) is received as

$$\mathbf{T^n}(\partial_x, \mathbf{n}(\mathbf{x}))\Gamma(\mathbf{x}-\mathbf{y}) = \Gamma_1(\mathbf{x}-\mathbf{y}),\ \ \Gamma_2(\mathbf{x}-\mathbf{y}) = [\mathbf{T^n}(\partial_y, \mathbf{n}(\mathbf{y}))\Gamma(\mathbf{y}-\mathbf{x})]'. \quad (8)$$

So, the vector function $\mathbf{u}_0(\mathbf{x})$ can be considered as a generalized simple layer elastic potential along the boundaries Ω

$$\mathbf{u}_0(\mathbf{x}) = \iint\limits_{\Omega} \Gamma(\mathbf{x} - \mathbf{y})\mathbf{p}_0(\mathbf{y})d\Omega_y, \quad (9)$$

where $\mathbf{p}_0(\mathbf{y})$ is an unknown density vector. The vector function $\mathbf{u}_0(\mathbf{x})$ satisfies the Lamé equations for any density $\mathbf{p}_0(\mathbf{y})$, integrable on Ω. Substituting (9) into boundary conditions (4), the integral equations are received

$$\iint\limits_{\Omega} \Gamma(\mathbf{x}-\mathbf{y})\mathbf{p}_0(\mathbf{y})d\Omega_y = \mathbf{u}_1(\mathbf{x}), \mathbf{x} \in \Omega_1, \mathbf{p}_0(\mathbf{x})$$

$$+ \iint\limits_{\Omega} \Gamma(\mathbf{x}-\mathbf{y})\mathbf{p}_0(\mathbf{y})d\Omega_y = \mathbf{X}(\mathbf{x}), \mathbf{x} \in \Omega_2 \quad (10)$$

Consider determining displacements $\mathbf{u}(\mathbf{x})$, treating them as a sum of the generalized elastic potentials of simple and double layers. The double layer potential is distributed over crack surfaces $S = \cup S_i$, whereas the simple layer potential is applied over the body surface boundary Ω. Therefore,

$$\mathbf{u}(\mathbf{x}) = \iint\limits_{S} \Gamma_2(\mathbf{x}-\mathbf{y})\boldsymbol{\alpha}(\mathbf{y})dS_y + \iint\limits_{\Omega} \Gamma(\mathbf{x}-\mathbf{y})\mathbf{p}(\mathbf{y})d\Omega_y. \quad (11)$$

Due to the properties of the simple and double layer generalized potentials, the vector function $\mathbf{u}(\mathbf{x})$ satisfies Eq. (1). Considering the boundary conditions for the function $\mathbf{u}(\mathbf{x})$ determining, we arrive to the system of singular integral equations.

3.2 Hypersingular Integral Equations

Equations obtained using presentation (11) are transformed to a system of hypersingular integral equations (HIE) relative to $\alpha(\mathbf{y})$. For the normal separation crack in an unbounded body, this system is reduced to the hypersingular equation

$$\frac{1}{4\pi} \iint\limits_{S} \frac{\alpha_3(\mathbf{y})}{|\mathbf{x} - y|^3} dS_y = Mq(\mathbf{x}), M = \frac{1 - \nu}{G}, \quad (12)$$

where function $q(\mathbf{x})$ characterizes the tensile load.

For testing numerical methods, it is convenient to use a penny-shaped crack in an unbounded elastic body under uniform tension. In this paper the penny-shaped crack is considered, and for estimating the stress-intensity factors (SIF), boundary and finite element method are applied. First, Eq. (12) has been considered for the penny-shaped crack as a circle of radius R, at the uniaxial uniform tension σ. So $q(\mathbf{x}) = \sigma$, and S is the circle in (12). The integration region S is divided into N flat triangular and quadrilateral elements, Fig. 1a). Consider the hypersingular integral over the approximated surface, and obtain the following system of linear algebraic equations:

$$\sum_{k=1}^{N} H_{kj}\alpha_{3k} = f\left(\mathbf{x}_{0j}\right), \quad j = 1, 2, \ldots N. \tag{13}$$

Here the matrix elements are evaluated by the formula

$$H_{kj} = \frac{1}{4\pi}\sum_{i=1}^{m}\frac{([\mathbf{l}_i \times \mathbf{r}_i]\cdot\mathbf{n})}{|[\boldsymbol{l}_i \times \boldsymbol{r}_i]|^2}\left[\frac{(\boldsymbol{l}_i \cdot \boldsymbol{r}_{i+1})}{r_{i+1}} - \frac{(\boldsymbol{l}_i \cdot \boldsymbol{r}_i)}{r_i}\right], \tag{14}$$

where m is the number of sides in the kth boundary element, and vectors $\mathbf{r}_k$, $\mathbf{l}_i$, $\mathbf{n}$ are

$$\mathbf{r}_k = (x_k - x_0, y_k - y_0, z_k - z_0), \mathbf{l}_i = (x_{i+1} - x_i, y_{i+1} - y_i, z_{i+1} - z_i), i = \overline{1, m}, \tag{15}$$

$$\mathbf{n} = (n_1, n_2, n_3), x_{m+1} = x_1, y_{m+1} = y_1, z_{m+1} = z_1, r_k = |\mathbf{r}_k|.$$

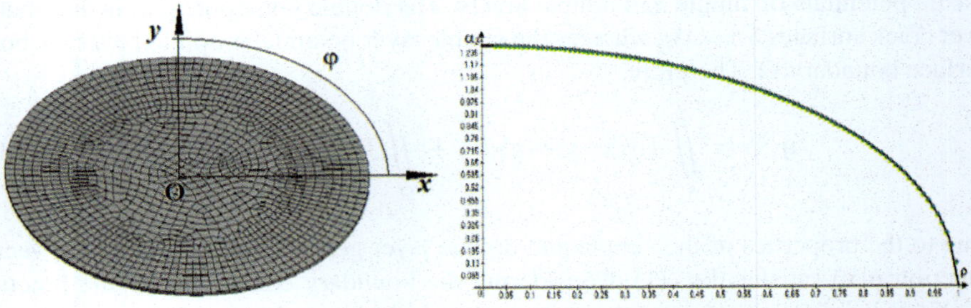

a) b)

Fig. 1. Discretization for penny-shaped cracks, a), and comparison of solutions for penny-shaped crack, b)

As the domain S in Eq. (12) is the circle, then the two-dimensional hypersingular equation is reduced to one-dimensional [13]. The cylindrical coordinate system is applied, so $x = \rho\cos\varphi$, $y = \rho\sin\varphi$, $x_0 = \rho_0\cos\varphi_0$, $y_0 = \rho_0\sin\theta_0$. Introduce the following notations $a = \rho^2 + \rho_0^2 + (z - z_0)^2$, $b = 2\rho\rho_0$, and obtain the next one-dimensional hypersingular integral equation:

$$\frac{1}{\pi}\int_0^{\pi}\alpha_3(\rho)\frac{\rho \mathrm{E}(\mathrm{k})d\rho}{(\rho + \rho_0)(\rho - \rho_0)^2} = \frac{1 - \nu}{\mathrm{G}}. \tag{16}$$

Here $\mathrm{E}(k)$ is the complete elliptic integral of the second kind with $k^2 = 2b/(a + b)$.

For numerical solution of Eq. (15) the one-dimensional BEM is applied [14]. Segment $[0, \pi]$ is divided into N_1 one-dimensional boundary elements. The method developed in [15] is used to calculate the inner elliptic integral. For comparison of results and their validation, the FEM is in use. The central penny-shaped crack with radius R in a homogeneous material is considered, Fig. 2a).

4 Results and Discussion

4.1 Benchmark Test

The penny-shaped crack of radius R is considered under uniaxial tension σ. Using BEM, the crack is considered in the unbounded 3D space. So, two formulations with hypersingular integral equations are involved.

The first one is concern with Eq. (12), whereas the second is relative to axially-symmetrical statement according to Eq. (15). At numerical simulations $R = 0.005$ m, and $\sigma = 1$MPa were chosen. Discretization of the S area on the boundary elements is shown in Fig. 2a). The total number of flat triangular and quadrilateral elements was $N = 9284$ for numerical solution of Eq. (16). Segment $[0, \pi]$ was divided into $N_1 = 100$ one-dimensional boundary elements. The analytical solution of Eq. (21) with right part $f(\rho_0) = -1$ has been obtained in [7] as $\alpha_3(\rho) = 4\sqrt{R^2 - \rho^2/\pi}$.

Figure 2b) shows comparison between analytical (solid line) and numerical solutions (green points) in in dimensionless coordinates $\rho = \rho/R, \alpha_3 = \alpha_3/R$. The accuracy $\varepsilon = 10^{-3}$ has been achieved at calculations.

The most important fracture parameters in linear elastic stress analysis are stress intensity factors (SIF). They characterize the stress and displacement fields near the crack tip and are successfully used in prediction the cracks propagation and structure failure.

In considered case of the penny-shaped cracks, the only SIF K_I is need to describe accurately the deformable state at the crack tip. Its analytical expression for the circle crack of radius R under uniaxial tension σ is $K_I = 2\sigma\sqrt{\pi R}/\pi$, [7]. This SIF in the theory of hypersingular equations [16] is expressed in terms of the displacement vector components as follows

$$K_1(\boldsymbol{x}) = \lim_{r \to 0} \sqrt{\pi/2r}\frac{G}{1-\nu}u_3(\mathbf{y}), \quad \mathbf{y} \in S, \tag{17}$$

where $r = |\boldsymbol{x} - \mathbf{y}|$ is the distance from an arbitrary point $\mathbf{y}$ near the crack to the point $\mathbf{x}$ of the crack contour, at that $\lim_{\mathbf{y} \to \mathbf{x}} u_3(\mathbf{y}) = \alpha_3(\mathbf{x})/2$.

Below, in Table 1 the numerical results of K_1 are presented using solutions of Eqs. (12) and (15) and compared with numerical FEM solution. In doing so, the penny-shaped crack in the rectangular specimen is considered, Fig. 2a). The finite-element grid consists of 43674 elements, and its half part is presented in Fig. 2b). The results presented in Table 1, testify the accuracy and credibility of FEM and BEM involved in estimations. Here the SIF is calculated in different contour points according to degrees φ, Fig. 1a).

It should be noted that the most accurate results are obtained with usage the one-dimensional BEM. But the drawback of this method is its limitation to circular and elliptical areas of integration only.

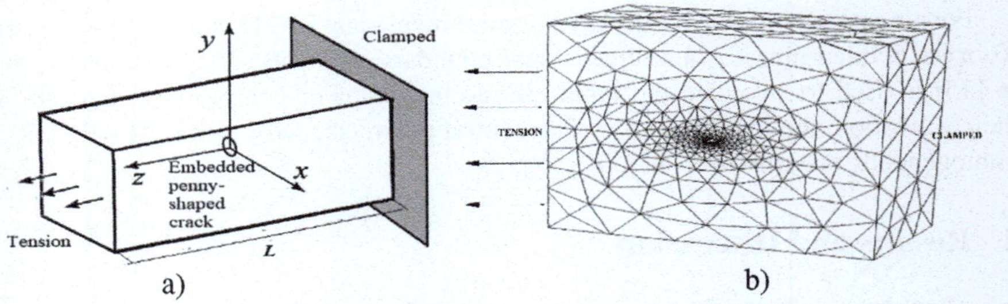

a) b)

Fig. 2. Penny-shaped crack in rectangular specimen

Table 1. Comparison of dimensionless $K_1/\sigma\sqrt{R}$ numerical evaluations.

Degrees	Analytical value	BEM, two-dimensional HIE, (16)	BEM, one-dimensional HIE, (16)	FEM
10	1.12827	1.13217	1.12988	1.14217
30	1.12827	1.12945	1.12837	1.14207
50	1.12827	1.12876	1.12830	1.14204
70	1.12827	1.12842	1.12828	1.14203
90	1.12827	1.12830	1.12828	1.14200

4.2 Crack Propagation in Fuel Tanks

Consider a horizontal fuel tank completely filled with propellant oxidizer. The 3D simplified model of the tank is presented as a cylinder with two bottoms, consisting two circle sectors of different radii. There are also two rings equidistant from the edges of the cylinder, for fastening. The geometric fuel tank model is shown in Fig. 3a). The absence of a free surface makes it possible to exclude sloshing from consideration as an important factor contributing to the resulting loads. The tank is very fragile, so and for its securely fixation two rings were installed of the same material as the tank, rigidly fixed along the contour. At transportation, stationary cyclic loads arise, associated with the speed of movement and the road quality of the railway track. These loads are non-stationary and have the "falling rain" distribution. However, this distribution can be obtained only by experiments. Since the experiment is not available, the stationary cyclic distribution is used as a simplified analogue, Fig. 3b).

The geometrical parameters of the fuel tank are following: diameter of the cylindrical part is1342 mm, the cylindrical part length is 4540 mm, small sector diameter is 500 mm, large sector diameter is1700 mm, tank wall thickness is 1.5 mm, the total model length is 5384 mm, and rings have rectangular sections of 150 x 30 mm. The tank material is Aluminum AMg6, its density is 2640 kg/m3, Young's modulus is 71 GPa, Poisson's ratio is 0.33, the yield strength is 280 MPa, and ultimate tensile strength is 310 Mpa.

The tetrahedral finite-element mesh containing 460000 elements is generated, Fig. 4a). The distribution of the structure equivalent Mises stress with maximum of

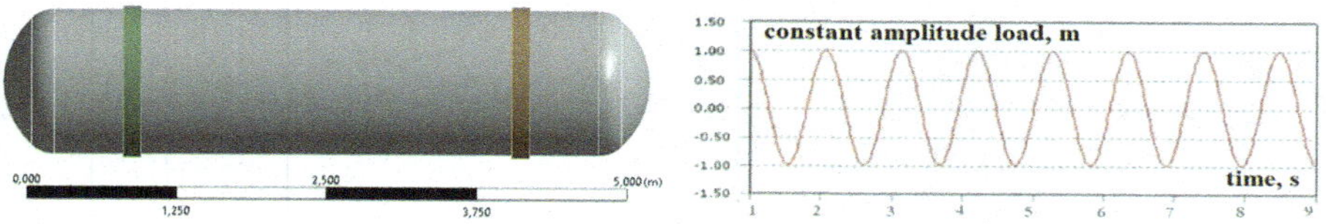

a) b)

Fig. 3. Geometrical model and loading distribution

19.2 MPa near structure inflection is obtained. The minimum is observed at the rings for securing the tank, Fig. 4b).

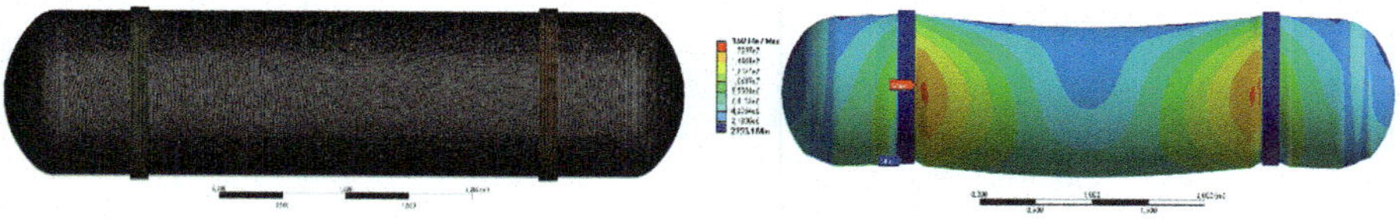

a) b)

Fig. 4. Finite-element model and equivalent Mises's stress.

The maximum stresses in the structure are much lower than the material yield point. This allows us to draw a conclusion about the sufficient strength of the tank, as well as its the long service life. The fatigue analysis of cyclic loading with an amplitude from -1 to 1 showed the service life with a number of cycles of more than 10^8. This allows the structure to be successfully transported over long distances.

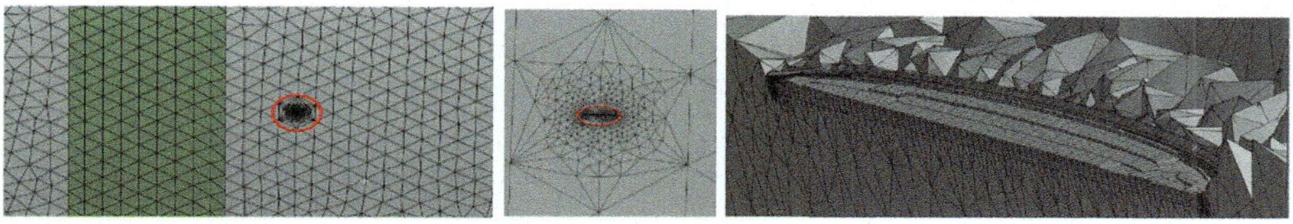

Fig. 5. Finite-element model of the cracked shell.

So, the structure without defects showed high strength and durability. But most structures contain high-quality welds, sometimes having greater strength than the structure itself. The situation becomes more complicated if the transverse weld not only passes through the places of greatest stress, but also contains a crack-like defect. To carry out a detailed strength analysis, a semi-elliptical crack with radii of 3 mm and 0.3 mm was placed in the zone of stress close to the maximum, Fig. 5.

The resulting distribution of the equivalent Mises stress in the structure visually almost coincides with the version of the structure without a crack, however, a significant surge with a maximum of 217.8 MPa appears in the vicinity of the crack. The level and location of the stress minimum have not changed. Figure 6a). Consider the crack

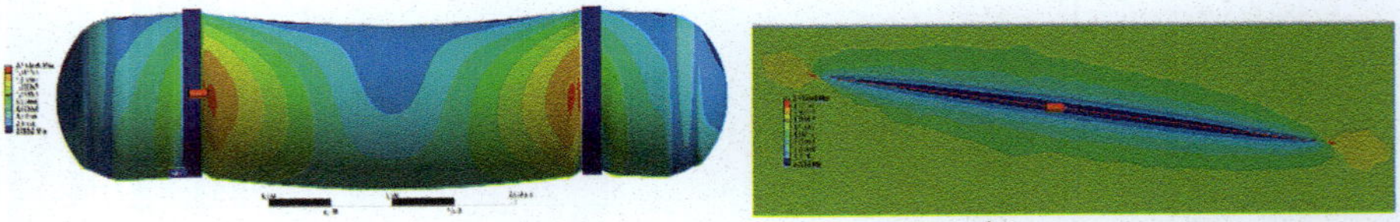

a) b)

Fig. 6. Distribution of equivalent Mises's stress

vicinity, Fig. 6b). Note that stress concentration is observed in the contact zone between the crack front and the fuel tank surface.

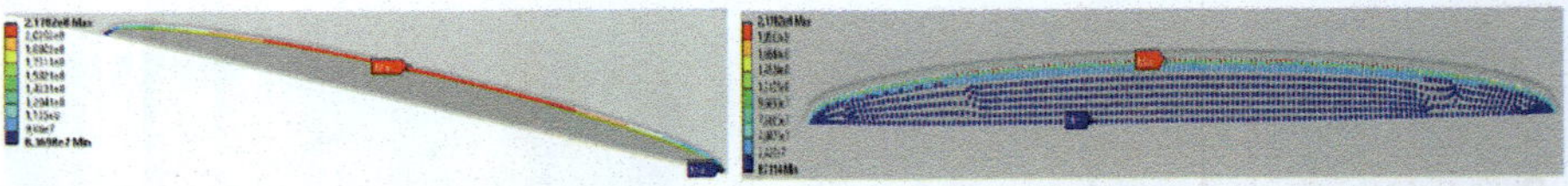

Fig. 7. Distribution of equivalent stresses along the crack growth front

The global stress maximum of 217.8 MPa is located almost in the middle of the crack front line, Fig. 7. The J-integral distribution of the crack front, that allows estimating the rate of the growth energy dissipation, is shown in Fig. 8a).

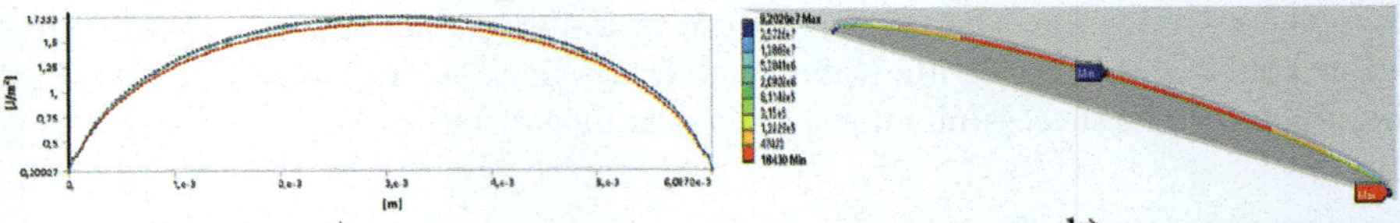

a) b)

Fig. 8. J-integral distribution and equivalent stresses along the crack growth front.

As follows from Fig. 8b), the number of cycles to failure is the lowest in the most loaded areas of the crack growth front, and it is only 18430. Certainly, this is enough to carry out the transportation over short distances, however, the long distance transportation can cause the cascade process of crack growth inside the shell followed by depressurization and destruction. Note that the uniform cyclic loading is applied, while in the case of non-stationary cycles, all residual strength and durability may be lost due to one cycle of increased load caused by a pothole in the roadway or sudden braking of vehicles [17].

As advantage of the proposed method, it would be mentioned the possibility to estimate durability of different structure elements with penny-shaped cracks. Although the BEM in the proposed formulation is limited by considering the only penny-shaped cracks, but due BEM benchmark test calculations, the FEM validity and accuracy have been proved.

5 Conclusion and Future Research

The boundary and finite element methods are applied for stress analysis of structures with crack-like defects. The boundary element formulation is based on using the hypersingular integral equations. The benchmark test calculations demonstrated high accuracy of the BEM implementing. Comparison of FEM and BEM calculations has proved the FEM validity and accuracy. The durability of the fuel tank is estimated at stationary cyclic loads. It has been found that transporting the defect-free tank can be reliable over long distances, because fatigue analysis showed the high service life of the structure with the number of cycles more than 10^8. After determining the highest stress concentration zones in the tank, the model penny-shaped crack was placed in this zone. For the cracked tank, the number of cycles to failure is only 18430. Further, scientific methods, criteria and software will be developed for estimating the resistance to fracture of rocket structures with various microdefects during transportation to the launch site.

Acknowledgments. The authors are very grateful to our foreign collaborator, professor Alexander Cheng for his continued support and cooperation.

References

1. Xu, B.B., Gao, X.W., Jiang, W.W., Cui, M., Jun, L.: Galerkin free element method and its application in Fracture Mechanics. Eng. Fract. Mech. **218**, 106575 (2019)
2. Bertolini, P., Eder, M.A., Taglialegne, L., Valvo, P.S.: Stresses in constant tapered beams with thin-walled rectangular and circular cross sections. Thin-Walled Struct. **137**, 527–540 (2019). https://doi.org/10.1016/j.tws.2019.01.008
3. Rokach, V.: Smoothed finite element method for stress intensity factor estimation: benefits and limitations, CoRR abs/1903.11401 (2019)
4. Serenza, E.N., Oscar, A.G., Suarez, O.A., Rossi, G.: A study about SIF estimation using XFEM, Proceedings of the XLI Ibero-Latin-American Congress on Computational Methods in Engineering, ABMEC Foz do Iguaçu/PR, Brazil (2020)
5. Simionato, F., Daros, C.H.: Boundary element method analysis for mode III linear fracture mechanics in anisotropic and nonhomogeneous media. Z Angew. Math. Mech. **99**, e201800211 (2019). https://doi.org/10.1002/zamm.201800211
6. Palladino, S., Esposito, L., Ferla, P., Minutolo, V.: Functionally graded plate fracture analysis using the field boundary element method. Appl. Sci. **11**, 8465 (2021)
7. Zhang, J., Xu, R., He, Y., Yang, W.: Direct Computation of 3-D stress intensity factors of straight and curved planar cracks with the P-version finite element method and contour integral method. Materials **14**(14), 3949 (2021). https://doi.org/10.3390/ma14143949
8. Yi, W., Rao, Q., Li, Z., Chun-Lin, C.: A New Method for predicting the crack propagation process of brittle rock under thermo-hydro-mechanical loading conditions. IEEE Access, 9 (2021). https://doi.org/10.1109/ACCESS.2021.3076001
9. Ren, Y., Dong, Y., Liu, B., Zhao, S., Yang, R.: Research on low stress high temperature precision shear separation process of metal bars. Heavy Mach. **05**, 37–41 (2021)
10. Zong, L., Shi, G.: Three-dimensional fatigue crack propagation analysis of welded steel beam based on global-local numerical model. Adv. Bridge Eng. **2**(4), 1–17 (2021)
11. Rusanov, A., Khorev, O., Agibalov, Y., Bykov, Y., Korotaiev, P.: Numerical and experimental research of radial-axial pump-turbine models with spliters in turbine mode. ICTM 2020, Lect. Notes Netw. Syst. **188** (2021). https://doi.org/10.1007/978-3-030-66717-7_36

12. Gnitko, V., Karaiev, A., Degtyariov, K., Strelnikova, E.: Singular boundary method in a free vibration analysis of compound liquid-filled shells. WIT Trans. Eng. Sci. **126**, 189–200, (2019), WIT Press. https://doi.org/10.2495/BE420171
13. Gnitko, V., Karaiev, A., Degtyariov, K., Vierushkin, I., Strelnikova, E.: Singular and hypersingular integral equations in fluid–structure interaction analysis. WIT Trans. Eng. Sci. **134**, 67–79 (2022). https://doi.org/10.2495/BE450061
14. Karaiev, A., Strelnikova, E.: Singular integrals in axisymmetric problems of elastostatics. Int. J. Model. Simul. Sci. Comput. **11**(1), 2050003 (2020). https://doi.org/10.1142/S1793962320500038
15. Karaiev, A., Strelnikova, E.: Axisymmetric polyharmonic spline approximation in the dual reciprocity method. Z. Angew. Math. Mech. **101**, e201800339 (2021). https://doi.org/10.1002/zamm.201800339
16. Obaiys, S.J., Ibrahim, R.W., Ahmad, A.F.: Hypersingular integrals in integral equations and inequalities: fundamental review study. In: Differential and integral inequalities. Springer Optimization and Its Applications, 151 (2019). https://doi.org/10.1007/978-3-030-27407-8_25
17. Zaitsev, B.P., Protasova, T.V., Smetankina, N.V., Klymenko, D.V., Larionov, I.F., Akimov, D.V.: Oscillations of the payload fairing body of the cyclone-4M launch vehicle during separation. Strength Mater. **52**(6), 849–863 (2020). https://doi.org/10.1007/s11223-021-00239-5

Structure Elements with Crack-Type Defects Durability Estimation

Olena Sierikova[1], Denis Kriutchenko[2](✉), Konstanin Vandyshev[3], and Ivan Vierushkin[2]

[1] National University of Civil Protection of Ukraine, Kharkiv, Ukraine
[2] Anatolii Pidhornyi Institute of Mechanical Engineering Problems NAS of Ukraine, Kharkiv, Ukraine
wollydenis@gmail.com
[3] Delft University of Technology, Delft, Netherlands

Abstract. The objective of this study consists in elaborating computational procedures to analyze durability of structural elements with crack-type defects under remote alternating loads. The novelty of the proposed method consists both in applying new benchmark tests for crack propagation analysis, and analyzing structure durability for structure elements with different defects. The problem of the stress intensity factor estimation for mode I loading is reduced to solution of a hypersingular integral equation. Analytical solutions of this equation with different right parts are obtained for penny-shape cracks. It is the basis for verifying boundary-type numerical methods. The problem of determining the number of cycles to failure for structural elements subjected to cyclic loading (tension-compression) with a given frequency and amplitude is solved. The Paris dependence for fatigue crack growth has been used to determine the critical number of cycles before failure. Numerical results characterizing the durability of structural elements with crack-like initial defects have been accomplished.

Keywords: Stress Intensity Factor · Hypersingular Integral Equations · Crack Propagation

1 Introduction

Due exhausted normative resource of hydro-turbine, petrochemical, power equipment at many industrial enterprises, the question arises of the possibility to extend the service life of both individual components, and complex parts of machinery. This is explained necessity to replace morally and physically obsolete equipment to ensure the operational reliability of the units in further operations whit fulfilling guarantees for power and efficiency. Since the specified equipment operates for a long time, its elements and assemblies are usually weakened by various kinds of microdefects. It should be noted, that crack propagation is a major mechanism affecting functionality and performance of structural elements. It is known that small cracks are usually not identified by visual inspection during repair works. But under alternating loads, such small cracks are

H. Altenbach et al. (Eds.): CAMPE 2023, pp. 301–311, LNME 2025.
https://doi.org/10.1007/978-3-031-82979-6_30

gradually developing, and this can lead to partial or complete destruction of the structural element. This explains the great interest in the study of fatigue cracks in the last decade and necessity of elaborating new advanced numerical methods for predicting the dangerous crack growth.

2 Literature Review

A lot of theoretical research is devoted to elaborating effective methods to analyze stress-strain states and destructive loads of responsible structural elements. Among them finite element methods [1], finite volume methods [2], and boundary element methods [3] should be mentioned. The presence of cracks in structural elements leads to a significant complication in numerical schemes and algorithms [4], as well as in experimental research [5]. The latest studies on the theory of cracks and inclusions, with forming clusters and chains, serve as the foundation to creation of modern technologies for determining the ultimate resource [6]. To date, a significant volume of experimental and theoretical studies of the crack growth patterns under the cyclically changing stress action has been accumulated. In [7] the fatigue crack growth model for metallic materials is proposed. Three-dimensional fatigue crack propagation analysis of welded steel beam was accomplished in [8]. New variants of boundary and finite element methods for 2D and 3D cracked structural elements were proposed in [9]. The problems to reduce the friction coefficient and the wear rate have been considered in [10]. In [11] the boundary element method (BEM) is used for stress intensity factors (SIF) calculation in functionally graded plates. It was found that apparatus of hypersingular integral equations is a powerful tool for SIF evaluations [12]. This makes it possible to analyze the durability of structural elements in the presence of cracks [13].

The literature analysis shows that the crack propagation problem remains the focus of attention of scientists and engineers. This requires developing advanced numerical methods to study crack propagation processes in real structures, estimate structures durability and residual life.

3 Research Methodology

3.1 Problem Formulation

Paper treats the durability of the hydraulic turbine structural elements. After determining the highest stress concentration zone, the growth problem of an internal isolated round crack, which initially had a given size (corresponding to control standards), has been considered under the load, obtained as a result of the stress-strain state analysis [14]. It has been found out for how many cycles the crack will grow to a size exceeding the allowable control standards (0.008 m).

The problem of determining the durability of a cracked structure is solved, following formulation of [14] and using hypersingular integral technique described in [15]. The task has been to determine the time (number of cycles $N = N^*$), when the crack grows to a critical size, and destruction occurs.

The Paris dependence is used in the form

$$dl/dN = N_G(\mathrm{D}R)^m, \tag{1}$$

Here l is the crack length, N is the cycles number, ΔR is a SIF amplitude value, N_G, and m are the material constants.

To determine the rate propagation of a fatigue crack, the following equality is used:

$$dl/dN = \Phi^{-1}(\mathrm{b}), \tag{2}$$

where $\Phi(\beta)$ is a characteristic function of the fatigue failure, the form of $\Phi(\beta)$ is established due experiments [6]. For example, for steels this function is approximated as follows

$$\Phi(\mathrm{b}) = N_G(1 - \mathrm{b})^m. \tag{3}$$

Parameter β here is determined by the next formula:

$$\mathrm{b} = 1 - k_1/K_{1c}. \tag{4}$$

Here K_{1c} is critical stress intensity factor; k_1 is the SIF of for mode I loading. To determine the critical crack size, the equation k1 $= K_{1c}$ is used. So

$$\Phi(\lambda)\frac{\partial l}{\partial N} = 1, \quad k_1 = K_{1c}. \tag{5}$$

The number of cycles before failure has been determined by the formula

$$\int_{l_*}^{l_0} \Phi\left(\sqrt{\frac{l}{l_*}}\right) dl = N, \tag{6}$$

where l_* is the critical crack length.

Integrating equality (6), and taking into account formula (3), it has been calculated the number of cycles before failure.

3.2 Hypersingular Integral Equations in 3D Crack Theory

Introduce the fundamental solution and stress operator in elasticity theory as in [16]:

$$\Gamma_{ij}(\mathbf{x}-\mathbf{y}) = \frac{\lambda+\mu}{8\pi\mu(\lambda+2\mu)}\left[\frac{\lambda+3\mu}{\lambda+\mu}\frac{\delta_{ij}}{|\boldsymbol{x}-\mathbf{y}|} + \frac{(x_i-y_i)(x_j-y_j)}{|\boldsymbol{x}-\mathbf{y}|^3}\right],$$

$$\mathbf{T}^{\mathbf{n}(x)}\mathbf{U} = 2\mu\frac{\partial \mathbf{u}}{\partial \boldsymbol{n}} + \lambda\mathbf{n}\mathrm{div}\mathbf{U} + \mu(\mathbf{n}\times \mathrm{rot}\mathbf{U}), \tag{7}$$

where $\mathbf{n}$ is a unit outward normal to the surface under consideration, λ and μ are Lame's coefficients, $\mathbf{U}$ is the displacement vector.

The next boundary value problem for 3D body bounded by the surface Ω and weakened by a system of cracks S_i ($i = 1,\ldots,n$) is formulated for defining **U**:

$$\mu\Delta U_j + (\lambda+\mu)\frac{\partial\vartheta}{\partial x_j},\ j=1,2,3,\ U_i(\boldsymbol{x}) = u_{i1}(\boldsymbol{x}),\ x\in\Omega_1,.$$

$$\boldsymbol{T}^{n(x)}\boldsymbol{U}(\boldsymbol{x}) = \boldsymbol{X}(\boldsymbol{x}),\ x\in\Omega_2, \tag{8}$$

Involve the following presentation for **U** based on Somigliana's identity

$$\boldsymbol{U}(\boldsymbol{x}) = \int\int_S \Gamma_2(\boldsymbol{x}-\boldsymbol{y})\boldsymbol{\alpha}(\boldsymbol{y})dS_y + \int\int_\Omega \Gamma(\boldsymbol{x}-\boldsymbol{y})\boldsymbol{p}(\boldsymbol{y})d\Omega_y,\quad S=\cup S_i, \tag{9}$$

where $\Gamma_2(\boldsymbol{x}-\mathbf{y}) = [\mathbf{T^n}(\partial_y,\mathbf{n}(\mathbf{y}))\Gamma(\mathbf{y}-\boldsymbol{x})]'$, and $\alpha(\mathbf{y})$, $\mathbf{p}(\mathbf{y})$ are unknown densities. Then the next system of singular integral equation is received to determine the unknowns:

$$\int\int_S \Gamma_2(\boldsymbol{x}-\boldsymbol{y})\boldsymbol{\alpha}(\boldsymbol{y})dS_y + \int\int_\Omega \Gamma(\boldsymbol{x}-\boldsymbol{y})\boldsymbol{p}(\boldsymbol{y})d\Omega_y = 0,\ \boldsymbol{x}\in\Omega_1, \tag{10}$$

$$\int\int_S \boldsymbol{T}^{n(x)}\Gamma_2(\boldsymbol{x}-\boldsymbol{y})\boldsymbol{\alpha}(\boldsymbol{y})dS_y - \boldsymbol{p}(\boldsymbol{x}) - \int\int_\Omega \Gamma_1(\boldsymbol{x}-\boldsymbol{y})\boldsymbol{p}(\boldsymbol{y})d\Omega_y = 0,\ \boldsymbol{x}\in\Omega_2,$$

$$\left(\boldsymbol{T}^{n(x)}\int\int_S \Gamma_2(\boldsymbol{x}-\boldsymbol{y})\boldsymbol{\alpha}(\boldsymbol{y})dS_y\cdot\mathbf{n}(\mathbf{x})\right) + \left(\int\int_\Omega \Gamma_1(\boldsymbol{x}-\boldsymbol{y})\boldsymbol{p}(\boldsymbol{y})d\Omega_y\cdot\mathbf{n}(\mathbf{x})\right) = N_{3i},\ \boldsymbol{x}\in S_i,$$

$$\left(\boldsymbol{T}^{n(x)}\int\int_S \Gamma_2(2(\boldsymbol{x}-\boldsymbol{y})\boldsymbol{\alpha}(\boldsymbol{y})dS_y\cdot\boldsymbol{\tau}_k(\boldsymbol{x})\right)$$
$$+\left(\int\int_\Omega \Gamma_1(\boldsymbol{x}-\boldsymbol{y})\boldsymbol{p}(\boldsymbol{y})d\Omega_y\cdot\boldsymbol{\tau}_k(\boldsymbol{x})\right) = N_{ki},\ \boldsymbol{x}\in S_i,\ k=1,2.$$

The boundary element method [17] is applied to numerical solution of system (10) in the case of penny-shaped cracks.

For the penny-shaped crack system (10) is reduced to the next hypersingular equation:

$$\frac{1}{4\pi}\iint_S \alpha(\mathbf{y})\frac{\partial^2}{\partial\mathbf{n}_x\partial\mathbf{n}_\xi}\frac{1}{|\mathbf{x}-\mathbf{y}|}\mathrm{dS} = \mathbf{f}(\mathbf{x}), \tag{11}$$

This is transformed as follows

$$\frac{1}{4\pi}\iint_S \frac{\gamma(y)}{|\boldsymbol{x}-\boldsymbol{y}|^3}dS_y = Mq(\boldsymbol{x}), M = \frac{1-\nu}{G}, \tag{12}$$

where G and ν are the shear modules and Poisson's ratio, and function $q(\boldsymbol{x})$ characterizes the applied load.

3.3 Analytical Solution of Hypersingular Integral Equation Over Circle

Equation (12) is considered in the next form:

$$\iint_S \frac{f(x,y)\Gamma(x,y)dxdy}{\sqrt{(x-x_0)^2+(y-y_0)^2}^3} = Mq(x_0,y_0),\quad \gamma(\xi) = f(x,y)\Gamma(x,y). \tag{13}$$

Let the domain S be a circle of radius R. Points lying on a circle whose coordinates satisfy the equation $x^2 + y^2 = R^2$ have been considered. Let (x_0, y_0) be an internal point of the circle S, that is, its coordinates satisfy the inequality $x_0^2 + y_0^2 < R^2$. The following integral has been considered.

$$I = \iint_S \frac{f(x, y)\Gamma(x, y)dxdy}{\sqrt{(x - x_0)^2 + (y - y_0)^2}^3}, \Gamma(x, y) = \sqrt{R^2 - x^2 - y^2}. \tag{14}$$

This integral must be understood as the following limit:

$$\mathrm{I} = \lim_{\varepsilon \to 0}(\iint_{S/S_\varepsilon} \frac{f(x, y)\Gamma(x, y)dxdy}{\sqrt{(x - x_0)^2 + (y - y_0)^2}^3} \frac{2\pi f(x_0, y_0)\Gamma(x_0, y_0)}{\varepsilon}). \tag{15}$$

In polar coordinates
$x - x_0 = \rho \cos \phi, y - y_0 = \rho \sin \phi, dxdy = \rho d \rho d \phi, x_0 = \rho_0 \cos \phi_0, y_0 = \rho_0 \sin \phi_0$

the integral limit value takes the form

$$\lim_{\varepsilon \to 0} \int_0^{2\pi} \int_\varepsilon^{\rho(\phi)} \frac{f \Gamma \rho d\rho}{\rho^3}. \tag{16}$$

Function $\rho(\phi)$ could be found by solving the corresponding quadratic equation as

$$\rho = -\rho_0 \cos(\phi - \phi_0) + \sqrt{R^2 - \rho_0^2 \sin^2(\phi - \phi_0)} \tag{17}$$

Let $f(x, y) \equiv 1$. After the transformations, substituting $u = \rho + \rho_0 \cos \phi$, according to [12], it has been established that the considered integral operator is equal to $-\pi^2$. By analogous reasoning, it has been obtained that the integral operator (14) transforms the function $f(x, y) = x + iy$ into $\frac{-3\pi^2}{2}(x_0 + iy_0)$.

Then the function $f(x, y) = z^n = (x + iy)^n$ has been taken. Because

$$z^n = \sum_{k=0}^{n} C_n^k z_0^{n-k} (z - z_0)^k,$$

the calculation reduces to considering integrals in the form

$$\iint_S \frac{(z - z_0)^{n+2} \sqrt{R^2 - x^2 - y^2}\, dxdy}{\sqrt{(x - x_0)^2 + (y - y_0)^2}^3}.$$

In the polar coordinates after transformations, it has been obtained

$$I = \lim_{\varepsilon \to 0} \int_0^{2\pi} \exp^{i(n+2)\phi} \int_\varepsilon^{-\rho_0 \cos(\phi - \phi_0) + \sqrt{R^2 - (\rho_0 \sin(\phi - \phi_0))^2}} \rho^{n+2} Q(\phi, \phi_0, \rho, \rho_0) \rho d \rho,$$

$$\mathrm{Q}(\phi, \phi_0, \rho, \rho_0) = \frac{\sqrt{R^2 - \rho_0^2 \sin^2(\phi - \phi_0) - (\rho + \rho_0 \cos(\phi - \phi_0))^2}}{\rho^3}. \tag{18}$$

So

$$I = \exp^{(n+2)i\phi_0} \int_0^{\pi} \exp^{(n+2)i\phi} d\phi \int_{-\sqrt{R^2-\rho_0^2\sin^2\phi}}^{\sqrt{R^2-\rho_0^2\sin^2\phi}} (u - \rho_0\cos\phi)^n \sqrt{R^2 - \rho_0^2\sin^2\phi - u^2 du} \tag{19}$$

Let now $a = \sqrt{R^2 - \rho_0 \sin^2\phi}$; $b = \rho_0 \cos\phi$. The next integral has been calculated:

$$J = \int_{-a}^{a} (u - b)^n \sqrt{a^2 - u^2} du.$$

Using the Newton binomial formula, it has been got

$$J = \sum\nolimits_{k=0}^{n} (-1)^n C_n^k b^k \int_{-a}^{a} u^{n-k} \sqrt{a^2 - u^2} du.$$

Since the integral of an odd function over a symmetric interval is equal to zero, there will be considered the only integrals with an even degree of the integrand. By changing the integration variable as $\frac{a}{u} = \sin t$, it will be obtained

$$\int_0^1 v^{2m} \sqrt{1 - v^2} dv = \frac{\pi(2m-1)!!}{(2m+2)2(2m)!!},$$

where the standard notation for the double factorial $n!! = n(n-2)(n-4)\ldots$ has been used. Finally, it has been gained

$$J = \sum\nolimits_{m=0}^{[\frac{n}{2}]} c(n, m) a^{2(m+1)} b^{n-2m},$$

where for $m = \frac{n-2k}{2}$ the coefficients $c(n, m)$ are follows

$$c(n, m) = C_n^{2k} \frac{(n-2k-1)!!}{(n-2k)!!} \frac{\pi}{n-2k+2},$$

and for $m = \frac{n-2k+1}{2}$ they are

$$c(n, m) = -C_n^{2k-1} \frac{(n-2k)!!}{(n-2k+1)!!} \frac{\pi}{n-2k+3},$$

Thus

$$\int_{-\sqrt{R^2-\rho_0^2\sin^2\phi}}^{\sqrt{R^2-\rho_0^2\sin^2\phi}} (u - \rho_0\cos\phi)^n \sqrt{R^2 - \rho_0\sin^2\phi - u^2 du} =$$

$$= \sum\nolimits_{m=0}^{[\frac{n}{2}]} c(n, m)(R^2 - \rho_0\sin^2\phi)^{m+1} (\rho_0\cos\phi)^{n-2m}. \tag{20}$$

Applying Newton's binomial formula, it is reduced to calculations of

$$\int_0^{\pi} \exp^{(n+2)i\phi} \sin^{2j}\phi \cos^{n-2m}\phi d\phi. \tag{21}$$

The real part of integral (21) has the form $\int_0^{\pi} \cos(n+2)\phi \sin^{2j}\phi \cos^{n-2m}\phi d\phi$. Taking into account the properties of Chebyshev's polynomials, the integral calculation is reduced to the trivial integral $\int_0^{\pi} d\phi$, which, obviously, is equal to π. The imaginary part of the integral from similar reasoning vanishes.

Thus, the integral under consideration has been found

$$\int_0^{\pi} \cos(n+2)\phi \sin^{2j}\phi \cos^{n-2m}\phi d\phi = \pi \sum_{t=0}^{n+2} \frac{T_{n+2}(1)_t j!!(2j-1)!!(n-2m+t-1)!!}{((n-2m+t)!!)^2}. \tag{22}$$

Substituting expression (22) into (19), the follow expression is received

$$\iint_S \frac{(z-z_0)^{n+2}\sqrt{R^2-x^2-y^2}dxdy}{\sqrt{(x-x_0)^2+(y-y_0)^2}^3}$$
$$= z_0^{n+2}\pi^2 \sum_{m=0}^{[\frac{n}{2}]} \sum_{j=0}^{m+1} (-1)^{n+j} C_n^{n-2m} C_{m+1}^j \frac{(2m-1)!!}{(2m+2)(2m)!!}\cdot$$
$$\cdot\left(\frac{R}{\rho_0}\right)^{2m-2j+2} \sum_{t=0}^{n+2} \frac{T_{n+2}(1)_t j!!(2j-1)!!(n-2m+t-1)!!}{((n-2m+t)!!)^2}. \tag{23}$$

So, analytical solutions for the hypersingular equation for circular cracks theory have been received for the right parts in the form of the real and image parts of function $f(x, y) = (x+iy)^n$.

3.4 Determination of the Stress Intensity Factor

For the problem of stress concentration in the body with the internal circular crack of radius R under constant load F is given, then according to the above, the follow equation is obtained

$$\Gamma(x, y) = \sqrt{R^2-x^2-y^2}F.$$

According to [12]

$$k_1(\boldsymbol{x}) = -\frac{\sqrt{2}}{\sqrt{\pi}} \frac{G}{1-\nu} \lim_{r\to 0} \frac{\sqrt{R^2-x^2-y^2-}}{\sqrt{r}} F, \tag{24}$$

where $r = |\boldsymbol{x}-\mathbf{y}|$ is the distance from an arbitrary point $\mathbf{y}$ near the crack to the point $\mathbf{x}$ of the crack contour.

Thus, the next formula for determining the critical crack size has been obtained

$$-2\sqrt{\frac{R_*}{\pi}} \frac{G}{1-\nu} F = K_{1c}. \tag{25}$$

The formula for the number of load cycles that are required for the crack growth to the size $R1$ takes the form

$$\int_{R_0}^{R_1} \Phi\left(\sqrt{\frac{R}{R_1}}\right) dR = N_* \tag{26}$$

Here R_0 is the initial crack size, R_1 is the allowable crack size.

4 Results and Discussion

The penny-shaped (circular) crack of radius R is considered under uniaxial tension-compression F. The number of cycles before the hydraulic turbine element's destruction has been determined. It has been assumed that the structural element under consideration is subject to the action of a cyclic load with an amplitude value σ and a frequency $\omega =$ 1.79 Hz. The calculation has been carried out for the initial dimensions of the defect, varying from 0.003 m to 0.0075 m. In Fig. 1 the results of numerical calculation of time t (in years) for the growth of internal isolated defects of various initial lengths to sizes that are unacceptable according to control standards for various loading levels σ has been presented. Data on loads and loading frequency have been taken from [18].

Knowing the total number of cycles to failure and the loading frequency, we determine the time until failure of the structural element.

In Fig. 1 numbers 1–7 indicate the curves corresponding to the following initial dsizes of the defect: 0.003, 0.0035, 0.0045, 0.0055, 0.0065, 0.007, 0.0075 m. Material with characteristics $N_G = 10^{11}$ has been considered; $m = 3.09$; $K_{1c} = 49$ MPa/m$^{1/2}$ (steel 40).

The maximum allowable defect size has been taken to be 0.008m. The most dangerous are cracks with a size of 0.0075m (the time to failure is about one year.)

The most dangerous are circular cracks with radius of 0.0075m with the time to failure about one year. Note that usually cracks with sizes larger than 0.005 m are considered unacceptable. That is, after scheduled inspections, structural elements with such cracks are subject to repair or modernization [19].

The proposed approach can be used not only to estimation of SIF for penny-shaped cracks. The obtained results especially concern with analytical solutions of the hypersingular integral equations will used for testing and validation of numerical methods. Note that the data obtained do not take into account the number of cyclic operating time before the formation of initial fatigue cracks. Therefore, it is topical to estimate the number of cycles before the appearance of fatigue cracks.

The method can be generalized to durability analysis of structure elements made from other materials. Especially topical is durability estimation of the structure elements made of composites and nanocomposites [20]. Here the additional possibility to prevent the crack propagation are arisen.

Conclusion and Future Research

The method has been developed for calculating the critical number of loading cycles for structural elements under the sign-variable cyclic load. It is assumed that in the zones

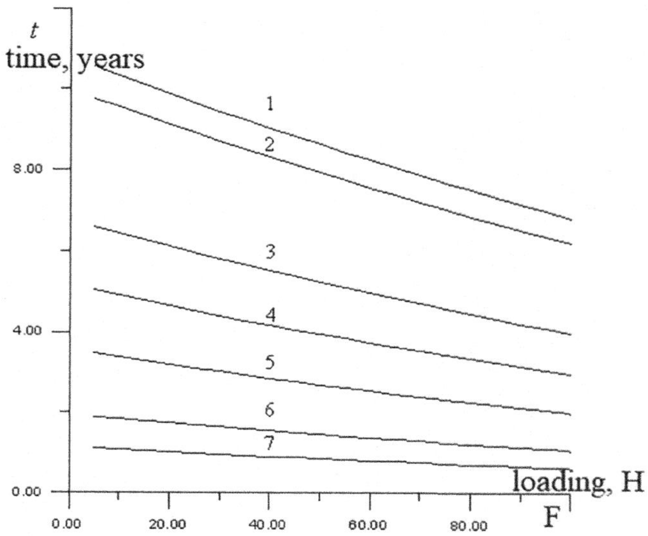

Fig. 1. Structure element durability.

of the highest stress concentration there may be defects such as cracks. The number of loading cycles, which leads to the growth of cracks to unacceptable sizes, is determined according to the Paris criterion. The calculation of SIF is accomplished using the hypersingular integral equation technique. The analytical solution is obtained for the circular crack in the unbound domain that will be used as benchmark test for validation of numerical methods, especially boundary-type ones. The penny-shape cracks are considered, and evaluation of the residual life is provided. The most dangerous crack size is estimated.

Further, scientific methods, criteria and software will be developed for estimating the resistance to fracture of rocket structures and turbine impellers under alternative loadings. The durability of structure elements made of composite and nanocomposite materials will be considered. The domain-type and boundary-type numerical methods will be involved into investigation.

Acknowledgments. The authors are very grateful to our foreign collaborators, professors Alexander Cheng and Stavros Syngellakis, Wessex Institute of Technology, for their constant support and cooperation.

References

1. Zaitsev, B.P., Protasova, T.V., Smetankina, N.V., Klymenko, D.V., Larionov, I.F., Akimov, D.V.: Oscillations of the payload fairing body of the cyclone-4M launch vehicle during separation. Strength Mater. **52**(6), 849–863 (2020). https://doi.org/10.1007/s11223-021-00239-5

2. Rusanov, A., Rusanov, R., Klonowicz, P., Lampart, P., Żywica, G., Borsukiewicz, A.: Development and experimental validation of real fluid models for CFD calculation of ORC and steam turbine flows. Materials **14**(22), 6879 (2021). https://doi.org/10.3390/ma14226879
3. Choudhary, N., Kumar, N., Strelnikova, E., Gnitko V., Kriutchenko, D., Degtyariov, K.: Liquid vibrations in cylindrical tanks with flexible membranes. Journal of King Saud University—Science **33**(8), 101589 (2021). https://doi.org/10.1016/j.jksus.2021.101589
4. Vojtov, V.A., Biekirov, A.S., Voitov, A.V., Tsymbal, B.M.: Running-in procedures and performance tests for Tribosystems. Journal of Friction and Wear **40**(5), 376–383 (2019). https://doi.org/10.3103/S1068366619050192
5. Kachanov, P., Lytviak, O., Derevyanko, O., Komar, S.: Development of an automated hydraulic brake control system for testing aircraft turboshaft gas turbine engines. Eastern-European Journal of Enterprise Technologies **6**(2–102), 52–57 (2019). https://doi.org/10.15587/1729-4061.2019.185539
6. Antunes, F.V., Ferreira, M.S., Branco, R., Prates, P., Gardin, C., Sarrazin-Baudoux, C.: Fatigue crack growth versus plastic CTOD in the 304L stainless steel. Eng. Fract. Mech. **214**, 487–503 (2019). https://doi.org/10.1016/j.engfracmech.2019.04.013
7. Li, H.F., Liu, Y.Q., Zhang, P., Z.F. Zhang, Z.F.: Full-stage fatigue crack growth model for metallic materials. Int. J. Fatigue **72**, 107662 (2023). https://doi.org/10.1016/j.ijfatigue.2023.107662
8. Zong, L. and Shi, G.: Three-dimensional fatigue crack propagation analysis of welded steel beam based on global-local numerical model. Adv. Bridg. Eng. **2**(4), 1–17 (2021) https://doi.org/10.1186/s43251-020-00029-0
9. Rokach, V.: Smoothed finite element method for stress intensity factor estimation: benefits and limitations. CoRR abs/1903.11401 (2019)
10. Vojtov, V.A., Kravtsov, A.G., Tsymbal, B.M.: Evaluation of tribotechnical characteristics for tribosystems in the presence of fullerenes in the lubricant. Journal of Friction and Wear **41**(6), 521–525 (2020). https://doi.org/10.3103/S1068366620060197
11. Palladino, S., Esposito, L., Ferla, P., Minutolo, V.: Functionally graded plate fracture analysis using the field boundary element method. Appl. Sci. **11**, 8465 (2021). https://doi.org/10.3390/app11188465
12. Obaiys, S.J., Ibrahim, R.W., Ahmad, A.F.: Hypersingular integrals in integral equations and inequalities: fundamental review study. In: Differential and Integral Inequalities. Springer Optim. Its Appl. **151** (2019). https://doi.org/10.1007/978-3-030-27407-8_25
13. Yi, W., Rao, Q., Li, Z., Chun-Lin, C.: A New Method for Predicting the Crack Propagation Process of Brittle Rock Under Thermo-Hydro-Mechanical Loading Conditions, IEEE Access, 9, (2021), https://doi.org/10.1109/ACCESS.2021.3076001
14. Moskalenko, R., Zaydenvarg, O., Strelnikova, O., Gnitko V.: Software development for the computational analysis of crack propagation and durability of structures, 2020 IEEE KhPI Week on Advanced Technology (KhPIWeek), 509–514 (2020). https://doi.org/10.1109/KhPIWeek51551.2020.9250089
15. Gnitko, V., Karaiev, A., Degtyariov, K., Vierushkin, I., Strelnikova, E.: Singular and hypersingular integral equations in fluid–structure interaction analysis. WIT Trans. Eng. Sci. **134**, 67–79 (2022). https://doi.org/10.2495/BE450061
16. Gnitko, V., Degtyariov, K., Karaiev, A., Strelnikova, E.: Multi-domain boundary element method for axisymmetric problems in potential theory and linear isotropic elasticity. WIT Trans. Eng. Sci. **122**, 13–25 (2019). https://doi.org/10.2495/BE410021
17. Karaiev, A., Strelnikova, E.: Axisymmetric polyharmonic spline approximation in the dual reciprocity method. Z. Angew. Math. Mech. **101**, e201800339 (2021). https://doi.org/10.1002/zamm.201800339

18. Rusanov, A., Subotin, V., Shvetsov, V., Rusanov, R., Palkov, I., Chugay, M.: Application of innovative solutions to improve the efficiency of the LPC flow part of the 220 MW NPP steam turbine. Archives of Thermodynamics **43**(1), 63–87 (2022). https://doi.org/10.24425/ather.2022.140925
19. Rusanov, A., Chugay, M., Rusanov, R.: Advanced computer technologies in the new flow part development for reactive type HPC steam turbine of T-100 series. LNME 55–63 (2023). https://doi.org/10.1007/978-3-031-18487-1_6
20. Sierikova O., Strelnikova, E., V. Gnitko, V., Degtyarev, K.: Boundary calculation models for elastic properties clarification of three-dimensional nanocomposites based on the combination of finite and boundary element methods, 2021 IEEE 2nd KhPI Week on Advanced Technology (KhPIWeek), Kharkiv, Ukraine, 351–356 (2021). https://doi.org/10.1109/KhPIWeek53812.2021.9570086

Effect of Alloying Elements in Weldability of API X70 HSLA Steels

Nabil Bensaid[1(✉)], Mohamed Farid Benlamnouar[1], Yazid Laib Dit Laksir[2], Tahar Saadi[1], and Riad Badji[1]

[1] Research Center, Industrial Technologies CRTI, 16014 Algiers, Algeria
nblbensaid@gmail.com

[2] Département de Génie Climatique, Université constantine1, Constantine 25000, Algérie

Abstract. This study aims to investigate and compare the evolution of microstructure and mechanical behavior in API X70 steels welded using SMAW techniques. A comprehensive analysis of the microstructure revealed that the weldability of low-carbon X70 steels is significantly influenced by the carbon content and microalloying elements. The microstructural evolution demonstrated that higher carbon content in X70 steel results in increased levels of martensite-austenite (M-A) compounds in the heat-affected zone (HAZ). Additionally, higher carbon content also leads to enhanced growth of ferritic grains in the HAZ. Furthermore, the rate of M-A compound formation in the HAZ increases with higher carbon content, but this rate is limited when a high proportion of niobium is introduced. Optimal mechanical properties, appropriate grain coarsening, and an acceptable ratio of M-A compounds throughout the HAZ were observed at a carbon content of 0.082%. These findings suggest that the weldability of HSLA-X70 steels can be improved by incorporating microalloying elements, without the need for complex metallurgical processes to enhance the microstructure. Fractographic examinations revealed a shift in the failure mode from low to high carbon ratios in X70 steels.

Keywords: Alloying elements · X70 HSLA Steel · Microstructure · SMAW Welding · Mechanical Behavior

1 Introduction

Despite advancements in high-strength low-alloy steel, X70 alloys remain extensively utilized in the oil and gas transportation sectors across various countries [1]. While there are multiple variants of X70 with distinct chemical compositions that provide superior strength, these compositions can influence various properties, particularly weldability, which remains a critical aspect in pipeline construction. Therefore, it is imperative to compare the welding characteristics among different variations of X70 to assess their behavior. The transportation of gas under extreme conditions necessitates the use of steels that are increasingly resistant to high pressure. To optimize the selection of steels based on their weldability, mechanical characteristics, and cost, the steel industry has developed new grades of high-strength low-alloy (HSLA) steels.

H. Altenbach et al. (Eds.): CAMPE 2023, pp. 312–323, LNME 2025.
https://doi.org/10.1007/978-3-031-82979-6_31

These steels exhibit improved mechanical performance while maintaining favorable formability and durability. By producing sheets with high-strength steels, it becomes possible to reduce the thickness of parts, leading to cost-effective products. The enhanced properties of high-yield strength steels are closely linked to the presence of small amounts of solid carbides or carbonitrides formed from microalloying elements such as Nb, V, and Ti. These elements facilitate precipitation and thermomechanical treatments during processing, resulting in grain refinement [2].

High-strength low-alloy (HSLA) steel, also known as microalloyed steel, is defined as an iron-based material with low carbon content, Mn, and small additions of Nb, V, and Ti [2, 3]. These additions impact precipitation hardening and grain size refinement. Among the alloying elements, carbon has the most pronounced effect on the properties of steel. Additionally, elements such as manganese and silicon, along with various microstructural variables like grain size, morphology, distribution of precipitation, nature of ferrite, and the type and amount of precipitates in ferrite, influence the mechanical properties of steel with a given chemical composition [1, 4, 5].

The investigation of multipass welding in steels has uncovered noteworthy microstructural changes between the heat-affected zone (HAZ) and the fusion zone (FZ). These changes significantly impact the mechanical behavior of welded joints [6, 7]. The process of arc welding subject's high-strength low-alloy (HSLA) steels to a series of thermal cycles, triggering intricate microstructural transformations. Within the heat-affected zone, the rapid alternation between heating and cooling leads to the formation of residual austenite-martensite compounds and the coarsening of ferritic grains [8]. Furthermore, even with similar yield strength values, variations in the chemical composition of the base material can result in different welding behaviors. Consequently, the mechanical properties of the weld can be affected by factors such as precipitation, the nature of phases, and the presence of austenite-martensite compounds [9].

In order to enhance the mechanical performance of high-strength low-alloy (HSLA) pipe steel, adjustments can be made to the alloying elements. Extensive research has been conducted since the 1980s to understand the relationship between metallurgical and mechanical properties of HSLA steels. However, limited information is available regarding the impact of base metal alloying elements on weld characteristics. Therefore, the primary objective of this study is to investigate the influence of carbon content and alloying elements on the weldability of low-carbon microalloyed pipeline steels, specifically focusing on X70 [10–12]. This research aims to assess the behavior of four X70-API pipeline steels used in gas pipeline projects. These steels, exhibiting high strength ranging from 70 Ksi (483 MPa) to 80 Ksi, have been developed to fulfill cost reduction requirements. They possess favorable properties, including a minimum yield strength that meets the API Grade X70 standards, making them suitable for the transportation of hydrocarbons such as condensate, crude oil, and natural gas liquids through pipelines.

The influence of alloying elements and carbide precipitation in the heat-affected zone has been the subject of some studies, but further investigation is still needed. Four major topics are particularly important: the presence of TiN and V(C, N) precipitates in ferritic phases during the cooling rate of the welding process due to the fraction conditions of N, V, and Ti [13–16]; the effect of welding parameters on HAZ characteristics such as grain size and microphase nature [17]; optimizing the influence of alloying elements on HAZ cracking, where the addition of elements like titanium can improve cracking behavior [18].

While extensive research has been conducted on the metallurgical aspects of niobium [19–21], its influence and interaction with carbon content during the welding process, particularly in the coarse-grained heat-affected zone (CGHAZ), require further investigation.

A pipeline is formed by connecting multiple pipes through welding, employing cost-effective and efficient welding techniques to ensure robust and dependable welds. In the oil industry, the SMAW manual arc welding process is widely favored due to its numerous advantages, and it can be executed in various positions to cater to the specific requirements of pipeline construction [22].

This study aims to examine the impact of alloying elements in X70 high-strength steel on the properties of welded joints. Experimental tests were conducted on various samples of X70 with different compositions using the shielded metal arc welding (SMAW) technique. The investigation focuses on evaluating the effects of different X70 compositions on key properties of the heat-affected zone (HAZ) and weld metal (WM), including hardness, Charpy impact resistance, tensile strength, and microstructure.

2 Experimental Procedure

The material of focus in this investigation was X70 steel. The samples were obtained in tube form, with a diameter of 28 inches (711.2 mm) and a wall thickness of 11.48 mm. The welding technique employed for the study was multipass Shielded Metal Arc Welding (SMAW), utilizing ER E9018 filler metal [23]. The chemical compositions of both the base material and the filler metal are outlined in Table 1 and Table 2, respectively. For specific information regarding the welding parameters used, please refer to Table 3.

Table 1. Chemical composition of X70 steels (in weight %).

Element	C	Si	Mn	Al	Cu	Cr	Mo	V	Nb	Ti
X70-A	0.06	0.34	1.65	0.03	0.02	0.02	0.3	0.08	0.02	0.02
X70-B	0.08	0.27	1.45	0.026	0.015	0.04	0.03	0.085	0.06	0.02
X70-C	0.11	0.35	1.50	0.038	0.013	0.03	0.005	0.090	0.02	0.05
X70-D	0.14	0.37	1.41	0.038	0.012	0.03	0.022	0.09	0.01	0.1

Optical and electron microscopy techniques (SEM) were utilized for microstructural analysis and fractographic examinations. Tensile specimens, prepared in accordance with ASTM E08 standards [24, 25], were tested using a servo-hydraulically controlled digital tensile testing machine. Three tensile specimens were prepared and evaluated for each welded steel sample. These tests were conducted to determine the mechanical properties, particularly the tensile strength of the joints.

Table 2. Chemical composition of filler material (in weight %).

Element	C	Si	Mn	Mo
(%)	0.12	0.80	0.90	0.5

Table 3. Welding parameters.

Number of passes	1	2	3	4	5
Electrode	G	D1	D1	D1	D1
Preheating (°C)	170	195	200	197	175
Average voltage (V)	27	25	24	26	23
Intensity (A)	83	138	144	161	143
Welding speed (mm/min)	42	96	71	71	120
Polarity	DC-RP	DC-RP	DC-RP	DC-RP	DC-RP

To assess Charpy impact resistance, the specimens, measuring 55 mm in total length and featuring a rectangular cross-sectional area of 10 mm × 7.5 mm, were tested following the standard methods for notch bar impact testing of metallic materials [6]. Specifically, a V-shaped notch with a flank angle of 45° and a depth of 2 mm (Fig. 3) was introduced exclusively in the Heat-Affected Zone (HAZ) [26].

Subsequently, SMAW welding tests were conducted on samples extracted from the tubes, with four different carbon content levels (0.060, 0.082, 0.11, and 0.014). These samples were then subjected to prestressed Post Weld Heat Treatment (PWHT). To investigate the mechanical behavior of the steel, both tensile tests and Charpy impact tests were performed on specimens taken perpendicular to the welding direction. Vickers hardness tests were conducted using a 200 g load.

3 Results and Discussions

3.1 Microstructures Evolution

Figures 1 (a, b, c, and d) depict the microstructures of the X70-A, X70-B, X70-C, and X70-D base materials, respectively. In the X70-A microstructure, acicular ferrite is observed along with some pearlite. This microstructure is achieved through accelerated cooling following the hot rolling of non-recrystallized austenite. In its as-received state, the X70-B microstructure consists of rows of ferrite (F) and pearlite (P), which are visible at lower magnifications. The microstructure of X70-C steel reveals alternating ferritic rows and pearlite, formed as a result of segregation during the solidification phase in its original delivery state. Lastly, the microstructure of X70-D is characterized by ferrite grains (F), including both polygonal ferrite and some acicular ferrite, along with regularly arranged rows of pearlite (P).

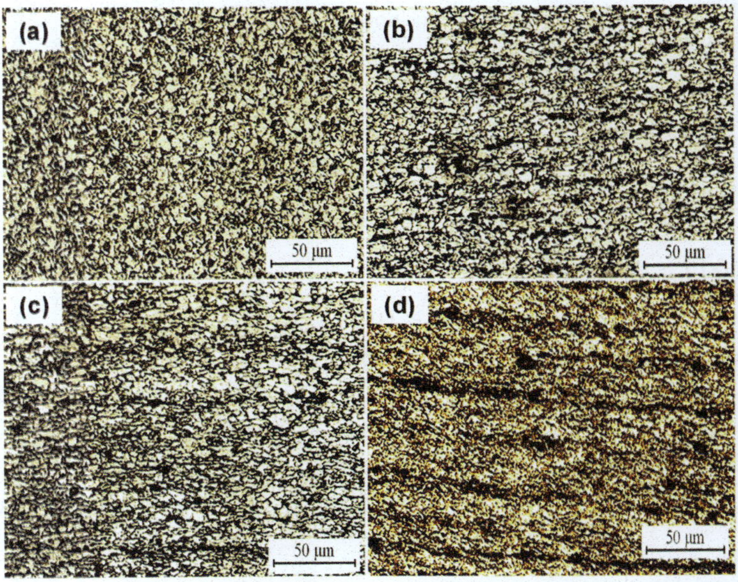

Fig. 1. Microstructures of BM: (a) X70-A, (b) X70-B, (c) X70-C, and (d) X70-D

The coarse heat-affected zone (HAZ) area refers to the initial microstructure of the base metal as it experiences higher temperatures due to the thermal cycles during welding (Fig. 2). At elevated temperatures (above 900 °C), Nb (C, N) precipitates form, refining the austenite grains. However, these precipitates completely dissolve due to the high welding energy provided by the SMAW process, resulting in the homogenization of austenite. The applied tension-compression during welding accelerates the coarsening kinetics of precipitates during cooling, unlike in cases without constraints. This phenomenon is attributed to the increased diffusivity of solute atoms caused by the higher dislocation density resulting from deformation. Furthermore, the phase transformation temperature ($\gamma \rightarrow \alpha$) decreases with increasing heating temperature. These two mechanisms, diffusivity and thermal cycles, lead to a mixed microstructure consisting of granular polygonal ferrite with dispersed small cementite particles.

In the CGHAZ of X70-A, the microstructure is characterized by granular bainite and coarse polygonal ferrite grains (Fig. 1(a)). In the CGHAZ of X70-B (Fig. 1(b)), the coarsening of ferrite grains is limited due to the presence of Nb (C, N) precipitates within the ferritic matrix, as indicated in Table 1, which hinder the coarsening of austenite grains. The highest coarsening ratio of ferrite grains is observed in the HAZ of X70-D (Fig. 1(d)).

The fusion zone (FZ) undergoes significant tempering, leading to an increased martensite content, which accounts for its relatively high hardness in this section (C). This area is formed by the rapid cooling of the molten filler metal and exhibits directional solidification patterns aligned with the cooling direction. The solidification of the fusion zone results in refined structures featuring regions such as proeutectoid ferrite, lamellar martensite, acicular ferrite, and granular bainite.

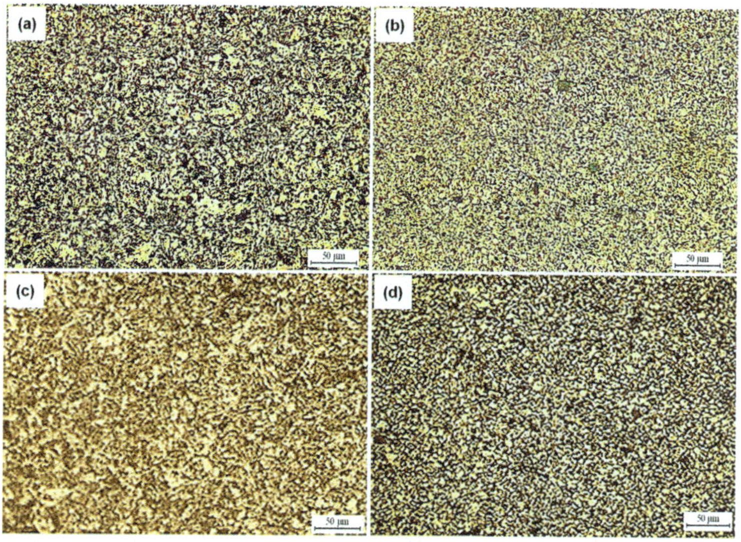

Fig. 2. Microstructures of HAZ: (a) X70-A, (b) X70-B, (c) X70-C, and (d) X70-D

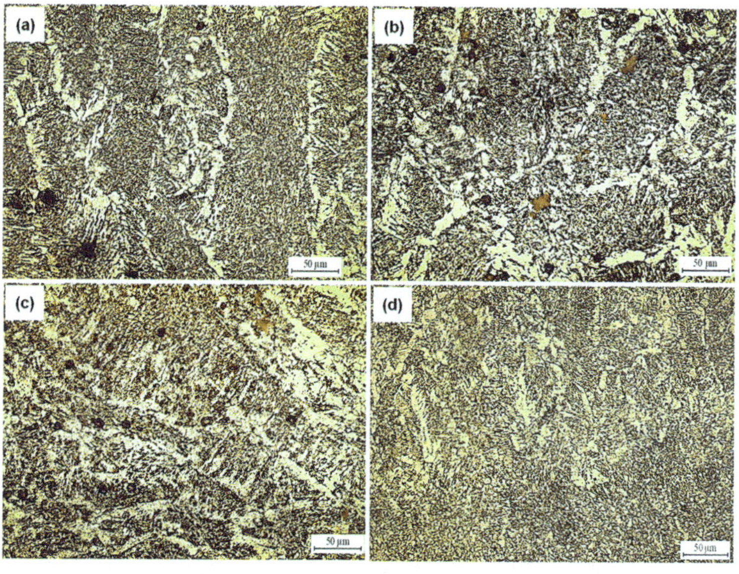

Fig. 3. Microstructures of FZ: (a) X70-A, (b) X70-B, (c) X70-C, and (d) X70-D

3.2 Mechanical Behavior

Figure 4 illustrates the variation in Vickers hardness measured across the three areas of the welded joints for the four different chemical compositions. These measurements were taken at mid-thickness of the cross-section of the welded joint. The hardness of the fusion zone (FZ) is consistently lower than that of the connecting region of the heat-affected zone (HAZ-FZ) in all samples. The appropriate hardness values observed in the weld metal can be attributed to the presence of acicular ferrite and polygonal ferrite, as well as the absence of hard phases such as bainite, martensite, and carbides. The increased hardness observed in the coarse-grained heat-affected zone (CGHAZ) can be attributed to two factors: first, the rapid cooling rate, which promotes the formation of hard phases (bainite, M-A compounds); and second, the presence of high residual stresses. These factors are consistent across the various X70 alloying element compositions.

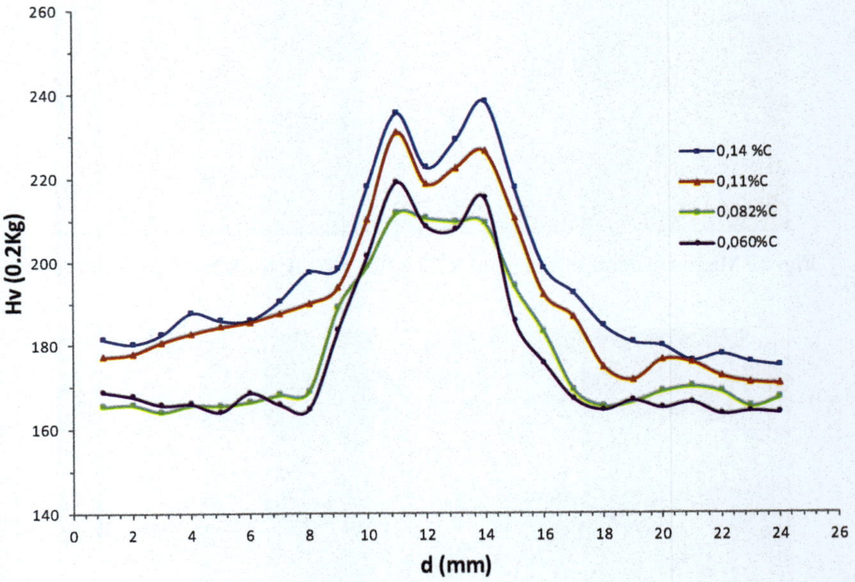

Fig. 4. Hardness of welded joints

In the heat-affected zone (HAZ), the presence of microalloying elements prevents carbide precipitation at the grain boundaries, instead promoting finely dispersed precipitation within the grains. The limited presence of hard compounds, such as martensite-austenite (M-A) constituents, which are sensitive to carbon content in X70-C, explains the decreased hardness in this sample. During deformation, these microalloying elements facilitate the enlargement of ferritic grain size [13–16] (Fig. 5).

The variation in tensile strength and deformation as a function of carbon content (and alloying elements) indicates that carbon significantly influences the tensile properties of the welds, as these factors are closely related to the microstructure of the welded joint. The influence of microalloying elements becomes more prominent with lower carbon content, as the heat-affected zone (HAZ) experiences alloying effects that overshadow the impact of microalloyed elements, particularly niobium (Table 1). It should be noted that deformation at the point of fracture decreases with an increase in niobium content.

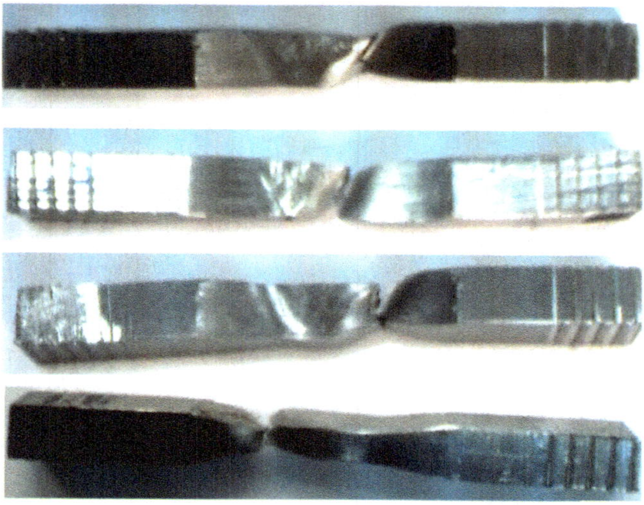

Fig. 5. Brocken tensile samples

Hardness measurements conducted in relation to the weld center reveal an increase in hardness in both the fusion lines and heat-affected zones (HAZ) of the four samples. The elevated hardness values observed in the fusion line can be attributed to the significant precipitation of carbides, as well as the presence of ferrite plates and martensite lattices in this specific region.

Table 4 presents the ultimate strength and deformations of X70-A, X70-B, X70-C, and X70-D specimens. It is evident that variations in the composition of the filler metal result in significant changes in the tensile properties. These differences can be attributed to variations in the microstructures of the fusion zones and the joining areas (Fig. 6).

The base material exhibits a higher energy absorption capacity before fracture compared to other welding zones. The high energy absorption of X70-A can be attributed to its microstructure, which consists of irregular acicular ferrite grains produced through controlled thermomechanical rolling. Conversely, the fusion zone shows lower energy absorption due to its microstructure, which includes areas of coarse solidification and the presence of hard phases. This microstructure leads to reduced resistance to crack propagation.

In the heat-affected zone of X70-B, excellent resilience is observed, a characteristic of X70 steel along with good yield strength. The lower energy absorption in the heat-affected zone compared to the base material can be attributed to several factors:

- The presence of polygonal ferrite, which is less resistant to crack propagation compared to acicular ferrite (ref),
- The occurrence of coarse regions that weaken resistance due to grain coarsening,
- The presence of brittle phases, such as bainite, hard martensite-austenite (M-A) compounds, and pearlite islands, which exhibit lower resistance to crack propagation.

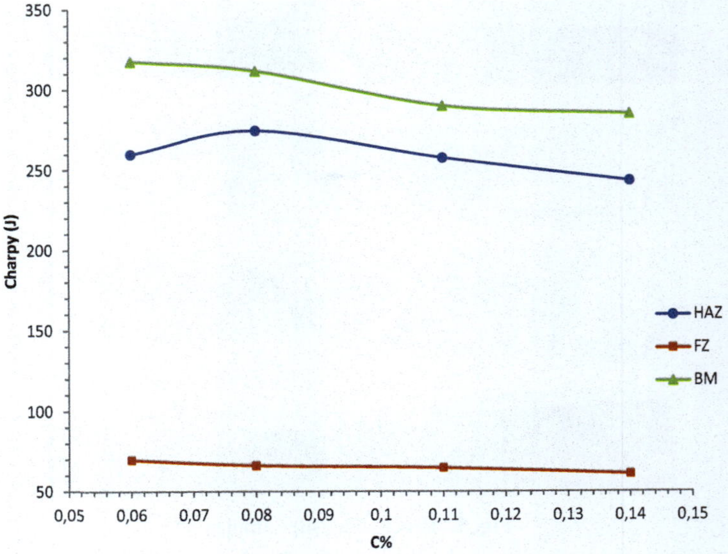

Fig. 6. Charpy test

Table 4. Tensile characteristics.

Samplers	X70-A	X70-B	X70-C	X70-D
R_m (MPa)	453.3	438.5	412.08	465.21
E_p (%)	2.77	2.24	2.1	1.17
ε_r (%)	8.66	9.26	7.96	6.36

3.3 Fractographic Evolution

The fracture behavior of X70 steels exhibits characteristics of plastic deformation. The fracture surface reveals isolated spherical cups or regions with fine dimples. The presence of these cupules and signs of plastic deformation increases with higher carbon content (Table 4). Conversely, higher carbon content improves weldability and resistance to brittle fracture, particularly in areas with increased necking. This phenomenon is attributed to the interaction of carbon with other interstitial atoms, forming slip bands known as 'Piobert-Lüders' bands that align along dislocation boundaries. This behavior is particularly pronounced in X70-D and X70-C steels, where rapid crack propagation initiation is observed (Fig. 7).

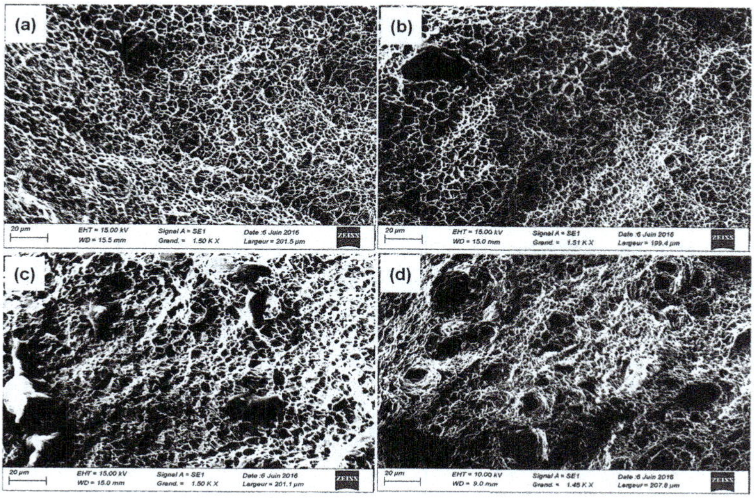

Fig. 7. SEM images of tensile fractures in base materials (a) X70-A, (b) X70-B, (c) X70-C, (d) X70-D.

4 Conclusions

This study highlights the influence of chemical composition, microstructure, and microalloying elements on the properties of microalloyed X70 steels. The optimal choice of composition, particularly low carbon content and the presence of Nb (C, N) precipitates, enables a superior combination of strength, ductility, and weldability in X70-B steel.

However, it should be noted that other factors, such as crack propagation resistance and energy absorption before fracture, should also be considered when selecting the material for specific applications. This research provides a solid foundation for optimizing the properties of microalloyed X70 steels and guiding the development of new alloys for the construction and pipeline industries. Further studies are needed to explore the relationships between composition, microstructure, and mechanical properties more deeply, with the aim of enhancing the performance of steels under specific service conditions.

References

1. Sohrabi, M.J., Mirzadeh, H., Mehranpour, M.S., Heydarinia, A., Razi, R.: Aging kinetics and mechanical properties of copper-bearing low-carbon HSLA-100 microalloyed steel. Arch. Civ. Mech. Eng. **19**(4), 1409–1418 (2019). https://doi.org/10.1016/j.acme.2019.09.001
2. Benlamnouar, M.F., Bensaid, N., Azzoug, M.O., Saadi, T., Zidani, M., Badji, R.: Optimization and evaluation of mechanical and electrochemical properties of ferritic stainless steel welding using taguchi design. Key Eng. Mater. **973**, 61–72 (2024). https://doi.org/10.4028/p-oqmgc5
3. Garcia, C.I.: High strength low alloyed (HSLA) steels. In: Rana, R., Singh, S.B. (eds.) Automotive Steels, pp. 145–167. Woodhead Publishing (2017). https://doi.org/10.1016/B978-0-08-100638-2.00006-7

4. Lee, C.-H., Shin, H.-S., Park, K.-T.: Evaluation of high strength TMCP steel weld for use in cold regions. J. Constr. Steel Res. **74**, 134–139 (2012)
5. Zhao, J., Jiang, Z., Lee, C.S.: Effects of tungsten on the hydrogen embrittlement behaviour of microalloyed steels. Corros. Sci. **82**, 380–391 (2014)
6. Yang, X., Xinjie, D., Liu, X., Dongpo, W.: Effects of heat input on microstructure and fracture toughness of simulated coarse-grained heat affected zone for HSLA steels. Mater Charact **155**, 109818 (2019)
7. Dornelas, P., Filho, J., Farias, F., Oliveira, V., Moraes, D., Júnior, P.: Influence of the interpass temperature on the microstructure and mechanical properties of the weld metal (AWS A5.18 ER70S-6) of a narrow gap welded API 5L X70 pipe joint. Int. J. Press. Vessel. Pip. **199**, 104690 (2022)
8. Poorhaydari, K., Ivey, D.G.: Precipitate alterations in the heat-affected zone of a Grade 100 microalloyed steel. J. Mater. Sci. **46**(14), 4953–4963 (2011)
9. Schroeder, N., Rhode, M., Kannengiesser, T.: Thermodynamic prediction of precipitation behaviour in HAZ of a gas metal arc welded S690QL with varying Ti and Nb content. Weld. World **67**, 2143–2152 (2023)
10. Wei, H., Liu, G.: Effect of Nb and C on the hot flow behavior of Nb microalloyed steels. Mater. Des. **56**, 437–444 (2014)
11. Sharma, L.: Advent of different grades of HSLA pipeline steels and their structural integrity issues related to welding – 'a review'. Mater. Today: Proc. **37**, Part 2, 3585–3587 (2021)
12. Nolan, T., Tracy, W.: Control of post-weld fracture toughness in friction stir processed X80 HSLA steel. Eng. Fract. Mech. **271**, 108586 (2022)
13. Bensaid, N., Hadji, M., Badji, R., Benlamnouar, M.F., Saadi, T., dit Laksir, Y.L., Senouci, S.: Microstructure and Mechanical Behavior of AISI 430 FSS Welds Produced with Different Elemental Metal Powder Addition. Solid State Phenom. **297**, 195–203 (2019). https://doi.org/10.4028/www.scientific.net/ssp.297.195
14. Saadi, T., Benlamnouar, M.F., Bensaid, N., Boutaghane, A., Soualili, M.A., Hachemi, H.: Optimization of automatic TIG welding parameters of AISI 304L ASS welds using response surface methodology. Defect Diffus. Forum **406**, 319–333 (2021). https://doi.org/10.4028/www.scientific.net/ddf.406.319
15. Benlamnouar, M.F., Bensaid, N., Saadi, T., Badji, R.: Multi-objective optimization and evolution of dissimilar welding process between Cr-Mo steel and austenitic stainless steel for power plant application. Mater. Res. Express **11**(2), 026507 (2024). https://doi.org/10.1088/2053-1591/ad28d2
16. Bose-Filho, W., Carvalho, A., Strangwood, M.: Effects of alloying elements on the microstructure and inclusion formation in HSLA multipass welds. Mater Charact **58**, 29–39 (2007)
17. Mohammadijoo, M., Valloton, J., Collins, L., Henein, H., Ivey, D.: Characterization of martensite-austenite constituents and micro-hardness in intercritical reheated and coarse-grained heat affected zones of API X70 HSLA steel. Mater Charact **142**, 321–331 (2018)
18. Oyyaravelu, R., Kuppan, P., Arivazhagan, N.: Comparative study on metallurgical and mechanical properties of laser and laser-arc-hybrid welding of HSLA steel. Mater. Today: Proc. **5**(5), Part 2, 12693–12705 (2018)
19. Cao, J.-C., Liu, Q.-Y., Yong, Q.-L., Sun, X.-J.: Effect of niobium on isothermal transformation of austenite to ferrite in HSLA low-carbon steel. J. Iron. Steel Res. Int. **14**(3), 51–55 (2007). https://doi.org/10.1016/S1006-706X(07)60043-0
20. Patterson, T., Lippold, J.C.: Effect of niobium on the microstructure and properties of submerged arc welds in HSLA steel. Weld. World **64**(6), 1089–1105 (2020)

21. Feng, C., Fang, H.-S., Zheng, Y.-K., Bai, B.-Z.: Mn-series low-carbon air-cooled bainitic steel containing niobium of 0.02%. J. Iron. Steel Res. Int. **17**(4), 53–58 (2010). https://doi.org/10.1016/S1006-706X(10)60086-6
22. Bensiali, H., Bidi, L., Cicala, E., Le Masson, P., Chibani, M.E.B., Boulahlib, M.S.: Effects of operating parameters on weld bead morphology with welding operations of API 5L X70 steel pipes by SMAW process. Weld. World **65**, 1119–1129 (2021)
23. ASME: Qualification standard for welding and brazing procedure, welders, brazers, and welding and brazing operators. Boiler & Pressure Vessel Code (2010)
24. Designation: E8/E8M-09: Standard test methods for tension testing of metallic materials. American National Standard (2010)
25. Benlamnouar, M.F., Saadi, T., Bensaid, N., Gousmine, M., Touggui, Y., Temmar, M.: Modelling and optimization of dissimilar welding between 304L and HSLA-X70 using response surface methodology. Lecture Notes in Mechanical Engineering, pp. 391–402 (2021). https://doi.org/10.1007/978-3-030-62784-3_39
26. ASTM E 23: Standard test methods for notched bar impact testing of metallic materials (2000)

Author Index

H. Altenbach et al. (Eds.): CAMPE 2023, pp. 325–326, LNME 2025.
https://doi.org/10.1007/978-3-031-82979-6

The manufacturer's authorised representative in the EU is Springer Nature Customer Service Centre GmbH, Europaplatz 3, 69115 Heidelberg, Germany. If you have any concerns regarding our products, please contact ProductSafety@springernature.com

Printed and bound by CPI Group (UK) Ltd, Croydon, CR0 4YY
01/10/2026
02225901-0004